# WATER DEFICITS AND PLANT GROWTH

## VOLUME V

Water and Plant Disease

# CONTRIBUTORS TO THIS VOLUME

PETER G. AYRES

C. M. CHRISTENSEN

R. D. DURBIN

D. W. FRENCH

D. M. GRIFFIN

C. T. INGOLD

DONALD F. SCHOENEWEISS

A. F. VAN DER WAL

C. E. YARWOOD

# WATER DEFICITS
# AND PLANT GROWTH

EDITED BY

## T. T. KOZLOWSKI

DEPARTMENT OF FORESTRY
THE UNIVERSITY OF WISCONSIN
MADISON, WISCONSIN

## VOLUME V

Water and Plant Disease

1978

ACADEMIC PRESS    New York    San Francisco    London

A Subsidiary of Harcourt Brace Jovanovich, Publishers

ACADEMIC PRESS, INC.
111 Fifth Avenue, New York, New York 10003

*United Kingdom Edition published by*
ACADEMIC PRESS, INC. (LONDON) LTD.
24/28 Oval Road, London NW1 7DX

Library of Congress Cataloging in Publication Data

Kozlowski, Theodore Thomas, Date
    Water deficits and plant growth.

    Includes bibliographies and index.
    CONTENTS: v. 1. Development, control and measurement.
—–v. 2. Plant water consumption and response. ––v. 3.
Plant responses and control of water balance.    [etc.]
    1. Plants––Water requirements.   2.  Growth (Plants)
3.  Plant–water relationships.   4.   Plant diseases.
I.  Title.
QK870.K69      582'.01'9212                   68–14658
ISBN  0–12–424155–7  (v. 5)

# CONTENTS

## 1. WATER RELATIONS OF DISEASED PLANTS
### PETER G. AYRES

## 2. WATER STRESS AS A PREDISPOSING FACTOR IN PLANT DISEASE
### DONALD F. SCHOENEWEISS

## 7. MOISTURE AND SEED DECAY

### C. M. CHRISTENSEN

## 8. MOISTURE AND DETERIORATION OF WOOD

### D. W. FRENCH AND C. M. CHRISTENSEN

## 9. MOISTURE AS A FACTOR IN EPIDEMIOLOGY AND FORECASTING

### A. F. VAN DER WAL

# LIST OF CONTRIBUTORS

Numbers in parentheses indicate the pages on which the authors' contributions begin.

PETER G. AYRES (1), Department of Biological Sciences, University of Lancaster, Bailrigg, Lancaster, England

C. M. CHRISTENSEN (199, 221), Department of Plant Pathology, University of Minnesota, St. Paul, Minnesota

R. D. DURBIN (101), Plant Disease Resistance Research Unit, ARS, USDA, and Department of Plant Pathology, University of Wisconsin, Madison, Wisconsin

D. W. FRENCH (221), Department of Plant Pathology, University of Minnesota, St. Paul, Minnesota

D. M. GRIFFIN (175), Department of Forestry, Australian National University, Canberra, Australia

C. T. INGOLD (119), Birkbeck College, University of London, London, England

DONALD F. SCHOENEWEISS (61), Illinois Natural History Survey, Urbana, Illinois

A. F. VAN DER WAL (253), Foundation for Agricultural Plant Breeding, Laboratory de Haaff, Wageningen, The Netherlands

C. E. YARWOOD (141), Department of Plant Pathology, University of California, Berkeley, California

# PREFACE

The philosophy of this volume departs somewhat from that of previous ones in this series because it does not restrict itself to water-stressed plants. Rather its purpose is to present a comprehensive treatment of the role of both water deficits and excesses in the plant disease complex.

The opening chapter, a broad overview of the water relations of diseased plants, deals separately with root diseases, vascular wilts, shoot diseases, foliar diseases, and the combined effects of water stress induced by disease and environment. The second chapter evaluates effects of water deficits on both pathogen and host, and also considers separately the degree and duration of water deficits as predisposing factors in plant disease. This is followed by a chapter on important abiotic diseases induced by water deficits as well as those induced by an excess of water. The fourth chapter discusses water in relation to both active and passive liberation of spores. The next chapter analyzes water in relation to the infection process. Particular attention is given to water in relation to spore germination, swelling and shrinkage of spores, and sources of moisture. The next chapter focuses on soil moisture in relation to spread and survival of pathogens. The seventh chapter, on water and seed decay, deals with seed longevity and appraises the activities of both field and storage fungi that affect seeds. The next chapter summarizes characteristics of deteriorated wood products, water in relation to wood deterioration, and prevention of deterioration. The final chapter is a comprehensive treatment of moisture as a factor in epidemiology and the forecasting of disease.

I thank the authors for their scholarly contributions as well as their patience and cooperation during the production phases. I especially want to express my deep appreciation to Dr. R. D. Durbin for calling my attention to the need for this volume and for his generous and valuable counsel during the planning and creation of the book. Mr. R. J. Norby and Mr. S. G. Pallardy assisted with preparation of the Subject Index.

T. T. KOZLOWSKI

# CONTENTS OF OTHER VOLUMES

CHAPTER 1

# WATER RELATIONS OF DISEASED PLANTS

Peter G. Ayres

DEPARTMENT OF BIOLOGICAL SCIENCES, UNIVERSITY OF LANCASTER, BAILRIGG,
LANCASTER, ENGLAND

## I. INTRODUCTION

Wilting, desiccation, and water-soaking are examples of the characteristic visible symptoms that arise when the water relations of a plant are
altered by a pathogenic microorganism. However, our increasing knowledge of the physiology of diseased plants shows us that in many diseases
host water relations are altered in a way that will seriously affect growth

1

and development, e.g., expansion of young tissue may be reduced, without the production of such characteristic visible symptoms. This chapter will consider diseases in which host water relations are altered as a primary event in pathogenesis, and not just diseases in which a disturbed water balance is obvious.

Physiological events in diseased plants may be studied and explained to a high level of understanding, but only if analogous processes in healthy plants are well understood can there be a proper appreciation of their importance to the plant. With the exception of those carrying seed-borne infections, plants begin life in a healthy state and disease is a progressive influence upon them; the progressive nature of disease is true also of seed-borne infection. The progress of disease depends upon the growth rate of the pathogen and the rate at which it releases factors such as enzymes and toxins. Depending upon factors such as the age and resistance of the host, and environmental conditions, the influence of disease may be mild or strong, slow or quick to develop, but its effects can only be judged by reference to the healthy plant. Thus, to appreciate fully the water relations of diseased plants the student should have a basic understanding of the water relations of healthy plants as described in Volumes I–IV of this series. (Kozlowski, 1968a,b, 1972, 1976).

Whereas the water relations of healthy plants may differ from species to species, each species may commonly be host to up to one-half dozen or more different pathogens, each attacking the plant in its peculiar way. Thus disease could potentially lead to an almost infinite variety of effects, but, fortunately for those interested in this subject, the number of ways in which a plant can respond to infection is limited, and similarities can be recognized between infections of remotely different origin. This is especially true of the response of plants to infection when examined at the cellular level. These changes are described first so that they may be recognized later as common factors when the water relations of apparently dissimilar diseases are examined.

Similarities between different diseases may also be recognized at the level of tissue and organ. For this reason, diseases will be studied in groups determined largely by the organ that is chiefly infected. Diseases of the root and vascular tissue will be examined before diseases of stem, leaf, and reproductive organs because this course follows the direction in which water moves in the plant and, thus, makes understanding more logical. It must always be borne in mind that organs affected by localized infections cannot be treated as though they were separate from the rest of the plant; each plant is a whole in which the parts interact, so that infection in one part of a plant will have repercussions on the remainder of the plant. Those who have studied root and vascular diseases probably are more aware of this than those who have studied diseases of the aerial parts of plants.

## II. SOLUTES AND CELL WATER RELATIONS

Although it is not the purpose of this chapter to reexamine the physiological and biochemical roles of water in the plant, or to reconsider the ways in which the plant regulates water uptake and loss by sophistication of growth habit, anatomy, or physiological process, one fundamental point will be restated because it is so important in disease: a cardinal feature which distinguishes the water relations of terrestrial plants from nonliving bodies is that plants are able to regulate their water content so that they are not subject to the extreme cycles of hydration and dehydration characteristic of nonliving objects. Plants achieve this regulation by selectively accumulating molecules of water-soluble substances behind membranes which are less permeable to the solutes than they are to water. Thus, the plant is characterized by the possession of "osmotic cells" whose water relations are described (given the additional presence of a semirigid cell wall) by the familiar equation:

$$\psi = \psi_s - \psi_p - \psi_m$$

where $\psi$ = water potential, and $\psi_s$, $\psi_p$, $\psi_m$ are the solute, pressure, and matric potentials, respectively (Slatyer, 1967). Any change in solute relations will affect the water relations of the cell and, probably, that of its neighbors. The plant maintains contact with liquid water in the soil and this water, obeying the laws of thermodynamics, moves down a gradient of (water) potential energy into the plant at a rate that is approximately equal to the rate of evaporation from the aerial parts of the plant.

Solutes can only be accumulated and maintained at a higher potential energy inside than outside the cell membrane by the expenditure of metabolic energy. The energy is needed to drive solute fluxes at the plasmalemma and to synthesize organic solutes. Energy and carbon skeletons are also needed to maintain the fabric of the membranes. When healthy plants senesce and their energy metabolism fails, widespread loss of membrane function occurs. Because infection advances senescence and an associated loss of membrane function in most plants, diseases must be grouped for our purposes according to whether permeability changes occur only when the death of the whole organ or plant is imminent, or whether they occur before, or are coincident with, the appearance of symptoms. In the latter group of diseases, which are of interest to us here, permeability changes may be localized or widespread and may arise from several different causes other than the general failure of energy metabolism. It is relevant to examine the more important causes because each has a different bearing on the subsequent development of disease.

## A. CAUSES OF ALTERED MEMBRANE PERMEABILITY

### 1. Enzymes

Phytopathogenic fungi and bacteria produce a range of polysaccharidases that attack plant cell walls and enable the pathogen to grow into or between cells of their host (Albersheim et al., 1969). Work with a range of pectic enzymes, purified to high specific activity, has shown that they are responsible for both cell death and tissue maceration in many diseases, the latter particularly in infections of seedlings or detached organs where an extensive soft, watery rot is produced. Cell injury caused by pectic enzymes is characterized by a rapid leakage of ions and water and a loss of the cell's ability either to retain vital stain or to plasmolyze in hypertonic solutions (Wood, 1972). However, although these observations indicate that the plasma membrane is damaged by pectic enzymes, isolated protoplasts in hyperosmotic media and cells plasmolyzed to the point of incipient plasmolysis before treatment with enzymes are not damaged (Basham and Bateman, 1975). Basham and Bateman suggest that membrane damage may result from the inability of enzyme-degraded cell walls to support the plasmalemma under conditions of osmotic stress. The effects of pathogen-produced cellulases and hemicellulases on host membrane permeability are not known. The role that pathogen-produced enzymes which attack the membrane itself, e.g., phosphatidases, have in changing its permeability is problematical. Various phytopathogenic fungi and bacteria have been shown to produce phosphatidases in culture (Tseng and Bateman, 1968), again the most active being those that produce soft watery rots in vivo, and phosphatidases have been detected in diseased but not in healthy tissues of bean (Phaseolus vulgaris) infected with Thielaviopsis basicola (Lumsden and Bateman, 1968) or Sclerotinia sclerotiorum (Lumsden, 1968) and tobacco leaves infected with Erwinia amylovora (Huang and Goodman, 1970). Measurements of membrane permeability in diseased tissues were not made, but the visible symptoms shown by the diseased tissues, e.g., loss of turgor and necrosis, suggested that permeability changes occurred. However, in an intensive study by Hoppe and Heitefuss of permeability and membrane lipid metabolism in P. vulgaris infected with rust (Uromyces phaseoli), it was found that although the pathogen caused permeability changes in the host 4 days after infection (Hoppe and Heitefuss, 1974a), significant increases in phospholipase activity and decreases in lipid concentration occurred mainly after the fourth day following inoculation (Hoppe and Heitefuss, 1974b). These investigators concluded that it was improbable that phospholipase activity caused the observed changes in permeability; in this and other diseases the role of phospholipases in causing permeability changes remains unproved.

## 2. Toxins

The extent to which toxins (nonenzymatic substances which interfere with metabolism of the host) may act, possibly in cooperation with enzymes, to cause changes in host membrane permeability is also uncertain. Many phytopathogenic microorganisms produce diffusible toxins that have the plasmalemma of the host as their primary site of action (Strobel, 1974) but often, e.g., in Victoria blight of oats caused by *Helminthosporium victoriae* (Wheeler and Black, 1963) and leaf spot and ear rot of corn caused by *Helminthosporium carbonum* (Scheffer and Samadder, 1970), altered water relations are not described among the characteristic symptoms of the disease, and there has been no detailed investigation of the water relations of the naturally infected host. However, there are some diseases in which the pathogen produces a toxin which alters both the permeability of membranes and the water relations of the host, e.g., bacterial wilt of potato caused by *Corynebacterium sepedonicum*, canker (and wilt) of peach *(Prunus persica)* and almond *(Prunus amygdalus)* caused by *Fusicoccum amygdali*, and angular leaf spot of cucumber caused by *Pseudomonas lachrymans* (Keen and Williams, 1971). All but the last of these diseases will be dealt with in detail in later sections. Toxin produced by *Helminthosporium maydis* race T, the cause of Southern corn leaf blight, causes solute leakage from roots of susceptible plants (Gracen *et al.*, 1972) and although it does not visibly alter the water relations of diseased plants, it may have important physiological effects because it induces rapid stomatal closure by inhibiting $K^+$ uptake (Arntzen *et al.*, 1973).

The toxin whose mechanism of action is probably best understood is helminthosporoside the disaccharide produced by *Helminthosporium sacchari*, the cause of "eyespot" of sugar cane (Strobel, 1975). The fungus is limited to eye-shaped lesions on the leaf but it releases a diffusible toxin that is carried up the leaf in the transpiration stream and causes appearance of a "runner." The leaf within the runner shows water-clearing, i.e., it becomes transparent, small droplets of water exude, and it finally becomes reddish-brown and necrotic. The runner may extend 10–20 in. away from the lesion. Such "action-at-a-distance" from the pathogen is typical of toxin action and similar runners are produced when healthy leaves are treated with the isolated toxin.

Helminthosporoside causes electrolyte leakage from treated tissues and from isolated protoplasts. Experiments with $^{14}$C-labeled toxin have shown that it binds to a protein in the plasmalemma of susceptible varieties. The protein has four identical subunits of molecular weight 12,000 and differs by only four amino acids in each subunit from a similar protein in resistant varieties. The binding protein is believed to be an ATPase stimulated by $K^+$ and $Mg^{2+}$ because the activity of such an enzyme is stimulated by the toxin and if leaves are heated to 50°C they lose both sensitivity to the toxin and the ATPase activity. The toxin would affect transport processes requiring ATP. Thus it would exert a selective effect on permeability because it would affect immediately only those transport processes that were actively driven. However, the observation that isolated protoplasts treated with toxin lose electrolytes, change shape, and eventually burst suggests that the toxin may be having other, more far-reaching effects on cell membranes. It is not clear how selective this, or other pathogens, can be in altering host membrane permeability, but the specificity of transmembrane carrier mechanisms in healthy cells and the fact that microbially produced antibiotics can alter the selectivity of membranes (Clarkson, 1974) both suggest that selective alterations occur.

### 3. Host Products

Alterations in membrane permeability after infection may be hastened by the activities of the host and may form part of a hypersensitive-type resistance reaction that serves to isolate the pathogen from living host cells. Phytoalexins are pathotoxic substances produced by many plants in localized responses to infection (Deverall, 1972). There is evidence that in peas infected with powdery mildew *(Erysiphe pisi)* (Shiraishi *et al.*, 1975) and in beans infected with rust *(Uromyces phaseoli)* (Elnaghy and Heitefuss, 1976) phytoalexins (pisatin and phaseolin, respectively) accumulated to concentrations that were injurious to the permeability of the host cell membranes. Both phytoalexins are lipophilic because they have a pterocarpan structure and may produce their effects by binding with membranes. It is interesting to note that in peas infected by *E. pisi* a common symptom is the premature desiccation of leaves (Ayres, 1976). Finally, if a plant cell contains lysosomes that release hydrolytic enzymes when the cell is attacked and dies [and this is a matter of contention (Wilson, 1973)], the enzymes released from these organelles may contribute to the breakdown of membranes in neighboring cells.

### B. METHODS FOR STUDYING PERMEABILITY CHANGES

The first investigators who studied permeability changes, notably Thatcher (1939), used plasmolytic methods, but apart from being extremely tedious these methods have the disadvantage of relying for their accuracy on the impermeability of the membrane to the osmoticum. Disease may make the membrane more permeable to the osmoticum, as well as to the cell sap. Other potential weaknesses of the plasmolytic method are discussed by Wheeler and Hanchey (1968). The method that is easiest and most widely used today involves immersing the tissue in deionized water and then measuring change in electrical conductivity of the water as electrolytes leak out. Tissues may be attached (Jones and Ayres, 1972) or detached, e.g., leaf disks (Ayres, 1977a), but in the latter case care must be taken to wash electrolytes from cut cells before measurement begins. The method only detects gross changes in permeability and assumes that electrolytes leak from tissue at the same rate as nonelectrolytes, which may not be true if, for example, the storage of different substances is compartmentalized within the cell, or if the membrane potentials are actively maintained by electrogenic pumps.

Altered membrane permeability has also been measured by studying the uptake or release of radioactively labeled tracers in healthy and diseased tissues (Van Etten and Bateman, 1971; Hancock, 1972; Jones and Ayres, 1972). Ideally, both fluxes should be studied for it is the net flux of the solute that is of interest, but in practice only one flux is usually studied. This is the influx because it is easier to study than the efflux, whose study requires equal preloading of healthy and diseased tissues with the isotope, which is difficult to arrange. Methods such as these detect changes in permeability with respect to the solutes under investigation. Again, the specificity of the changes is not known and can only be demonstrated if it is shown that simultaneous movement of other solutes is unaffected.

## C. METHODS FOR STUDYING THE WATER RELATIONS OF DISEASED PLANTS

The methods that have been used to study the water relations of diseased plants are much the same as those used in the study of healthy plants and described by Barrs (1968). It will be pointed out below where these methods are unsatisfactory for the study of diseased plants and the few additional methods used for diseased plants will be described briefly.

### 1. Permeability Changes Absent

If diseased tissue is water-stressed compared with healthy tissue held under similar conditions, but there is no evidence that permeability changes take place early in the infection cycle, then the cause of stress may lie in a decreased supply of water to the tissue or an increased loss of water from it. A decreased supply may result from a failure of roots to take up water, or of the vascular system to conduct water. As will be mentioned again later, root function in diseased plants is a virtually unexplored subject. Blockages in the vascular system are common, particularly in diseases caused by vascular wilt pathogens (see Section IV), and have either been observed microscopically or detected experimentally as an increased resistance to flow (decreased hydraulic conductance) when water is forced through the vascular system under pressure or suction. There are objections to the latter method because the pressures induced in the system may be sufficient to break down resistances to flow, such as those caused by aggregations of the pathogen's cells. As proof of this, resistance in stems of tomato infected by the wilt pathogen *Fusarium oxysporum lycopersici* decreased markedly as water flow was caused by an experimentally applied pressure (Duniway, 1971a). The same investigator (Duniway, 1971b) compared resistances to water flow without applying external suction or pressure. He measured the transpiration rate of healthy and *Fusarium*-infected tomato cuttings under conditions controlled to keep leaf water content constant (as measured with a $\beta$-gage). If the assumption is made that the changes in illumination, used to control leaf water content, have no effect on stomatal diffusion resistances, which is dubious, then

$$\text{Transpirational flux } (g\ s^{-1}) = \Delta\psi/R = \Delta C/r$$

where $\Delta\psi$ is the water potential gradient across the system in the fluid phase (bars), $R$ is the resistance in the fluid phase, $\Delta C$ is the vapor pressure concentration gradient between leaf and bulk air, and $r$ is the resistance in the vapor phase. $R$ can be calculated without reference to $r$, which includes stomatal resistance.

If the sole factor causing stress in leaves is a reduced supply of water to the tissue then the relationship between stomatal diffusion resistance and either relative water content (RWC),

$$\text{RWC} = (\text{fresh weight} - \text{dry weight})/(\text{turgid weight} - \text{dry weight})$$

or leaf water potential, which is characteristic of each species, will be unaltered by disease. The relationship will be altered if the pathogen interferes with the solute relations of the tissue. Stomatal resistances are sometimes difficult to measure in diseased plants. In healthy plants they are commonly measured using a diffusion porometer with a water-sensitive element and are derived from the difference between the water lost in light and darkness. It is assumed that stomata close in the dark and that the cuticular resistance is not affected by illumination. In diseased plants the cuticular resistance may not be affected by illumination but is very often affected by the pathogen which may directly invade its host via the epidermis or rupture the epidermis to sporulate. Moreover, stomatal closure in the dark is often incomplete in diseased plants. In these cases changes in stomatal resistance have been followed by direct measurement of stomatal apertures (Duniway and Durbin, 1971a), or by use of a resistance porometer which is not affected by the cuticular resistance (Ayres, 1972). Changes in cuticular resistance can only be observed when stomatal apertures are at a fixed value, usually fully closed.

## 2. Permeability Changes Present

The first sign of impaired membrane permeability is that water-stressed tissue from the diseased plant, unlike tissue from its healthy counterpart, is unable to regain full turgor when freely supplied with water under conditions of minimal transpiration. Turgor may be judged visually or measured as the weight of water in the tissue per unit dry weight, or if the tissue is a leaf, as the fresh weight per unit area (Duniway, 1973). Thus, when tobacco plants were infiltrated with cells of the bacteria *Pseudomonas fluorescens* or *P. syringae,* or distilled water (control) and allowed to wilt before leaf disks were taken and floated on water, only disks infiltrated with *P. fluorescens* regained the same turgor as the controls (Duniway, 1973) (Fig. 1). Confirmation that *P. syringae* caused permeability changes in the host was obtained when measurements of leaf

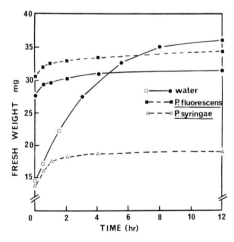

Fig. 1. Recovery from wilting in leaf disks taken from tobacco plants infected by *Pseudomonas fluorescens* or *P. syringae.* Healthy plants served as controls. Open and closed symbols represent wilted and turgid disks, respectively. (From Duniway, 1973.)

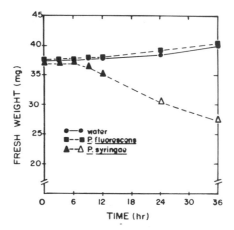

Fig. 2. Fresh weights of floating tobacco disks at various times after infection with water, *Pseudomonas fluorescens* or *P. syringae*. Intercellular spaces remained filled with water or bacterial suspensions throughout the experiment. Open and closed symbols represent wilted and turgid disks, respectively. (From Duniway, 1973.)

solute potential were made after disks had been floated on water. The potential of disks infiltrated by *P. fluorescens* was −8.4 bars, and close to that of water controls, whereas that of *P. syringae* was −0.8 bars and very close to that of healthy disks in which the semipermeability of membranes had been destroyed by freezing.

For the physiologist used to working with healthy plants a study of the water relations of diseased plants holds some surprises for wilting can occur when the water potential of a tissue and the surrounding medium are zero. Disks cut from tobacco leaves immediately following inoculation with *P. syringae* were floated on water in a saturated atmosphere; nevertheless, they showed significant decreases in fresh weight and turgor (Fig. 2) (Duniway, 1973).

Although studies of the turgor relations of plants, as indicated above, may suggest that permeability changes have taken place in diseased tissue, the evidence is only indirect and proof must rely on direct measurement of those changes.

## III. ROOT DISEASES

The microorganisms that cause root diseases may actively penetrate the root of their host or may enter at the site of wounds; sometimes these wounds are caused by vectors of the microorganism such as nematodes or earthworms. Root-infecting microorganisms, with few exceptions, live in the soil when the host is absent but the infections they produce may

extend on the foot, or basal, region of the stem and beyond, for the
pathogen often grows within the vascular system of the host. One group,
the vascular wilt pathogens, have become highly specialized toward growing in this situation. They spread extensively within the vascular system
of the stem and sometimes even into the veins of the leaves. The aerial
parts of plants are sometimes directly infected by typical root-infecting
pathogens when seedlings emerge from the soil.

The pathogens that cause diseases of seedling or senescent tissues,
e.g., *Pythium, Phytophthora, Rhizoctonia* spp., are commonly regarded as
primitive or unspecialized, for the more extreme species attack a wide
range of hosts, destroy tissues with a powerful enzymatic attack, and,
having killed or "damped-off" the host, return to a saprophytic life in the
soil (Garrett, 1970). Recent studies of root diseases indicate that infection
causes progressive water stress in the aerial parts of the plant by increasing stem and root resistances to the conduction of water. In Phytophthora
root rot of safflower (Duniway, 1975) and "black rot" of sugar beet
caused by *Aphanomyces cochliodes* (Safir and Schneider, 1976) leaf diffusive resistances increased with decreasing water potential just as in
healthy plants. Five days after infection of sugar beet the diffusive resistance of a highly and moderately susceptible variety in the light were 4.5
and 2.6 sec cm$^{-1}$, respectively, compared with control values of 1.6 sec
cm$^{-1}$, indicating that water stress was developing in infected leaves (Safir
and Schneider, 1976). The whole-plant resistance to water flow was calculated from steady-state transpiration data and was ten times greater in
infected plants than in controls 9 days after inoculation. Death of seedlings resulted from the progressive decline in water supply to the leaves.
Phytophthora rot of safflower caused a marked increase in resistance to
water uptake through the root system that was independent of the reduction (43–56%) in root fresh weight that was also caused. Stem xylem resistance was increased, but the presence of fungal cells in the xylem could
not be wholly blamed for the increase since it extended beyond the highest
point colonized by the fungus. In a woody species, *Pinus ponderosa*, infection of 1-yr-old seedings by *Verticladiella wagenerii* (root stain disease)
similarly caused decreased leaf water content, transpiration rate, and
stomatal aperture (Helms *et al.,* 1971). Leaf water potential fell from
between −12 and −15 bars in healthy plants to −18 bars at 1 month, and
−30 bars at 6 weeks, after infection. Most of the infected seedlings died
soon afterward. The fungus is confined to the underground parts of the
plant and the evidence is consistent with there being blockages in the
vascular system of the root that prevent normal conduction of water to the
aerial parts of the plant. It is known that *V. wagenerii* grows within xylem
tracheids and this may well cause blockages. Wilting of tobacco affected

by black shank disease (Powers, 1954) and of rhododendron infected by *Phytophthora* spp. (Deroo, 1969) have also been attributed to blockage of vascular elements by the pathogen.

In no root disease where the conduction of water is interrupted has there been a separate examination of factors affecting absorption, probably because of the technical difficulty of resolving the two interdependent factors. However, it is possible that widespread permeability changes occur in root cells of many plants with root diseases. These would reduce the ability of cells of the cortex and pericycle to absorb and transport ions into the xylem, and so establish the normal osmotic gradient along which water can flow. In roots of tabasco peppers infected by tobacco-etch virus, a change in permeability, indicated by a leakage of sodium and potassium ions, was detected 24–48 hr before wilt symptoms appeared (Ghabrial and Pirone, 1967). The permeability changes also preceded a decrease in respiration rate by 12–24 hr and histological changes by 24–48 hr, suggesting that altered permeability was the key event in pathogenesis. No permeability changes occurred in systemically infected leaves, and wilted plants recovered when root tips were excised and roots placed in water, indicating that wilt resulted from the failure of water to reach the xylem rather than from any blockage of water transport in the xylem itself. In hypocotyls of sunflower above lesions caused by *Sclerotinia sclerotiorum,* cells had a reduced permeability to water, ions, and urea and this peculiar effect on the movement of water into cells was considered to contribute to the inability of infected sunflowers to cope with water stress (Hancock, 1972).

The death of root cells does not have an immediate effect on water uptake which may continue for some time through dead cells, but, because of loss of selectivity in absorption, serious ionic imbalance can arise in tissues (Subramanian and Saraswathi-Devi, 1959). Studies of the specialized root parasite *Gaeumannomyces (Ophiobolus) graminis,* the cause of "take-all" of cereals, showed that invasion of living phloem cells is a particularly significant event in root infection (Clarkson *et al.,* 1975). Invasion reduced the movement of $^{86}$Rb toward the apical meristem of roots. Roots ceased to elongate and within 2 to 3 days transport of ions from the external solution into the xylem ceased. Both changes probably were the result of carbohydrate starvation, for lesions are known to interrupt the movement of $^{14}$C-labeled carbohydrate toward root apices (Manners and Myers, 1975).

Desiccation is an obvious symptom of plants severely infected with take-all in the field; when plants were grown in sand culture in the laboratory, infection reduced both shoot growth and water content (Asher, 1972a). The seminal root system was severely attacked by the fungus and its growth was retarded, but this was somewhat compensated by the fact that inoculated plants translocated a greater proportion of their total as-similates basipetally and produced an extensive system of adventitious roots. In addition to altering root growth, infection reduced expansion of young leaves. This pattern of changes in growth was not apparent if plants were continuously irrigated (Asher, 1972b) but was similar to changes that occurred in healthy plants of *Lolium temulentum* subjected to water stress (Wardlaw, 1969) suggesting that the distinctive growth pattern of infected plants was the result of pathogen-induced water stress. Studies of take-all disease show not only how infection may reduce growth by inducing water stress, but also remind us that plants may partially compensate for the destruction of existing roots by producing fresh growth; they illustrate perfectly how root–shoot interactions should always be considered when water relations of diseased plants are explored. It may be that the reduced growth shown by diseased plants in general is often due to pathogen-induced water stress, for in healthy plants leaf expansion is one of the processes most sensitive to water stress (Boyer, 1976).

## IV. VASCULAR WILTS

By the strictest definition, vascular wilts are diseases in which the pathogen, such as species of *Verticillium* or *Fusarium,* is confined to the vascular system of the host for all but the beginning and end of the infection cycle. However, a looser, but more useful, definition of vascular wilts is usually employed so that the group includes diseases such as Dutch elm disease (caused by *Ceratocystis ulmi*) in which the pathogen may grow into xylem parenchyma and ray cells, as well as the vascular system of the stem early in the infection cycle. Because of their high level of specialization the group is treated separately, although vascular occlusions resulting in water stress in leaves and stem are undoubtedly common in a wide range of diseases cuased by less-specialized root-infecting pathogens, as seen in the previous section.

Vascular pathogens multiply and spread within the xylem by hyphal growth or by production of numerous bud cells that are small enough to be carried along with the transpiration stream. Lateral spread is less common and relatively slow, but it is important in perennial hosts for it enables the pathogen to invade new tissues. Upward movement of the pathogen is stopped when its bud cells reach a point at which they are broader than the diameter of the path ahead. This commonly happens at imperforate end walls of xylem vessels, at tracheids near the extremities of veins, or at constrictions resulting from infection. Narrowing of vessel lumina to pro-duce such constrictions may result from activities of the pathogen or host. Many wilt pathogens produce pectic enzymes that attack host cell walls, leading to the formation of aggregates of cell wall fragments or even collapse of cell walls. Infected plants often produce tyloses, gums, or gels which stop the spread of the pathogen. These changes check the movement of water through the plant, as well as movement of the pathogen. Dimond (1967, 1970) and Talboys (1968) discussed such changes in detail along with the nature of further restrictions on water movement, such as increased viscosity of tracheal fluid due to production of polysaccharide slimes by bacterial pathogens.

For the best part of a century plant pathologists have argued whether wilting in these diseases results from occlusion of vascular elements due to accretion of pathogen's cells plus host material, or from increased water loss from the aerial parts of the plant, in particular leaves, resulting from damage to host membranes caused by toxic products of the fungus. The arguments for and against these theories were explored in depth by Talboys in 1968; consequently only work published since that date will be considered here. It is heartening to record that much recent work has remedied a weakness pointed out by Talboys, namely, that "few investigators . . . have taken adequate account of stomatal movements during the course of their experiments."

Recent work, including information about stomatal behavior and transpiration in plants with vascular wilt diseases, largely supports Talboys' view that, with the exception of those vascular infections with which a general wilting is not associated (see Section IV,C), vascular occlusions, and not toxins, are the cause of wilting. Dimond (1970, 1972) concluded that, with the exception of certain glycopeptides, toxins were not responsible for production of wilt symptoms *in vivo*. Wilt toxins, such as fusaric acid and lycomarasmine produced by cultures of *Fusarium*, may induce wilt symptoms on cuttings of host plants, but *in vivo* many critical aspects of the disease syndrome are lacking. Thus, the permeability of parenchymatous cells of cuttings treated with fusaric acid and lycomarasmine first increased then decreased (Gaumann, 1958), whereas in naturally infected plants parenchymatous cells showed normal osmotic functioning and wilted tissues regained turgor when water was made available to them (Dimond, 1967; Dimond and Waggoner, 1953). Furthermore, lycomarasmine has not been detected in naturally infected plants. In only a very few diseases is there evidence that membrane-damaging toxins are involved in the vascular wilt syndrome.

A. VASCULAR OCCLUSION

*1. Verticillium Wilt*

The fungus *Verticillium dahliae* causes classic wilt symptoms in chrysanthemum; the first foliar symptom is slight flaccidity at the tip of

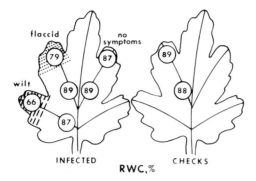

Fig. 3. Relative water content (RWC) within the tips and bases of uninfected chrysanthemum lobes and *Verticillium dahliae*-infected lobes with symptomless, flaccid, or wilted tips. [From MacHardy *et al.* (1976). Reproduced by permission of the National Research Council of Canada from the *Can. J. Bot.* **54**, 1023–1034, 1976.]

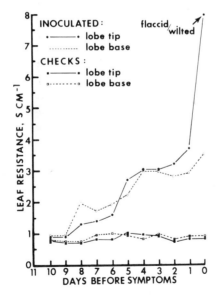

Fig. 4. Relationship between the development of internal water stress (indicated by increased resistance to water vapor diffusion) and the appearance of water-stress symptoms within the tips and bases of chrysanthemum leaf lobes infected with *Verticillium dahliae*. Diffusive resistance of uninfected leaves during a similar period served as checks. [From MacHardy *et al.* (1976). Reproduced by permission of the National Research Council of Canada from the *Can. J. Bot.* **54**, 1023–1034, 1976.]

the leaf lobes, followed by more extensive wilting and, eventually, by chlorosis and necrosis (Alexander and Hall, 1974). All available evidence indicates that occlusion of vascular elements causes the wilt. The fungus colonizes the leaf before symptoms occur and is then localized, but reversible wilt develops. At this stage symptom development is most clearly related to the patterns and degree of colonization of the leaf by the fungus (Fig. 3). Colonization increases as chlorosis and, finally, necrosis develop and progress to the rest of the leaf.

Early in infection, when symptom development is still localized in the distal regions of infected leaves, the noninfected regions of infected leaves have stomatal diffusion resistances (Hall *et al.*, 1975) and a relative water content (MacHardy *et al.*, 1974) similar to those of leaves on healthy plants. Only immediately before the development of symptoms does stomatal resistance increase and relative water content decline (Fig. 4) (MacHardy *et al.*, 1976). There is no evidence that diffusion resistance decreases (transpiration increase) at any time prior to symptom development. Comparison of the vascular movements of the water-soluble dye basic fuchsin, or of $^{14}$C-mannitol, showed that both solutes moved freely in healthy leaves and the turgid areas of infected leaves but did not move into areas of infected leaves that displayed wilt symptoms (Hall *et al.*, 1975; MacHardy *et al.*, 1976). No evidence could be found that the fungus altered the permeability of host membranes in advance of symptom production; conductivity measurements showed similar rates of electrolyte leakage from the turgid areas of infected leaves and from healthy leaves whether whole leaves (Hall and Busch, 1971) or leaf disks (MacHardy *et al.*, 1974) were used. It seems clear that *V. dahliae* causes wilt in chrysanthemum by occluding vascular elements, thereby cutting off the supply of water to distal parts of the plant.

In sunflowers infected by *V. dahliae* wilting is not a significant symptom; instead, tiny chlorotic spots appear on the leaves and these increase in diameter, eventually fusing to form large patches which are chlorotic at first and later necrotic. The fungus causes similar ultrastructural changes in sunflower (Robb *et al.*, 1975b) and chrysanthemum (Robb *et al.*, 1975a). In both there is wall breakdown and occlusion of vessels and in both there is degeneration of the mesophyll tissues, but the timing of the phenomena is different in the two hosts. In sunflower, vessel blockage is not significant until chlorosis is well advanced, but in chrysanthemum vessel blockage is well advanced before wilt symptoms develop and these precede chlorosis and degenerative changes in the mesophyll. These results have led investigators at Guelph, who have been responsible for so much research into *Verticillium* wilts, to consider the possibility that in wilt of sunflower a toxin initiates the development of foliar symptoms; however, if a toxin is involved it must have a limited range of action because visible symptoms do not extend more than a millimeter beyond the extent of the colonized area of the leaf. Thus, the behavior of the toxin is very different from that of toxins described previously, e.g., *Helminthosporium sacchari* toxin. The characteristic visible and ultra-structural changes seen in *Verticillium* wilt of sunflower closely resemble those associated with senescence; it may be that the fungus, possibly through the action of a toxin, alters the hormonal balance of the plant in infected areas to induce localized acceleration of senescence (a possibility that will be examined in more detail in Section IV,C).

In other host plants infected by *Verticillium* species an interruption in the supply of water to the leaves seems to be the cause of wilting. Potato plants infected by *V. albo-atrum* had leaf relative water contents lower than those of healthy plants and which fell to a midday value of 80% late in the growing season (Harrison, 1970). The transpiration rate of infected leaves was lower than that of healthy leaves and was attributed to their lower relative water content, for in both healthy and diseased leaves there was the same linear relationship between transpiration and relative water contents of 80–98%. At full turgor detached diseased leaves transpired faster through both stomata and cuticle than did detached healthy leaves because drought caused diseased leaves to develop a different anatomy from that of healthy leaves, e.g., in diseased leaves there were more stomata per unit area than in healthy leaves (Harrison, 1971). In *Verticillium*-infected cotton, just as in *Verticillium*-infected potato (and *Fusarium*-infected tomato, as seen later) the vascular wilt pathogen does not alter the relationship between water stress and transpiration that is characteristic of the healthy host, or, apparently, the solute relations of the host, in any way that could lead to desiccation. Thus Duniway (1973) found that the diffusion resistances of healthy and *Verticillium*-infected cotton leaves were equal to or higher than the resistances of healthy leaves over the range 60–100% relative water content.

## 2. Fusarium Wilt

In tomato plants infected by *Fusarium oxysporum lycopersici* wilting clearly results from vascular occlusion. Healthy and diseased leaves wilt at the same water potentials (Fig. 5) and in both the wilting is fully revers-ible over the same range of leaf water potentials (Duniway, 1971a). The solute potential of sap expressed from disks cut from healthy and diseased leaves and floated on water to become fully turgid (thus eliminating differ-ences in leaf water content) was −7.2 bars for healthy leaves and −8.6 bars for diseased leaves, which shows that infection leads to some solute accumulation rather than any increased leakiness of cell membranes. The

Peter G. Ayres

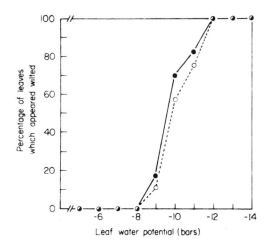

Fig. 5. Percentage of tomato leaves which appeared wilted as a function of decreasing leaf water potential in healthy (●——●) and *Fusarium*-infected (○----○) plants. Water potentials were rounded off to the nearest bar and each point represents at least 20 determinations. (From Duniway, 1971a.)

water potential of the leaves of which the solute potential was measured was not stated, but since the relationship between wilting and leaf water potential was similar in healthy and infected leaves it would seem that diseased leaves are more turgid than healthy leaves at water potentials that precede wilting.

Excessive transpiration can be ruled out as a cause of wilting because leaves on infected plants transpire less than those on healthy plants. The primary reason for this may be that diseased leaves have stomatal diffusion resistances equal to, or often slightly higher than, those of healthy leaves over the range of leaf water potentials −5 to −13 bars (Fig. 6). The cause of the reduced stomatal opening is not known.

The xylem sap of diseased plants has the same viscosity as the xylem sap of healthy plants (Waggoner and Dimond, 1954). Wilting occurs because the supply of water to the leaf is reduced as resistances to flow increase at several points in the path along which water moves. Measurements of water flow through excised root systems to which a pressure of 2 bars was applied showed that infection did not increase the resistance of roots to water flow as judged by the volume of exudate collected from the cut stump (Duniway, 1971a). The resistance of stem and petiole sections was measured on cuttings taken 10–16 days after infection when the first symptoms appear (Duniway, 1971b). Plants were cut, with the stem under water, until all that remained was a leaflet plus petiole and stem. The leaflet was enclosed in a chamber that incorporated a β-gage to measure leaf water content and a differential psychrometer to measure transpiration. By adjustment of temperature and illumination, steady transpiration rate and leaf water content were attained. The petiole was then excised close to the stem under water and the conditions were altered until the original water content and a new, steady transpiration rate were obtained. The petiole was then excised close to the leaflet under water and the procedure was repeated. Finally, the leaf was removed from the chamber and its water potential was measured. Resistances (R) to water flow through the different parts of the cuttings could then be calculated using the relationship described previously (Section II,C,1).

In infected plants with turgid leaves, stem resistance (52.7 bars $gm^{-1} \times 10^3$) was 500 times greater than that in healthy stems but petiole (0.1 bars $gm^{-1} \times 10^3$) and leaflet (52.4 bars $gm^{-1} \times 10^3$) resistances were normal. The increase in stem R is not serious for the plant since stem resistance is low in healthy plants compared with other resistances, and

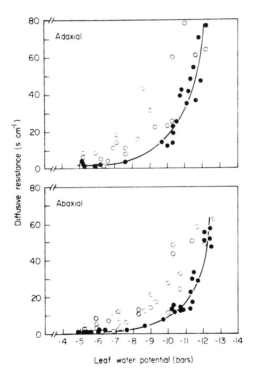

Fig. 6. Diffusive resistance of the adaxial and abaxial surfaces of tomato leaves plotted as functions of decreasing leaf water potential. Each point represents one porometer and pressure chamber determination on a leaflet from a healthy (●——●) or *Fusarium*-infected ( - ○ - ) plant. (From Duniway, 1971a.)

the 500-fold increase in diseased stems only doubled the total resistance between ground level and the experimental leaf. When measurements were made on wilted plants the stem resistance rose to 111.6 bars $gm^{-1} \times 10^3$, but the petiole resistance was very high and in some plants approached infinity because the water content would only remain constant if there was no transpiration. Water was only taken up by these leaflets when the petiole was cut near the chamber. Leaflet resistance was high where the petiole resistance was very high, probably because the leaflet measurements included xylem resistance within the leaflet and the remaining petiolar stump. Pieces of wilted leaf tissue from healthy and diseased plants absorbed water with equal speed when floated on water.

Fusarium wilt of tomato illustrates well a general point about vascular diseases. The resistance of stem bundles to water flow is relatively large compared with that of petioles and, more particularly, leaf veins (Dimond, 1970). The cross-sectional area available for conduction of water becomes ever less as the terminus of the conductive system is reached. In plants such as tomato, the large bundles of the stem form an interconnecting network, whereas small stem bundles and petiolar bundles are independent of each other (Dimond, 1966). Hence an occlusion of the same size will have a greater effect on distal tissues when it occurs in a petiolar bundle or a leaf vein than when it occurs in a stem.

### 3. Oak Wilt and Dutch Elm Disease

There is good evidence that in both oak wilt and Dutch elm diseases the pathogen causes occlusion of vascular elements in the host. Gregory (1971) found that the resistance to water flow of 1-in. sections cut from the lower stem of 1- to 2-yr-old seedlings of red oak *(Quercus borealis)* increased 8 days after inoculation with *C. fagacearum.* As time after inoculation increased, the resistance of lower stem sections increased and this increase traveled up the plant into the petioles. There was a good correlation between increased resistance to water flow in stem and petiole sections and the first appearance of the fungus in that particular part of the plant. At the time the leaves first wilted the fungus was present throughout the stem and also in some petioles (though not always those showing wilt symptoms). Final proof that vascular occlusion causes oak wilt must await a full examination of the water relations of leaves on infected plants.

Just such a comprehensive examination has been made of elms infected by Dutch elm disease, the disease that has periodically made such devastating attacks on the elms of northwest Europe and North America in the last half century. Infection of trees of a susceptible clone reduced the hydraulic conductance of internodal segments taken from 3- to 4-yr-old branches by 66% within 11 days of infection, compared with a reduction of 27% in segments taken from branches of trees of a resistant clone (Melching and Sinclair, 1975). When the transpirational behavior of single leaves on young plants was examined under controlled environmental conditions, it was found that the appearance of foliar symptoms was accompanied, or followed, by a decrease in transpiration rate (MacHardy and Beckman, 1973). When infected plants were transferred into the light, transpiration first increased as stomata opened, but then showed a series of alternate decreases and increases of diminishing amplitude. The reason for the initial opening being as large, or almost as large, as in the healthy plant was that the leaf had increased its water content during the dark period. Soon after opening, water loss exceeded uptake and some stomata closed. Then uptake exceeded loss until stomata began to reopen. The cycle was repeated until a balance between uptake and loss was established (Fig. 7). A similar phenomenon occurs in the transpirational behavior of banana plants infected with *Pseudomonas solanacearum* (Beckman *et al.,* 1962). It indicates that stomata function normally in both diseases; closure is a response to water stress, not a direct response to infection. This contrasts with the stomatal closure of rust- and powdery mildew-infected plants (see Sections VI,C and D) which results directly from infection.

In Dutch elm disease the relationship between the vascular anatomy of the host, the position of occlusions, and wilting has been examined. When movement of the water-soluble dye "light green" was followed after feeding to the cut end of infected stems it was found that the inten-

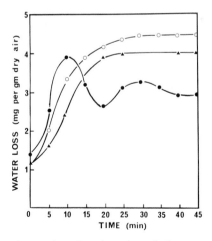

Fig. 7. The cause of water loss from branches of elm exposed to continuous light (beginning at time 0); after 9 hr of darkness. Non-inoculated control (○———○) *Ceratocystis ulmi*-infected branch 1 day (▲———▲) and 23 days after inoculation (●———●). (From MacHardy and Beckman, 1973.)

sity of the dye progressively diminished toward the shoot tip, whereas in healthy cuttings it was uniformly distributed. Major blockages to the movement of dye appeared to be located at the nodes of leaves, 1-yr-old twigs, and green shoots; these are exactly the sites at which constrictions in the vascular pathway caused radioactively labeled spores of *C. ulmi* to accumulate (Pomerleau and Mehran, 1966). As pointed out previously (Section IV,A,2), vascular occlusions are most serious when they occur nearest the leaf. However, leaves rarely wilted when the vascular elements of current shoots were healthy, even if the water-conducting elements of the older twig sections were extensively blocked, and studies of dye movement indicated that this was because the latewood vessels often remained functional in segments that were otherwise blocked (MacHardy and Beckman, 1973). In elm and other perennials there is the possibility of escape from disease by the annual production of new healthy wood.

Several studies have indicated that resistance to vascular disease, i.e., lack of symptom expression, may be associated with the particular anatomy of the vascular system of the host. Thus, Elgersma (1970), McNabb *et al.* (1970), and Sinclair *et al.* (1972) all found that Dutch elm disease-resistant trees of both *Ulmus americana* and *U. hollandica* possessed vessels of smaller mean diameter, fewer earlywood vessels, fewer contiguous vessels in transverse section, and a larger proportion of relatively short vessels than were possessed by susceptible trees. The hydraulic conductivity of stem segments was lower in resistant plants than in susceptible plants. It seems that the vascular anatomy of the resistant plants restricts the movement of both water and the pathogen and, in doing so, keeps the pathogen well away from the leaves.

Differences in hydraulic conductivity also exist among clones of sugarcane of different susceptibility to the bacterial disease "ratoon stunting." When water was sucked under vacuum through single nodes of the cane, an immune clone allowed a flow rate of $1.4 \text{ cm}^3 \text{ min}^{-1}$, three tolerant clones allowed rates of $2.6–8.4 \text{ cm}^3 \text{ min}^{-1}$, and susceptible clones allowed rates of $13.0–19.6 \text{ cm}^3 \text{ min}^{-1}$ (Teakle *et al.*, 1975). The number of vessels that are continuous across nodes is smaller in resistant than in susceptible clones and this probably limits the movement of both water and the pathogen in the resistant clones. It is interesting that clone CP29-116, which is tolerant of "ratoon stunt," is also resistant to "red rot" disease caused by *Physalospora tucumanensis*. As long ago as 1938 Atkinson attributed the red rot resistance of this clone to the presence of discontinuous vessels at the node which limited the spread of the pathogen. In wilt of alfalfa caused by *Corynebacterium insidiosum* resistance to the pathogen has also been attributed to the vascular anatomy of the host (Van Alfen and Turner, 1975b). The roots and stems of resistant varieties have

fewer vascular bundles and shorter vessel elements than those of susceptible varieties (Cho *et al.*, 1973).

To return to Dutch elm disease, the blocking of xylem elements is
probably caused not only by the cells of the pathogen but also by toxins
that it produces. In culture, *C. ulmi* produces at least two toxins: one is a
high-molecular-weight carbohydrate that is insoluble in water (Takai,
1974), the other is a glycopeptide of molecular weight between $0.5 \times 10^6$
and $2 \times 10^6$, containing 5% w/w protein (Van Alfen and Turner, 1975a).
The wilt-inducing properties of the latter have been examined by Turner
and Van Alfen, who found that while the transpiration rate of elm cuttings
was reduced by 22% if cuttings were placed for 3 hr in a solution of the
toxin containing 400 $\mu$g cm$^3$, a small, measurable decrease in stem and
petiole conductance occurred when only 0.01 cm$^3$, i.e., 4 $\mu$g of toxin, had
entered the stem. At the same concentration of toxin, cumulative
transpiration over 12 hr was reduced by 45%, compared with water controls; higher concentrations of toxin would reduce transpiration more
rapidly. Leaf water potential was also reduced by the toxin. No evidence
could be found that the toxin affected the permeability of cell membranes.
Electrolyte loss was the same from toxin- and water-treated leaf disks,
and the relationship between leaf water potential and stomatal diffusive
resistance was the same in both water- and toxin-treated cuttings. Thus,
no metabolic role for these toxins has been demonstrated, and they should
be regarded as toxins *sensu lato* and as distinct from the toxins described
earlier.

It was found that high-molecular-weight dextrans, substances which
it is believed cannot become metabolically integrated into elm cuttings,
mimicked the action of the toxin on stem and petiole conductance, and
their ability to do so was correlated with their molecular weight. Clearly,
if the toxin is produced *in vivo*, its large molecule can contribute to wilting
by blocking the pathways of water movement without impairing the permeability of membranes. The production of high-molecular-weight compounds by pathogens causing wilt diseases is quite common and the list of
pathogens that have this potential includes some of the species that have
already been described as causing vascular occlusion, e.g., *Ceratocystis
fagacearum* (McWain and Gregory, 1972) and *Verticillium albo-atrum*
(Keen and Long, 1972), but there has been no detailed examination of the
way in which these substances produce their effects.

## 4. Bacteria

There are about sixteen species of bacteria that may be regarded as
causing wilt diseases. These grade from primarily systemic vascular invaders to those pathogens that mainly attack parenchymatous tissue and

only cause wilt indirectly (Buddenhagen and Kelman, 1964). All the wilt-inducing bacteria, with the exception of *Corynebacterium sepedonicum* (on potato) and *C. michiganense* (on tomato), cause wilt by interfering with water supply to leaves of infected plants. They do this principally by multiplying within the vascular system of the host until it is blocked and, apart from their extra speed of multiplication, in this respect they resemble the fungal pathogens that have just been discussed. However, a number of pseudomonads and xanthomonads produce a sheath of slime around their cells which may become detached so as either to block the vascular system or to increase significantly the viscosity of the xylem sap and, thus, its rate of flow.

Strains of *Pseudomonas solanacearum* produce wilt disease on a wide variety of hosts including economically important crops, e.g., banana and members of the Solanaceae. The first clear demonstration that slime produced by a bacterium could contribute to wilt disease was made by Husain and Kelman (1958), who found that the hydraulic conductivity of tomato stems was reduced by infection but that wilted leaves would recover full turgor if they were excised and placed in water. A heat-stable, wilt-inducing polysaccharide of high molecular weight could be isolated from culture filtrates and was apparently similar to a substance that could be isolated from xylem vessels of infected plants. Only virulent strains of the pathogen produced the polysaccharide in culture. A detailed study of the water relations of infected or polysaccharide-treated tomatoes was not made, but in bananas infected by the same organism an examination of the pattern of transpiration of infected plants prior to wilting showed fluctuations, but no general increase in transpiration rate (Beckman *et al.*, 1962). The fluctuations were similar to those described before (see Section IV,A,3) in Dutch elm disease and also occurred in healthy plants if water was withheld. No adverse effect of the pathogen on guard cell function could be detected and it was concluded that wilting was due to the presence of bacterial cells and slime in the xylem.

Black rot disease of crucifers, caused by *Xanthomonas campestris*, differs from those vascular diseases discussed previously because the pathogen normally enters the plant via hydathodes at the margins of leaves rather than through roots or lower stems. However, it spreads from the hydathodes through the veins of leaves and causes V-shaped lesions delimited by the major leaf veins. An extracellular polysaccharide is produced by the bacterium in culture and a similar compound can be isolated from infected plants. It contains 0.04–0.08% organic nitrogen, and thus may be a glycopeptide. The material plugs the xylem vessels, as shown by the limited movements of eosin and $^{32}PO^{3-}$, and restricts the movement of water, causing localized reductions in relative water content and wilting (Sutton and Williams, 1970). Water stress appears to lead to chlorosis, the collapse of host cells, and loss of electrolytes; if water stress is prevented by keeping infected leaves in a saturated atmosphere, leaves remain green and turgid and maintain their electrolyte balance. In water-stressed infected leaves the bacteria multiply until they rupture the xylem vessels and are released into the parenchyma of the leaf.

*Corynebacterium insidiosum* causes wilt of alfalfa *(Medicago sativa)*. In culture the bacterium secretes a glycopeptide of molecular weight $5 \times 10^6$ and a similar substance can be isolated from diseased plants in amounts sufficient to induce wilting when reapplied to healthy plants (Ries and Strobel, 1972). When toxin or water was forced under 2 bars pressure through alfalfa stems the hydraulic conductance of the stems increased with the passage of toxin, but remained relatively constant with the passage of water (Van Alfen and Turner, 1975b). Toxin treatment did not increase electrolyte leakage from stem sections. When cuttings that had previously been in water were transferred to 200 $\mu$g cm$^{-3}$ of toxin they showed a decrease in transpiration and stomatal conductance compared with controls remaining in water. After 2.5 hr in the toxin, leaves were wilted, with a water potential of $-14.5 \pm 1.7$ bars as compared with $-7.9 \pm 1.0$ bars in turgid leaves of cuttings stood in water; this pattern of water relations, in which wilting is associated with reduced transpiration and normal stomatal functioning without gross changes in cell permeability, is by now the familiar characteristic of plants whose vascular system is occluded. Although Van Alfen and Turner found that a dextran of molecular weight $5 \times 10^5$ did not affect stem conductance in alfalfa, as it had in their studies of elm (Van Alfen and Turner, 1975a), they speculated that the toxin of *C. insidiosum* plugs the pores of bordered pit membranes between vessels, particularly where one vessel ends adjacent to another. Gradual plugging of these membranes would not only interfere with lateral transport of water, but also, because of the finite length of vessels, the upward flow of water through the stems would be reduced and embolism might result.

## B. WILT TOXINS

The suggestion of Dimond (1970) that some glycopeptides might prove the exception to the rule that vascular wilts are simply caused by occlusion of vascular elements has proved to be true in the case of some bacterial diseases. Glycopeptides that can cause wilt without blocking vascular elements have been isolated from cultures and plants infected by

*Corynebacterium sepedonicum* (Strobel, 1970) and *C. michiganense* (Rai and Strobel, 1969). The toxin from *C. sepedonicum* has a single highly branched oligosaccharide chain attached to a single peptide of 7–9 amino acids (Strobel *et al.*, 1972). The toxin, which has a molecular weight of 21,400, contains about 10% 2-keto-3-deoxygluconic acid, which appears to be vital for its biological activity since methyl esterification of these groups renders the toxin inactive (Johnson and Strobel, 1970). Studies by Strobel and Hess (1968) and Hess and Strobel (1970) found that the pattern of wilting produced by this toxin is recognizably different from that produced by toxins, organisms, or high-molecular-weight substances, e.g., dextrans, causing vascular occlusion because stem flaccidity is apparent before wilting of the leaves, a sequence of events that does not occur with vascular occlusion. Water movement throughout the plant, as indicated by movement of water-soluble dye, was unrestricted and no evidence of blockages could be found when the vascular system was examined in the electron microscope. The toxin appeared to have its effect upon the membrane system of the host for cells from infected plants lost the ability to plasmolyze and leaked electrolytes. Electron micrographs of tissues taken from infected plants showed that both cellular and organelle membrane systems bore evidence of disruption, often in advance of spread of the bacterium. The breakdown of membrane structure and function would be sufficient to prevent affected cells from absorbing and retaining water. This would probably cause wilting, though altered transpiration, which is an unexplored possibility, may be a contributory factor. The widespread nature of the wilt probably results from the ability of the toxin to move freely in advance of the bacterium. When [14]C-labeled toxin was supplied to test plants it was evenly distributed throughout the plant, forming an isolatable complex with a cell component of high molecular weight (200,000 or more), possibly a constituent of cell membranes.

Cultures of *C. michiganense* produce three phytotoxic glycopeptides, of molecular weight 200,000, 130,000, and 30,000; all produce wilt symptoms in tomato cuttings and immunological studies have shown the presence of similar glycopeptides in infected plants (Rai and Strobel, 1969; Strobel, 1974). Water-soluble dye moves freely through toxin-treated plants and [14]C-labeled toxin becomes generally distributed through the plant prior to wilting, though some concentration occurs at the leaf margin where the first wilt symptoms appear. This evidence, taken with the failure of microscopic investigation to find any blockages in the vascular system of wilted plants, strongly indicates that the toxin causes wilting by disturbing the normal structure and functioning of membranes, as does the toxin of *C. sepedonicum*.

## C. ACCELERATED SENESCENCE

A vascular pathogen that produces distinct wilt symptoms in one host may produce a very different set of symptoms, of which wilting forms only a small part, in another host. An example of this phenomenon has been seen already in the case of *V. dahliae* which causes wilting in chrysanthemum but development of spreading chlorotic flecks in sunflower. Talboys (1972) noted that other hosts of *Verticillium*, e.g., hop and raspberry, show a minimum of wilting and he noted that the two most consistent symptoms in a wide range of vascular diseases are chlorosis and the abscission of leaves. Since the pattern of symptom development parallels the normal pattern of senescence in these plants—it is sequential, and acropetal in direction—Talboys suggested that the induction of senescence might be fundamental to the wilt syndrome, though, presumably, not readily apparent in hosts showing marked wilting. Senescence is characterized by a decline in tissue protein and RNA levels as well as a loss of chlorophyll. It is associated with widespread permeability changes and is controlled by a complex series of interdependent changes in the levels of a number of hormones. Enhanced levels of ethylene production are involved in the changes but the exact role of ethylene is uncertain (Abeles, 1973; Lieberman, 1975). The timing of ethylene production appears to be critical; in several phenomena, such as abscission and fruit ripening, senescence appears to be initiated when small bursts in ethylene production occur, but the bulk of ethylene production does not occur until senescent events are well advanced. Thus, although Talboys (1972) found that ethylene production by stems of hops and petioles of strawberry, infected by *V. albo-atrum* and *V. dahliae*, respectively, was highest in segments showing disease symptoms, the ethylene cannot be regarded as the cause of senescence since no evidence was presented to show that enhanced levels of ethylene were present in tissues immediately before they senesced. The accumulation of ethylene in this case, and the accumulation of auxins that occurs in the tissue of many plants infected by vascular wilts may be effects rather than causes of the disease syndrome, for, as Sequeira (1973) has pointed out, hormonal changes almost invariably accompany the response of plant cells to pathogens.

MacHardy *et al.* (1974) have argued that the similarity between wilt-induced and natural senescence may be only superficial, arising out of the inability of any plant to express more than a narrow range of symptoms in response to a wide range of stimuli. In *Verticillium*-infected hops and strawberries there was no evidence that the events leading up to the appearance of chlorosis were the same as those occurring in natural senescence. In Verticillium wilt of chrysanthemum MacHardy *et al.*

(1974) did not find any evidence of changes in RNA and protein levels before wilting occurred. However, in their examination of Verticillium wilt of sunflower Robb *et al.* (1975b) found that chlorosis was well advanced before vessel blockage occurred, but they did not determine whether chlorosis arose from events resembling those of natural senescence.

The relationship between pathological wilt and senescence is hardly clarified if the levels of other hormones involved in natural senescence are examined. The senescence of leaves is generally retarded by cytokinins (Kende, 1971) but in tracheal fluid and leaf extracts of *Verticillium*- infected cotton plants normal levels of cytokinins were present when symptoms first became apparent. Levels of cytokinins only declined after chlorosis developed (Misaghi *et al.*, 1972). Abscisic acid (ABA) accumulates in healthy plants subjected to water stress (Vaadia, 1976), and in cotton infected by a defoliating strain (T9) of *V. albo-atrum*, ABA began to accumulate as defoliation began and continued to accumulate as defoliation spread upward within cuttings. Ethylene production by infected plants increased at the same time as ABA, and treatment of healthy plants with as little as 0.2 $\mu$l of ethylene per liter of air caused abscission similar to that caused by *V. albo-atrum* (Weise and DeVay, 1970). Here, there is at least correlative evidence linking altered hormone levels and senescence-type symptoms in a disease caused by a vascular pathogen, but the work may be faulted on the grounds that ethylene and ABA production were measured on whole plants enclosed in bags or jars. The enclosure of the plants may have altered ethylene levels and the soil in the pots probably contributed to the total ethylene present. Also, it is known that the hormonal relations of two sides of the zone of abscission differ significantly from each other and from those of the remainder of the plant. These objections have been largely overcome in the work of Pegg and Cronshaw (1976), who enclosed only attached leaves and stem internodes of tomato plants in ventilated Perspex chambers and measured ethylene production over short periods after inoculation with *V. albo-atrum*. However, Pegg and Cronshaw's results did little to resolve the problem of whether wilt fungi exert their fundamental effect by accelerating senescence. In support of the theory they found that when plants were inoculated with strain T179 (originating from tomato) susceptible plants showed a peak of ethylene production immediately following, or coinciding with, chlorosis and abscission, and resistant plants showed a smaller peak that also coincided with the more limited development of symptoms. Against the theory, they found that when the same plants were inoculated with strain H93P (originating from hop), to which all the plants were susceptible, there was no clear relationship between ethylene production and symptom development. Moreover, in susceptible plants infected by T179, ethylene production declined after symptoms began to appear, a pattern opposite to that occurring in senescent tissues (Abeles, 1973). Healthy plants did not produce ethylene when subjected to water stress until after they had passed the point of irreversible wilting and Pegg and Cronshaw concluded that the ethylene measured in infected plants resulted from damage to host cells.

When detached tomato leaves were preloaded with [86]Rb and exposed to ethylene before treatment with pectic enzymes or a polysaccharide fraction isolated from culture filtrates of *V. albo-atrum,* increased tissue permeability and drastic wilting followed. Since ethylene itself had no effect on tissue permeability, Cronshaw and Pegg (1976) suggested that ethylene acts as a toxin *(sensu lato)* synergist in Verticillium wilt of tomato. More information about permeability changes *in vivo* would be valuable but ethylene may well have an *in vivo* role in certain wilt diseases as Talboys suggested. However, there is little evidence that the complete senescence syndrome is fundamental to pathological wilting.

## V. SHOOT DISEASES

This section will discuss one disease, canker of peach and almond, the fungus *(Fusicoccum amygdali)* that causes the disease, and the host nonspecific toxin (fusicoccin) produced by the fungus. Attention is focused on this subject because the water relations of *Fusicoccum*-infected and fusicoccin-treated plants have been extensively studied in recent years and are probably better understood than those of most other diseases. Examination of this disease naturally follows the studies of vascular wilts, for although stem canker is the named symptom of the disease, the pathogen infects the vascular tissue of the host's stem and causes wilting in leaves. Thus it has much in common with vascular wilts. It will be seen that much is known about the way fusicoccin affects stomatal behavior to cause wilting of leaves; discussion of this information will provide a useful basis for comparisons when the subject of alterations in stomatal behavior caused by foliar pathogens is discussed in Section VI.

*Fusicoccum amygdali* infects hosts through buds or leaf scars. Spread of the fungus is very slow but within a few days wilting of leaves up to 40 cm from the infection site is seen. At this stage both bark and xylem are free of gums, which are soon produced in tissues adjacent to the infection area; hence wilting must result from a diffusible product of the fungus (Graniti, 1964). The rate of spread of the pathogen and its toxin slows down when gummosis sets in and the pathogen is localized. Cankers form when corky layers build up around the lesion. It has been known for several years that when the fungus is grown in submerged culture it produces several related compounds; the principal one is fusicoccin (Ballio *et al.*, 1964), which when applied to healthy host or nonhost plants induced wilt symptoms (Graniti, 1964). Only recently an active toxin was isolated from tissues of peach and almond infected with *F. amygdali* and shown to be fusicoccin (Ballio *et al.*, 1976).

From studies of the action of fusiccocin on several nonhost species by Turner and Graniti (1969), and of the action of the fungus or its toxin on almond (Turner and Graniti, 1976), there is agreement that the toxin increases stomatal opening in light and darkness and thus increases transpiration to such an extent that the leaves wilt because the transpired water is not replaced fast enough. Turner and Graniti (1969) simultaneously measured leaf diffusion resistances and water potentials in cuttings of dogwood *(Cornus florida)* that had been stood in distilled water overnight to allow full hydration of the tissues and then transferred to a solution of $10^{-5}M$ fusicoccin or water. The diffusion resistance of the abaxial surface of leaves showed an initial increase, but after 1 hr in fusicoccin it was lower than that of water controls and continued to decrease with time. It was concluded that fusicoccin stimulated stomatal opening since the diffusion resistance of the astomatous adaxial surface of the leaf was relatively constant and differed little from the resistance of the adaxial surface of control leaves. Silicone rubber impressions taken from leaf surfaces confirmed that fusicoccin stimulated stomatal opening. Leaf water potentials of cuttings stood in water varied little over 6 hr but the potentials of fusicoccin-treated leaves decreased from $-2.1$ bars to $-11.4$ bars over the same period; fusicoccin had induced a transpiration rate that exceeded the rate at which cuttings could take up water. When cuttings of tomato plants were treated with fusicoccin (ca. 2 $\mu$g/gm fresh weight), increased stomatal opening, accelerated transpiration, and wilting also resulted (Chain *et al.*, 1972). No evidence of impaired vascular flow could be found in fusicoccin-treated tomatoes; living cells were still apparently healthy, for they showed normal protoplasmic streaming, and the semipermeability of membranes appeared to be unaffected by the toxin, for there was no increase in solute leakage from cells after treatment. The effect of fusicoccin appeared to be limited to the stomata.

Turner (1972a) investigated the effects of fusicoccin on the movement of $K^+$ ions into and out of guard cells. In healthy plants the massive movement of $K^+$ into and out of guard cells that takes place when they are respectively illuminated or darkened, appears to play a central role in the change in turgor of the guard cells that alters their shape and the dimensions of the pores between them (Meidner and Willmer, 1975). Fusicoccin was found to stimulate stomatal opening and $K^+$ uptake by guard cells when strips of detached epidermis, taken from plants equilibrated in light or darkness, were incubated on a 10 $\mu M$ solution of fusicoccin, or control solution, and then examined immediately under the light microscope either for stomatal aperture or, after histochemical staining, for potassium (Fig. 8). Similar conclusions were reached when the epidermis was stripped from fusicoccin-treated or control plants, frozen immediately, and analyzed quantitatively in the electron microscope with an X-ray microprobe attached. In fusicoccin-treated and control leaves the concentration of $K^+$ in the guard cells was related by the same linear regression to stomatal aperture (Fig. 9). It is not known exactly how fusicoccin affects $K^+$ movement, but the observation that fusicoccin-stimulated stomatal opening was twice as fast in epidermal strips taken from plants held in the light as in strips taken from plants held in the dark, and that the effects of the toxin on stomatal opening could be reversed by substances such as 2,4-dinitrophenol, which interfere with energy metabolism of the cell, suggest that fusicoccin may act by interfering with the production of energy required to drive $K^+$ transport.

There is little evidence in the literature to suggest that fusicoccin may stimulate water uptake as well as transpiration. Lado *et al.* (1972) found that toxin-treated leaf disks of tomato, clover, and tobacco took up more water than control disks. Since the osmotic pressure values of treated tissues were lower than those of controls, the authors concluded that fusicoccin acts to irreversibly extend the cell wall. Ballio *et al.* (1971) found that fusicoccin and a number of closely related derivatives stimulated water uptake in cuttings of etiolated pea seedlings. Structural alterations of the molecule, such as deacetylation, did not destroy this property, whereas they did destroy the ability to cause wilt, which appears to be distinct from the ability to increase water uptake.

## VI. FOLIAR DISEASES

Wherever the water relations of plants infected by foliar pathogens have been investigated, an effect on transpiration has been found. It is reasonable to expect that the main effect of the pathogen will be localized about its site of infection. Probably for that reason most investigators have not looked beyond the leaf in their attempts to explain the alterations in water relations; the effects of foliar infections on root growth and, more importantly, root activity have largely been ignored, as have root–shoot hormonal interactions. Changes in transpiration have usually been explained in terms of altered stomatal behavior, often coupled with a change in the nonstomatal evaporative pathway as the pathogen grows through or on the surface of the leaf in order to sporulate. Fungi growing at the leaf surface also lose water and contribute to the total water lost by the diseased leaf. Thus, the changes that occur are temporally related to

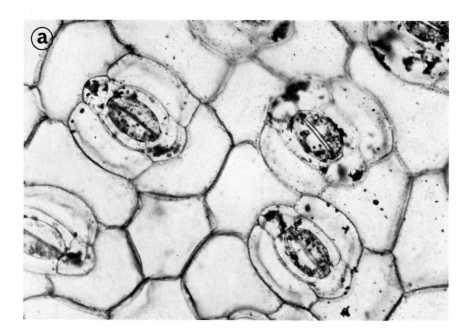

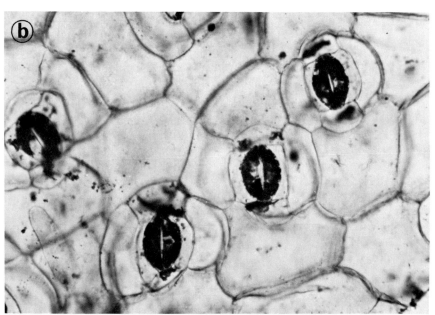

Fig. 8. Fusicoccin-stimulated $K^+$ uptake in stomatal guard cells. Epidermal strips of *Commelina communis* were incubated for 3 hr in a medium containing 1 m$M$ $K^+$, (a) without and (b) with 10 $\mu M$ fusicoccin, and then stained for potassium. (By courtesy of Dr. T. A. Mansfield.)

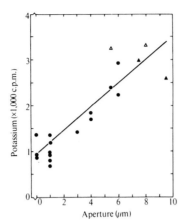

Fig. 9. Relation between concentration of potassium (cpm) in guard cells and stomatal aperture for leaves painted with 10 $\mu M$ fusicoccin in 0.02% v/v aqueous methanol and 0.05% v/v Triton X-100 in either light ($\triangle$) or dark ($\blacktriangle$), or 0.02% v/v methanol and 0.05% v/v Triton X-100 in both light and dark ($\bullet$). The concentration of potassium was measured as c./10s with an electron microprobe. (From Turner, 1972a.)

development of the pathogen: in fungal diseases the most critical change often occurs when the pathogen commences sporulation; e.g., in rust diseases transpiration is often reduced early in the infection cycle but increases after the pathogen ruptures the epidermis to release its spores. Changes in water relations are also spatially related to development of the pathogen since most infections are localized, at least initially, and the pathogen grows outward from a single infection site. Thus, tissue at different distances from the infection site will be in different stages of disease development. The last point is often overlooked by physiological pathologists, or avoided by the use of "blanket" inoculations that produce artificially uniform infections over the leaf surface. Because the general principle that altered water relations are intimately related to the growth pattern of the pathogen is particularly evident when foliar diseases are considered, the subject of foliar diseases will be dealt with by examining in groups those diseases that are caused by closely related pathogens. It will then be seen that similarities may be recognized between a variety of plants infected by pathogens that are closely related to each other.

## A. Peronosporales

*Phytophthora infestans,* the cause of potato blight, and *Peronospora tabacina,* the cause of blue mold of tobacco, are closely related members of the Peronosporales (Phycomycetes). When *P. infestans* attacks plants

via leaves (it may also spread from tubers) it grows intracellularly producing a rapid and drastic disorganization of host tissues. In colonized tissue
the fungus causes abnormal opening of stomata in the light, and failure of
stomata to close in the dark (Farrell *et al.*, 1969). The affected area, which
may be up to 7 mm wide, extends from the necrotic tissue at the center of
the growing lesion to within 1–2 mm of the advancing periphery of the
mycelium (Fig. 10). Although guard cells were not directly infected by the
fungus, their osmotic (solute) potentials were higher than normal. No
measurements were made of the osmotic pressure of the sap of epidermal
cells, but clearly the turgor relations between guard and epidermal cells
were altered. Other aspects of water relations of infected plants were not
examined. Increased stomatal opening would lead to increased transpiration unless the diffusion pathway through stomatal pores was sufficiently
blocked by the sporangiophores that emerge through them. Since blight is
favored by cool, wet conditions it seems unlikely that the altered stomatal
behavior would impose any significant water stress on the host, but it may
produce an increase in humidity close to the leaf surface sufficient to
stimulate sporulation, particularly at night. Cruickshank and Rider (1961)
suggested that a similar change takes place around tobacco leaves infected by *P. tabacina* as a result of the pathogen-induced increase in dark
transpiration.

   *Peronospora tabacina* is a less destructive pathogen than *P. infestans*
for it grows intercellularly and only haustoria invade living cells. The

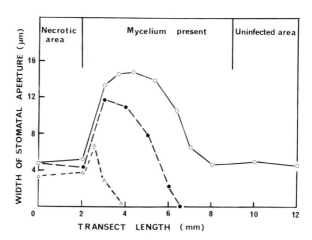

Fig. 10. The effect of increasing water deficit on width of stomatal apertures, measured
in belt transects through infections of *Phytophthora infestans* on three potato leaflets.
○——○, 0% water deficit; ●——●, 11% deficit; △——△, 38% deficit. (From Farrell *et al.*,
1969).

finding (Cruickshank and Rider, 1961) that in the presporulation phase *P. tabacina* causes a small increase in transpiration in the light and almost doubles the rate of dark transpiration (measured gravimetrically) strongly suggests that it produces an effect in stomata similar to that caused by *P. infestans*. Unfortunately, no measurements of stomatal aperture were made. At the end of the presporulation phase, and in the postsporulation phase of infection, transpiration from the diseased tissue decreased to less than that of healthy tissues, the exact ratio being controlled by the environment. The cause of the decrease is unknown, but it may have been the blocking of stomatal pores by the emerging sporangiophores or collapse of the tissue. No measurements of tissue water content were made but, since leaf expansion depends inter alia on the uptake of water and it was noted that leaf growth was almost zero in diseased plants in the presporulation phase, significant water stress may well have developed within diseased plants at this stage.

The work of Cruickshank and Rider provides a unique analysis of the energy balances of healthy and diseased leaves. Leaf and air temperatures were monitored by a series of fine wire thermocouples and it was found that the energy requirement necessary to sustain the observed transpiration rate from healthy leaves was provided in the main by radiative exchange; there was a small contribution of heat directly from the air, and probably a little from respiratory processes. In the diseased leaf only about half the energy required to sustain measured rates of transpiration in the presporulation phase could be supplied by radiative exchange. It was calculated that about 10% of the excess energy requirement could come from respiration, which as a general rule is more active in diseased than in healthy plants, but the origin of the remaining 40% of the excess energy requirement remains a mystery.

## B. Subcuticular Parasites

Some foliar pathogens grow initially in a subcuticular position, between the cuticle and the outermost layers of the epidermal cell wall of their host. The epidermis and underlying tissues are only colonized when an extensive subcuticular mycelium has developed. *Rhynchosporium secalis*, the cause of barley leaf blotch, is an imperfect fungus with a subcuticular phase during which it causes stomata within the infected area of the leaf to open more widely than normal in the light and to close incompletely in the dark or in high concentrations of $CO_2$. Changes in stomatal behavior become more acute as infection progresses and are associated with altered rates of transpiration (Fig. 11) (Ayres, 1972; Ayres and Jones, 1975). Freezing point determinations showed that as infection develops,

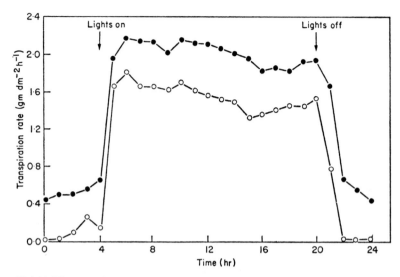

Fig. 11. The transpiration of barley plants, healthy (○) and infected 6 days with *Rhynchosporium secalis* (●), over 24 hr including a photoperiod of 16 hr. (From Ayres and Jones, 1975.)

the turgor pressure of guard cells and, more particularly, epidermal cells of diseased leaves fell below that of similar cells in healthy leaves. The drop could lead to increased stomatal opening. Although the turgor pressure surplus of guard cells over epidermal cells was not increased, Glinka (1971) has shown that at different levels of turgor pressure in epidermal cells a fixed surplus can lead to different stomatal apertures, with a maximum response when the epidermal cells are at about incipient plasmolysis. Solutes were probably lost from epidermal cells because of a change in their membrane permeability, for when attached leaves were bathed in deionized water electrolyte leakage was greater from diseased than from healthy plants (Jones and Ayres, 1972). A study of diseased tissue by electron microscopy revealed one likely cause of altered permeability; the normal close association between plasmalemma and cell wall was often disrupted in regions adjacent to subcuticular hyphae (Jones and Ayres, 1974).

Since barley leaf blotch, like potato blight discussed previously, is favored by cool, wet conditions it is not obvious how significant altered stomatal behavior would be in inducing any degree of water stress under field conditions, but the increased transpiration in diseased leaves led to an accumulation of root-absorbed solute, $^{86}$Rb, at the infection site. Increased rates of water loss in diseased plants could well have the important role of facilitating the accumulation of solutes at infection sites, that occurs in many diseases (Durbin, 1967). Finally, on the subject of sub-cuticular parasites, it has been reported that *Spilocaea (Cycloconium) oleagina,* the cause of olive leaf spot, causes increased transpiration in the host (Larcher, 1963). In view of the Mediterranean distribution of the host it would be interesting to know more about the mechanism of this in-crease and whether it leads to water stress.

## C. Rust and Smut Diseases

Rust diseases occur worldwide on a variety of angiosperms, gymno-sperms, and pteridophytes, often with serious economic consequences. Rust fungi belong to the order Uredinales (Basidiomycetes), are obli-gately biotrophic (*sensu* Lewis, 1973), and frequently have complex life cycles involving up to five different spore types and sometimes two unre-lated hosts. Studies of the water relations of rusted plants have dealt with the uredospore stage, for the uredospore is the asexually produced "re-peating" spore that gives rise to epidemics among crop plants. Uredo-spores of many rusts, e.g., *Puccinia graminis,* f.sp. *tritici* (stem rust), form germ tubes that penetrate the leaf via stomatal pores. An intercellular mycelium with intracellular haustoria is established, and at a suitable time the colony ruptures the host's epidermis to release more reddish/brown uredospores (or teleutospores at the end of the growing season) (Fig. 12).

The water relations of rusted plants have been intensively investi-gated since the beginning of the century; the results of the major investi-gations are summarized in Table I. There is widespread agreement that rust diseases eventually cause an increase in transpiration per unit area of leaf. However, rust diseases also reduce leaf area development and dry matter accumulation to a degree that depends on the growth stage at which the host is first infected; this reduction often causes the total amount of water taken up or lost by rusted plants to be less than that taken up or lost by healthy plants, whether calculated on a daily or (see Fig. 13) a lifetime basis (Johnston and Miller, 1934, 1940; Murphy, 1935; Bever, 1937). The increase in transpiration per unit area is due to nonstomatal water loss because it only occurs after the rust pustule erupts, is more marked with reference to healthy plants at night than in the day (Johnston and Miller, 1940), and can occur when stomatal apertures are smaller (by a factor of two) than in healthy leaves (Duniway and Durbin, 1971a). The contribution that the fungus makes to total transpiration in sporulating leaves is unknown. It is a remarkable fact that only one paper (Yarwood, 1947) of any significance has been published on the subject of transpira-tion from loosely organized, i.e., nonstromatic, fungal mycelia. Yarwood

## TABLE I
### Major Investigations into the Water Relations of Rust-Infected Plants, 1900–1976

| Host plant/rust | Effects on water relations and associated factors | Reference |
| --- | --- | --- |
| *Rubus/Gymnoconia interstitialis* | Heavily infected shoots took up twice as much water as healthy shoots, yet wilted while healthy shoots were turgid | Blodgett (1901) |
| Clematis, violet, rose of Sharon, rye, and rose/unspecified rusts | Rust infected twigs and shoots lost more water than comparable healthy tissues | Montemartini (1911) |
| Apple/*Gymnosporangium juniperi-virginianum* | *In situ* measurements showed that rusted leaves lost less water than healthy tissues in all but the very latest stage of infection | Reed and Cooley (1913) |
| Wheat, rye, barley, oats, corn/stem and leaf rusts | Rusted cereals in sealed pots lost more water per unit leaf area than healthy plants | Weaver (1916) |
| *Xanthium, Helianthus, Dianthus*/unspecified rusts | Rusted dicotyledonous plants only lost more water than healthy plants after sporulation commenced | Weaver (1916) |
| Wheat/*Puccinia graminis* f.sp. *tritici* (stem rust) Wheat/*Puccinia triticina = recondita* (leaf rust) | Stem rust increased total amount of water lost by plants in sealed pots and reduced dry weight of grain; leaf rust had no effects | Weiss (1924) |
| Wheat/*P. graminis tritici* | Water-soluble light green dye accumulated where pustules ruptured the host epidermis, indicating increased transpiration in that area | Harvey (1930) |
| Wheat/*P. triticina = recondita* | Plants grown in pots; rust increased water loss per unit dry weight of leaf or grain, but total water loss per plant was slightly reduced due to reduction in plant growth; root growth was most affected by rust; Severity of effects decreased as plant age-at-infection increased | Johnston and Miller (1934) |

38

| Host/Pathogen | Observations | Reference |
|---|---|---|
| Oats/*P. coronata avenae* (crown rust) | Plants grown under field conditions; results as for wheat/*P. triticina* [Johnston and Miller (1934)] | Murphy (1935) |
| Wheat/*P. glumarum* = *striiformis* (stripe or yellow rust) | Heavy infection increased water loss | Gassner and Goeze (1936) |
| Wheat and barley/*P. glumarum* = *striiformis*; Wheat/*P. triticina* = *recondita* | As wheat/*P. triticina* [Johnston and Miller (1934)]; Increased transpiration caused by rust was greater in the dark than in the light | Bever (1937); Johnston and Miller (1940) |
| Bean/*Uromyces phaseoli* | Transpiration per unit leaf area reduced by rust until sporulation, when it increased, particularly in the dark; stomata over unopened pustules were closed in the light | Yarwood (1947) |
| Bean/*U. phaseoli* | Transpiration per plant and per unit dry weight of leaf reduced by rust; transpiration increased to almost healthy rates when sporulation began; root growth inhibited by rust | Gerwitz and Durbin (1965) |
| Oats/*P. coronata avenae* | Seedlings used; rust increased water loss after sporulation began; greatest increases were during dark periods; root growth strongly inhibited by rust | Amatya and Jones (1966) |
| Bean/*U. appendiculatus* | Transpiration per gram water content of plants was reduced until sporulation began when it exceeded transpiration in healthy plants; stomatal movements gradually diminished, aperture remains partly open | Sempio *et al.* (1966) |
| Wheat/*P. recondita* f.sp. *tritici* | Plants in a potometer; the difference between transpiration and absorption was greater in rusted than in healthy plants | Parodi and Bitzer (1969) |
| Bean/*U. phaseoli* | Stomatal aperture and transpiration in light reduced before sporulation; transpiration increased, particularly in dark, after sporulation; root growth reduced by rust | Duniway and Durbin (1971a,b) |
| Wheat/*P. recondita* | As wheat/*P. triticina* [Johnston and Miller (1934)] | Van der Wal and Cowan (1974); Van der Wal *et al.* (1975) |

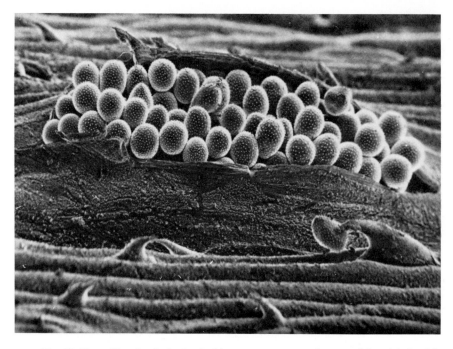

Fig. 12. The epidermis of a barley leaf is torn open as a uredosorus of *Puccinia hordei* (brown rust) erupts. Surface view of fresh frozen specimen. ×:600. (By courtesy of Rothamsted Experimental Station.)

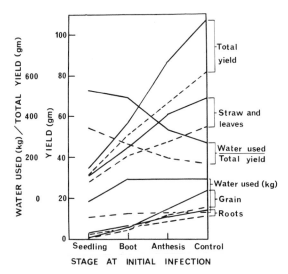

Fig. 13. Yield and water relations of oats grown at 85% (————) and 50% ( - - - - - - - - - - - - - ) soil moisture content and initially infected with crown rust at seedling, boot, or anthesis stage. (Modified from Murphy, 1935.)

concluded *tentatively* that "there is relatively little water loss from the fungus itself." This serious lack of information about the transpirational characteristics of phytopathogenic fungi applies in all diseases where nonstromatic mycelium grows on the surface of the host. Many experiments with cereal rusts have examined water relations over long periods and under conditions enabling several generations of the pathogen to be completed (e.g., Johnston and Miller, 1934; Murphy, 1935), with the result that the effects of the early stages of rust development on water relations have been masked by the presence of old infections. Recent research on bean rust has studied water relations over a single cycle from infection to sporulation. It has been found that infection initially causes a decrease in stomatal opening and transpiration (Figs. 14 and 15). The cause of altered stomatal behavior is unknown, but a small contributory factor to reduced transpiration may be the blocking of stomatal pores by fungal infection structures (Wynn, 1976). The reduction in transpiration from apple caused by *Gymnosporangium juniperi-virginianum* was attributed to a collapse of the mesophyll and air spaces within the leaf resulting from infection (Reed and Cooley, 1913).

Inspection of Table I shows that root growth is reduced by many rust diseases. There has been no quantitative investigation of the effect that this reaction has on the ability of roots to take up water, but it would seem

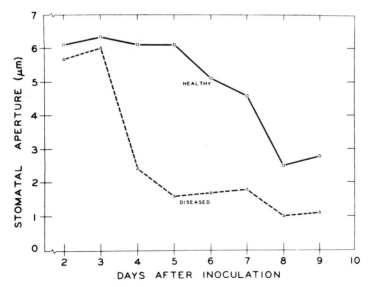

Fig. 14. Average stomatal apertures of healthy and rusted bean leaf disks in the light on different days after inoculation. A difference of $1.1\mu$ between apertures indicates significance at the 1% level. (From Duniway and Durbin, 1971a.)

Peter G. Ayres

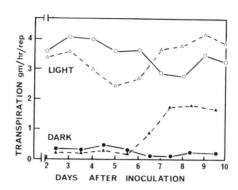

Fig. 15. Transpiration rates of healthy (———) and rusted ( - - - - - - - ) bean leaves on different days after inoculation. (Modified from Duniway and Durbin, 1971a).

to be seriously affected; Bushnell and Rowell (1968) found that severe infection of wheat by *P. graminis tritici* caused a decline in growth and $CO_2$ evolution per unit volume of root system and was associated with the premature death of adult plants. Severe infection at the heading stage caused abrupt desiccation and death of shoots about 10 days before ripening time whether plants were grown in soil or solution culture. In the latter medium death was preceded by a slower decline in the rate of $CO_2$ evolution. Rust of moderate severity did not lead to premature death but did keep the rate of $CO_2$ evolution by roots from increasing at the same rate as that of roots of healthy plants. The decline of roots in plants infected by rusts and other foliar diseases is probably due to carbohydrate starvation similar in effect to that suffered by roots infected by "take-all" lesions (see Section III).

A decline in root growth and activity directly reduces the capability of the root system to absorb water, but it may affect the water relations of plants in other indirect ways. The ability of roots to absorb the inorganic solutes that are needed for osmoregulation may be impaired. Cytokinins are synthesized in roots (Kende, 1971), and these and other hormones such as abscisic acid mediate root–shoot water relations (Collins, 1974). The decline of roots may be a contributory cause of the altered cytokinin levels found in rust-infected beans (Király *et al.,* 1967).

Although smut fungi, which belong to the order Ustilaginales, are closely related to rusts and, like rusts, are obligate biotrophs causing economically serious diseases of a wide range of Angiosperms, all that is known of their effects on host water relations is contained in two old reports (Kourssanow, 1928; Nicolas, 1930) which show that smuts cause increased transpiration on various hosts.

## D. POWDERY MILDEWS

Powdery mildew fungi belong to the order Erysiphales (Ascomycetes). They have certain similarities to rust and smut fungi as they are obligate biotrophs causing economically serious diseases worldwide on a variety of hosts. However, they are unlike other phytopathogenic fungi because they grow entirely on the surface of host tissues, except for the haustoria that they insert into host epidermal cells. Asexual conidiospores are produced rapidly, often within 3 to 4 days of infection, and in such large numbers that infected tissues take on a powdery appearance (Fig. 16). The water relations of plants attacked by these fungi are especially important, for powdery mildews appear to be favored by warm, dry conditions (Boughey, 1949).

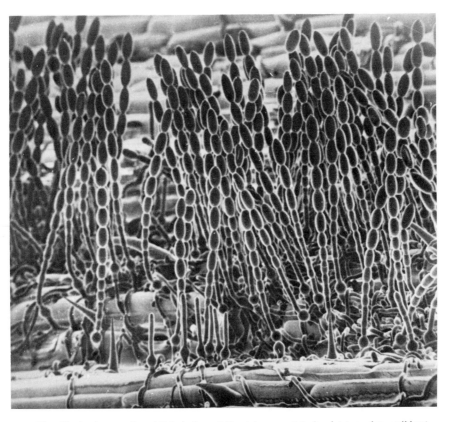

Fig. 16. A cluster of conidial chains of *Erysiphe graminis hordei* (powdery mildew) extends about 0.25 mm above the epidermis of a barley leaf. Younger chains and surface hyphae may be seen in the foreground. Surface view of fresh frozen specimen. ×:500. (By courtesy of Rothamsted Experimental Station.)

The two powdery mildews whose water relations have been most closely studied are pea mildew, caused by *Erysiphe pisi*, and barley mildew, caused by *E. graminis* f. sp. *hordei*, and infection seems to cause similar changes in both hosts. Although Graf-Marin (1934) reported that powdery mildews caused increased stomatal opening in the light, as measured by direct observation of pieces of detached epidermis, continuous recordings of stomatal movements, made on attached leaves with a resistance porometer, show that as the primary infection cycle progresses there is a reduction in the ability of stomata to open in the light and close in the dark (Ayres, 1976; Majernik, 1971). Eventually stomata remain fixed in a partially open state (Fig. 17a). In pea, and also in one of the barley varieties (Dvoran) used by Majernik, diminution of stomatal movements was preceded by a transient increase in the opening response to light. The cause of altered stomatal behavior is unknown. Martin *et al.* (1975) reported that in wheat infected by *E. graminis* f. sp. *tritici* stomatal opening and also transpiration were reduced within 6, and possibly within 3 hr of inoculation; they suggested that a volatile product of the fungus was responsible for the effect.

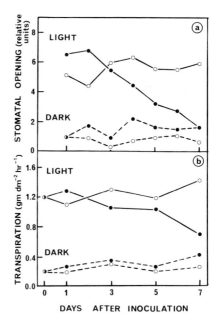

Fig. 17. Stomatal opening (a) and transpiration (b) in pea leaves in light or darkness following infection by powdery mildew. Each point is the mean of four replicates. (O) Healthy, (●) infected. (Modified from Ayres, 1976.)

Transpiration per unit of leaf fresh weight or leaf area follows closely the pattern of stomatal behavior in both pea (Fig. 17b) (Ayres, 1976) and barley (Paulech and Haspelova-Horvatovicova, 1970). Transpiration decreases in the light and increases in the dark with the result that total transpiration is reduced in a 24-hr cycle of 16 light and 8 dark hr. Reduced transpiration in the light results principally from reduced stomatal opening, though a contributory factor is the increase in the diffusion resistance of the leaf boundary layer caused by the presence of the network of mycelium over the leaf surface, for when the mycelium was removed from pea leaves there was an immediate increase in transpiration of 12.37% (Ayres, 1976). It is noteworthy that as long ago as 1895, Müller-Thurgeau suggested that reduction in transpiration of grape infected by powdery mildew *(Peronospora viticola)* was due to blockage of stomatal pores by the fungus. Increased transpiration in the dark results in part from incomplete stomatal closure [Paulech and Hespelova-Horvatovicova (1970) ignored this factor and, incorrectly, called dark transpiration "cuticular" transpiration] and partly from the mycelium itself. It is unlikely that the truly transcuticular pathway of transpiration is increased since there is no electron microscopic evidence to suggest that the structure of the cuticle outside penetration points is altered by powdery mildew fungi.

The transpirational behavior of oak *(Quercus robur)* infected by powdery mildew *(Microsphaera alphitoides)* differs from that of mildewed pea and barley. Although mildew reduced the mean area of stomatal pores in oak leaves from 4.65 to 3.01 $\mu m^2$ 6 days after infection, infected leaves transpired at a faster rate than healthy leaves in both light and darkness (Hewitt and Ayres, 1975). Two reasons for the difference between oak and other mildews may be that, first, oak has a comparatively low ratio of stomatal to cuticular transpiration, which means that total transpiration would be less affected in oak than in other species by factors altering stomatal opening. Second, the proportion of the total water loss from mildewed leaves that occurs through the fungus probably is greater in oak than in other species since removal of the mycelium reduced transpiration in light and darkness to control levels, whereas in pea it increased transpiration, as noted earlier. The relative water content of mildewed oak leaves fell below that of healthy leaves by 5.7% and 10.8% at 4 and 6 days after infection, respectively, but the effects of increased transpiration on leaf water status were somewhat moderated, since 6 days after infection diseased leaves had slightly higher water potentials than healthy leaves at any value of relative turgidity. This shift in the relationship between relative water content and leaf water potential suggested that infected tissues began to lose solutes about 6 days after infection. Measurements were not taken beyond 6 days, but the observation that desiccation was not a characteristic symptom of mildew on fully expanded oak leaves suggested that altered water relations were not a primary cause of their premature death.

Premature desiccation of leaves is a characteristic symptom of pea mildew in spite of the reduction in transpiration caused by the pathogen. Seven days after inoculation diseased leaves had a higher water potential ($-2.7 \pm 0.2$ bars) and relative water content ($91.56 \pm 1.73\%$) than healthy leaves ($-3.9 \pm 0.3$ bars, $86.25 \pm 0.95\%$, respectively), but then they rapidly wilted and desiccated, even when held in a saturated atmosphere (Ayres, 1977a). The cause of wilting seemed to be a loss of solutes from within cell membranes since wilting was coincident with a rapid increase in the tissues' permeability to electrolytes. Earlier, more localized changes in tissue permeability might result in altered turgor relations between guard cells and epidermal cells and be a contributory cause of altered stomatal behavior. The phytoalexin, pisatin, may have a key role in altering tissue permeability since, within 4 days of infection, it accumulates to a concentration, $300\ \mu g\ gm^{-1}$ fresh weight, that damages the plasmamembrane of isolated pea protoplasts.

In spite of the exhaustive attention of researchers at the Slovak Academy of Sciences, the water status of mildewed cereal tissues is far from clear. Certain results do suggest, however, that changes in mildewed barley resemble those in mildewed pea. During 7 days of the primary infection cycle the amount of water taken up per gram of dry matter accumulated by seedlings in potometers was 18% greater in diseased leaves than in healthy ones (Priehradny, 1971, Table 3), suggesting that the relative water content and water potential of diseased leaves increased above those of healthy leaves. Majernik (1965), Paulech and Haspelova-Horvatovicova (1970), and Priehradny (1975; see this paper for similar publications by the same author) all presented evidence to show that transpiration rates per plant, or grams of water in the plant, increased sharply above those of healthy plants 8 days or more after infection. The increase was attributed by Majernik to increased cuticular transpiration, but it is difficult to see why this should suddenly increase when the fungus has been sporulating since the third day after infection. An alternative explanation is that the permeability of infected tissues to solutes and water increased at this time. In barley, discrete pustules are formed, unlike the more uniform infection of pea leaves, and the healthy tissues around the pustules may be able to supply enough water to sustain a short burst of increased transpiration. The absence of obvious wilting in mildewed barley, in contrast to mildewed pea, probably also reflects the greater structural strength of the cereal leaf. Nevertheless, Majernik noted that "increased water output characterizes the end of the disease process as it causes irreversible drying."

Decreased root growth can contribute to the reduction in transpiration and eventual irreversible drying of mildewed tissues. At the beginning of sporulation, 4 days after infection, barley roots showed reduced elongation, inhibition of lateral root growth and a reduction in the number and size of xylem vessels (Vizárová and Minarćic, 1974). At the same time mitotic divisions at root tips were reduced in number (Minarćic and Paulech, 1975), respiration declined, and [14]C-labeled photoassimilates leaked from the roots (Frič, 1975). Undoubtedly these changes would reduce the capacity of roots to take up and transport water and solutes to the leaves. Changes in root structure and function are attributed by the authors to the increased level of cytokinins that may be detected in mildewed plants (Vizárová, 1974), although, since cytokinins are synthesized in roots, the altered levels might be an effect of the changes as much as a cause. Changes in levels of hormones such as cytokinins and abscisic acid, which have a role in regulating the water relations of healthy plants, e.g., by regulating stomatal opening or permeability of membranes to water and ions (Vaadia, 1976) and the effect that these might have in mildewed, rusted, and other diseased plants, is a subject ripe for exploration.

E. VIRUSES

Although the majority of virus infections have a systemic element, wilting is rarely a primary symptom of virus disease (Holmes, 1964). A notable exception is wilt of tabasco peppers which, as described previously, results from an alteration in permeability of root cells. Where viruses have been observed to alter the water relations of their host, transpiration usually was reduced by infection, e.g., in potato leaf roll (Muller, 1932), bean mosaic (Harrison, 1935), tobacco mosaic (Gondo, 1935), and barley yellow dwarf (Orlob and Arny, 1961) virus diseases. Selman (1945) reported that tobacco mosaic and spotted wilt viruses increased transpiration from the host by altering the structure of the cuticle. The results were based on experiments that used the cobalt chloride paper method of measuring transpiration, which the author admitted was prone to error, and it is important to note that in the same paper (and this has often been overlooked) gravimetric experiments are reported that showed that infection reduced transpiration in both light and dark during the first three days after infection. At this stage the virus could have had little effect on leaf area. Schuster (1957) reported that several mosaic-type viruses increased host transpiration but the mechanisms responsible for the increase were not detailed.

It is common for viruses to induce curling or twisting of leaves or reduction in leaf size, shape, or attitude (Holmes, 1964); all these factors may reduce water loss from the plant, but in diseases such as potato leaf roll (Muller, 1932) and maize dwarf mosaic virus (Lindsey and Gadauskas, 1975) transpiration is also reduced because infection reduces stomatal opening in the light. Reduced stomatal opening in maize was associated with the inability of guard cells to accumulate $K^+$ ions on illumination. Lindsey and Gadauskas attributed this to the failure of photosynthesis to supply enough ATP to drive active transport of $K^+$. Similar changes affecting stomatal behavior were noted previously (see Section V) in leaves treated with the fungal toxin fusicoccin. The onset of reduced transpiration in maize was coincidental with the first appearance of symptoms and infected plants had higher water potentials ($-5$ and $-4$ bars in fourth and fifth leaves, respectively) than healthy leaves ($-9$ and $-6$ bars in fourth and fifth leaves, respectively) when stress was induced by the cessation of watering.

Leaves of tomato infected by aspermy virus, which causes dark-green mottling and distorted growth, also had higher water potentials ($-10$ bars) and relative water contents (83.5%) than healthy controls ($-11.1$ bars and 80.9%) when plants were in a nonsaturated atmosphere (Tinklin, 1970). However, when transferred to a saturated atmosphere, the water potential of healthy leaves increased to $-2.9$ bars and relative water content rose to 96.7% while the water potential and relative water content of infected leaves only rose to $-9.3$ bars and 89.2%, respectively. Tinklin suggested that the virus caused permeability changes in the leaf which prevented water uptake; however, since stomatal behavior was not affected by infection, it is difficult to see why infected plants did not have a much lower water potential than healthy plants when water uptake was impaired. The explanation may be that water potential was measured by the Shardakov method (see Barrs, 1968) which would give erroneous values when solute leakage occurred from diseased tissues as the result of permeability changes. The possibility remains that some viruses cause permeability changes in their host.

## VII. COMBINED EFFECTS OF WATER STRESS INDUCED BY DISEASE AND ENVIRONMENT

The studies described so far have dealt mainly with well-watered plants maintained under laboratory conditions. Many aspects of water relations have been investigated but associated measurements of the water status of the tissue are often absent. Fewer investigators have measured water stress under simulated, or actual field conditions, or considered the possible combined effects of environmentally imposed and disease-imposed water stress. A picture of such combined effects has to be compiled by inference from laboratory studies and the meager direct information that is available.

In powdery mildew and rust diseases and in potato blight, stomatal movements were totally inhibited as infection progressed. If stomata were fixed in an open position, then, under conditions of environmental stress, the plant would be unable to check water loss through this pathway by closing its stomata in the normal way. Stomata around potato blight lesions were unable to close in the normal way in response to leaf water deficits of 11 and 38% (see Fig. 10) (Farrell *et al.*, 1969). Stomata of barley infected with leaf blotch (Ayres, 1972) and tomato infected with leaf mold, caused by *Cladosporium fulvum*, also showed increased opening and stomata of tomato failed to close when leaves were exposed to high temperatures that induced closure in healthy leaves (Dvoretskaya *et al.*, 1959). The presence of permanently open stomata in potato blight and barley leaf blotch lesions may not adversely affect the water relations of the whole plant since these diseases occur under cool, wet conditions when soil water would be readily available. Failure of stomatal closure would be much more serious for powdery mildew diseases that occur under hot dry conditions. Mildewed peas transferred from Hoagland's solution to a rooting medium with an osmotic potential of $-3.0$ bars (Hoagland's solution plus polyethylene glycol 4000) 4 days after inoculation were desiccated by the seventh day after inoculation, whereas plants remaining at the original water potential of $-0.5$ bars (Hoagland's solution) did not desiccate until the tenth day (Ayres, 1977a). Thus, environmental stress can exacerbate the effects of pea mildew. However, environmentally imposed water stress, measured as a decline in leaf water potential, also checks growth of the pathogen (Ayres, 1977b). Clearly, the combined effects of mildew and environmental stress will be complex in the field.

Duniway and Durbin (1971a) found that bean rust totally inhibited the movements of stomata which became fixed in a closed position. Nevertheless, infected plants were unusually susceptible to drought after sporulation because they could not control water loss through the ruptured cuticle. Such plants wilted (leaf water potential, $-14.9$ bars) when the water potential of the vermiculite rooting medium was below $-1$ bar, whereas healthy plants (leaf water potential of $-7.8$ bars at a root-medium potential of $-1$ bar) did not wilt until the potential of the rooting medium fell below $-2.5$ bars. The authors considered that reduction of the root–shoot ratio also upset the water economy of rusted beans under conditions of even mild drought.

The investigation by Murphy (1935) of the effects of crown rust on oats growing at two soil moisture levels showed that the rust had proportionally less effect on host physiology at the lower than at the higher soil moisture level, but that disease and water stress had additive effects (Fig. 13). Recent work at Wageningen in The Netherlands has shown that in leaf rust *(Puccinia recondita)* of wheat the relationship is complex between stresses imposed on the plant by disease and by a lack of soil moisture. Plants were grown and infected at three soil water potentials, $-10.25$ bars (dry), $-4.25$ bars (medium), and $-2.50$ bars (wet). At the lowest water potential plant development was faster, growth rate was lower, and transpiration per plant was less than at higher potentials (Van der Wal *et al.*, 1975), which agrees with field observations that in hot, dry years rusted plants show a drought-induced early maturity (Greenall, 1956; Hartill, 1961). Yield in healthy cereals is somewhat more protected from the effects of water stress than other processes such as leaf expansion and photosynthetic activity (Wardlaw, 1967; Boyer, 1976) and the same appears to be true of rusted plants, although there were considerable losses of both total dry weight and yield in rusted plants grown at $-10.25$ bars. Considerable losses also occurred at $-2.50$ bars, but at $-4.25$ bars the growth and yield of rusted plants was only slightly less than that of uninfected controls. It was concluded that this was because host resistance was related to soil water potential and was most fully expressed at $-4.25$ bars. Thus, at the highest and the lowest soil water potentials rapid development of the fungus and severe water stress were, respectively, the overriding factors affecting the plant. Leaf water potentials, obtained indirectly by calculation from other measurements, decreased with decreasing soil water potential and also with time after heading in both rusted and healthy plants (Cowan and Van der Wal, 1975). Rusted leaves had slightly lower water potentials than leaves on healthy plants after heading but the differences were small ($-11.40$ bars for rusted and $-10.25$ bars for healthy, both at a soil water potential of $-4.25$ bars). The only significant difference between plants at preheading stages occurred in plants grown at a soil water potential of $-2.10$ bars; rusted leaves had a water potential of $-24.8$ bars compared with $-5.80$ bars in healthy leaves. The large difference in this instance was attributed to the particularly luxuriant development of the rust under conditions of high soil water potential. The effects of *Puccinia recondita* on root growth were very small in the Wageningen experiments, probably because the plants were grown in containers that limited root growth.

     Murphy (1935) emphasized the importance of the root-to-shoot ratio under field conditions: "Heavy epiphytotics of crown rust are usually preceded by a period of wet weather with attendant high humidity and soil moisture. These conditions, while highly favourable for rust, also favour top growth and root development suffers. Then as rust appears and increases, the ratio of roots to tops is still further decreased by the effect of the fungus." Such a plant is at a great disadvantage when, as is often the case later in the season, deficient soil moisture becomes a limiting factor. The lack of sufficient root system directly reduces the ability of the plant to obtain moisture, while at the same time its water requirement is greatly increased as a result of the rust infection.

It may be seen from the examples of wheat rust and pea mildew that the combined effects of pathogen-induced and environment-induced water stress are confounded by the effects of environment on development of the pathogen. These effects must always be taken into account when the water relations of diseased plants are considered. The effect of environmental conditions on predisposition to disease, and the subsequent development of disease are discussed in Chapters 2 and 5, respectively, of this volume.

## VIII. CONCLUSIONS

Disease alters the water relations of plants in a direction that can result (and in many cases has been demonstrated to result) in water stress in part or all of the aerial parts of the plant. Exceptions to this rule appear to be powdery mildew diseases in the early stages of infection and several virus-induced diseases. In these diseases water loss from the plant is reduced and there are few data to show that, at least temporarily, hydration of tissues increases. Another exception is blue mold disease of tobacco in which water loss was reduced in the later stages of disease. It would be interesting to know whether such changes are characteristic of downy mildew diseases. Water stress may be brought about in one or a combination of ways: water loss may be accelerated; uptake may be reduced without an associated change of the same proportion in transpiration; or the ability of cells to retain solutes and water may be destroyed.

Water stress and disease both reduce plant growth. Many of the changes shown by water-stressed and by diseased plants appear to be similar, e.g., reduced expansion of young tissues, reduced photosynthetic activity, increased photo- and dark respiration, altered hormone levels. Research to date has made some progress in showing that the similarity does not arise by chance, but it is still a major task for the future to determine whether these phenomena in diseased plants are in part, or wholly, the results of water stress.

We also need to know whether certain phenomena that are peculiar to diseased plants, such as accumulation of solutes, including growth regulators, at sites of infection by biotrophic pathogens are influenced by altered water regulations in the host. It is often forgotten that the vascular system of higher plants has a circulatory element; the phloem is supplied with water by the xylem and vice versa. Little is known about the effects of altered water relations on phloem transport in healthy or diseased plants. The relationship between pathogen-induced water stress, reduced leaf expansion, and altered assimilate distribution in take-all of cereals has been discussed. The use of controlled infections by highly specialized parasites could be a useful tool for physiologists as they attempt to unravel the complexities of translocation. Whitbread and Bhatti (1976) have shown that when dwarf beans are systemically infected by *Xanthomonas phaseoli*, a necrotrophic pathogen causing blight, the distribution of $^{14}$C-labeled assimilate was altered to a pattern similar to that found in healthy plants subjected to water stress. As in take-all disease, the modification in diseased plants probably arose because water stress reduced leaf expansion and thus altered normal source–sink relationships. Assimilate movement in dwarf beans was also affected when cells of *Pseudomonas phaseolicola* entered petioles in the late stages of halo-blight infections; transport of assimilate out of the leaf was reduced (Hale and Whitbread, 1973).

Questions about the consequences of water stress in diseased plants can only be answered meaningfully when we know more about the degree of water stress that plants suffer as a result of infection, particularly under field conditions where, unfortunately, studies are further complicated by direct effects of water stress on development of the pathogen. If water stress is widespread among diseased plants, then drought resistance should be regarded as a factor vital to the success of any selective breeding program aimed at improving field (major gene, horizontal) resistance in plants.

## REFERENCES

Abeles, F. B. (1973). "Ethylene in Plant Biology." Academic Press, New York.
Albersheim, P., Jones, T. M., and English, P. D. (1969). Biochemistry of the cell wall in relation to infective processes. *Annu. Rev. Phytopathol.* 7, 171–194.
Alexander, S. J., and Hall, R. (1974). *Verticillium* wilt of chrysanthenum: Anatomical observations on colonization of roots, stem and leaves. *Can. J. Bot.* 52, 783–789.
Amatya, P., and Jones, J. P. (1966). Transpiration changes in oat plants infected with crown rust. *Proc. Arkansas Acad. Sci.* 20, 59–67.
Arntzen, C. J., Haugh, M. F., and Bobick, S. (1973). Induction of stomatal closure by *Helminthosporium maydis* pathotoxin. *Plant Physiol.* 52, 569–574.
Asher, M. J. C. (1972a). Effect of *Ophiobolus graminis* infection on growth of wheat and barley. *Ann. Appl. Biol.* 70, 215–223.
Asher, M. J. C. (1972b). Effect of *Ophiobolus graminis* infection on the assimilation and distribution of $^{14}$C in wheat. *Ann. Appl. Biol.* 72, 161–167.
Atkinson, R. E. (1938). On the nature of resistance of sugar cane to red rot. *In Proc. 6th Int. Cong. Soc. Sugar Cane Technol.*, Baton Rouge. pp. 684–692.

Ayres, P. G. (1972). Abnormal behaviour of stomata in barley leaves infected with *Rhynchosporium secalis* (Oudem.) J. J. Davis. *J. Exp. Bot.* **23**, 683–691.

Ayres, P. G. (1976). Patterns of stomatal behaviour, transpiration and $CO_2$ exchange in pea following infection by powdery mildew *(Erysiphe pisi)*. *J. Exp. Bot.* **27**, 354–363.

Ayres, P. G. (1977a). Effects of powdery mildew *(Erysiphe pisi)* and water stress upon the water relations of pea. *Physiol. Plant Pathol.* **10**, 139–146.

Ayres, P. G. (1977b). Effects of leaf water potential on sporulation of *Erysiphe pisi* (pea mildew). *Trans. Br. Mycol. Soc.* **68**, 97–100.

Ayres, P. G., and Jones, P. (1975). Increased transpiration and the accumulation of root absorbed $^{86}R6$ in barley leaves infected by *Rhynchosporium secalis* (leaf blotch). *Physiol. Plant Pathol.* **7**, 49–58.

Ballio, A., Chain, E. D., De Leo, P., Erlanger, B. F., Mauri, M., and Tonolo, A. (1964). Fusicoccin: A new wilting toxin produced by *Fusicoccum amygdali* Del. *Nature (London)* **203**, 297.

Ballio, A., Pocchiari, F., Russi, S., and Silano, V. (1971). Effects of fusicoccin and some related compounds on etiolated pea tissues. *Physiol. Plant Pathol.* **1**, 95–103.

Ballio, A., D'Alessio, V., Randazzo, G., Bottalico, A., Graniti, A., Sparapano, L., Bosnar, B., Casinovi, C. G., and Gribanovski-Sassu, O. (1976). Occurrence of fusicoccin in plant tissues infected by *Fusicoccum amygdali* Del. *Physiol Plant Pathol.* **8**, 163–169.

Barrs, H. D. (1968). Determination of water deficits in plant tissues. *In* "Water Deficits and Plant Growth" (T. T. Kozlowski, ed.), Vol. 1, pp. 235–368. Academic Press, New York.

Basham, H. G., and Bateman, D. F. (1975). Relationship of cell death in plant tissue treated with a homogenous endopectate lyase to cell wall degradation. *Physiol. Plant Pathol.* **5**, 249–262.

Beckman, C. H., Brun, W. A., and Buddenhagen, I. W. (1962). Water relations in banana plants infected with *Pseudomonas solanacearum*. *Phytopathology* **52**, 1144–1148.

Bever, W. M. (1937). Influence of stripe rust on growth, water economy and yield of wheat and barley. *J. Agric. Res.* **54**, 375–385.

Blodgett, F. H. (1901). Transpiration of rust-infected *Rubus*. *Torreya* **1**, 34–35.

Boughey, A. S. (1949). The ecology of fungi which cause economic plant diseases. *Trans. Br. Mycol. Soc.* **32**, 179–189.

Boyer, J. S. (1976). Photosynthesis at low water potentials. *Philos. Trans. R. Soc. London, Ser. B* **273**, 501–512.

Buddenhagen, I., and Kelman, A. (1964). Biological and physiological aspects of bacterial wilt caused by *Pseudomonas solanacearum*. *Annu. Rev. Phytopathol.* **2**, 203–230.

Bushnell, W. R., and Rowell, J. B. (1968). Premature death of adult rusted wheat plants in relation to carbon dioxide evolution by root systems. *Phytopathology* **58**, 651–658.

Chain, E., Mantle, P. G., and Milborrow, B. V. (1972). Investigations on the phytotoxicity of fusicoccins. *In* "Phytotoxins in Plant Diseases" (R. K. S. Wood, A. Ballio, and A. Graniti, eds.), pp. 395–398. Academic Press, New York.

Cho, Y. S., Wilcoxson, R. D., and Frosheiser, F. I. (1973). Differences in anatomy, plant extracts and movement of bacteria in plants of bacterial wilt resistant and susceptible varieties of alfalfa. *Phytopathology* **63**, 760–763.

Clarkson, D. T. (1974). "Ion Transport and Cell Structure in Plants." McGraw-Hill, New York.

Clarkson, D. T., Drew, M. C., Ferguson, I. B., and Sanderson, J. (1975). The effect of the take-all fungus, *Gaeumannomyces graminis*, on the transport of ions by wheat plants. *Physiol. Plant Pathol.* **6**, 75–84.

Collins, J. C. (1974). Hormonal control of ion and water transport in excised maize root. *In* "Membrane Transport in Plants" (U. Zimmerman and J. Dainty, eds.), pp. 441–443. Springer-Verlag, Berlin and New York.

Cowan, M. C., and Van der Wal, A. F. (1975). An ecophysiological approach to crop losses exemplified in the system wheat, leaf rust and glume blotch. IV. Water flow and leaf-water potential of uninfected wheat plants and plants infected with *Puccinia recondita* f.sp. *triticina*. *Neth. J. Plant. Pathol.* **81**, 49–57.

Cronshaw, K. D., and Pegg, G. F. (1976). Ethylene as a toxin synergist in *Verticillium* wilt of tomato. *Physiol. Plant Pathol.* **9**, 33–44.

Cruickshank, I. A. M., and Rider, N. E. (1961). *Peronospora tabacina* in tobacco: Transpiration, growth and related energy considerations. *Aust. J. Biol. Sci.* **14**, 45–57.

Deroo, H. C. (1969). Sap stress and water uptake in detached shoots of wilt-diseased and normal rhododendrons. *Hortic. Sci.* **4**, 51–52.

Deverall, B. J. (1972). Phytoalexins and disease resistance. *Proc. R. Soc. London, Ser. B* **181**, 233–246.

Dimond, A. E. (1966). Pressure and flow relations in vascular bundles of the tomato plant. *Plant Physiol.* **41**, 119–131.

Dimond, A. E. (1967). Physiology of wilt disease. *In* "The Dynamic Role of Molecular Constituents in Plant-Parasite Interaction" (C. J. Mirocha and I. Uritani, eds.), pp. 100–120. Am. Phytopathol. Soc., St. Paul, Minnesota.

Dimond, A. E. (1970). Biophysics and biochemistry of vascular wilt syndrome. *Annu. Rev. Phytopathol.* **8**, 301–322.

Dimond, A. E. (1972). Symptoms of vascular wilt disease. *In* "Phytotoxins in Plant Disease" (R. K. S. Wood, A. Ballio, and A. Graniti, eds.), pp. 289–309. Academic Press, New York.

Dimond, A. E., and Waggoner, P. E. (1953). On the nature and role of vivotoxins in plant disease. *Phytopathology* **43**, 229–235.

Duniway, J. M. (1971a). Water relations of *Fusarium* wilt in tomato. *Physiol. Plant Pathol.* **1**, 537–546.

Duniway, J. M. (1971b). Resistance to water movement in tomato plants infected with *Fusarium*. *Nature (London)* **230**, 252–253.

Duniway, J. M. (1973). Pathogen-induced changes in host water relations. *Phytopathology* **63**, 458–466.

Duniway, J. M. (1975). Water relations in safflower during wilting induced by Phytophthora root rot. *Phytopathology* **65**, 886–891.

Duniway, J. M., and Durbin, R. D. (1971a). Some effects of *Uromyces phaseoli* on the transpiration rate and stomatal response of bean leaves. *Phytopathology* **61**, 114–119.

Duniway, J. M., and Durbin, R. D. (1971b). Detrimental effect of rust infection on the water relations of bean. *Plant Physiol.* **48**, 69–72.

Durbin, R. D. (1967). Obligate parasites: Effect on the movement of solutes and water. *In* "The Dynamic Role of Molecular Constituents in Plant-Parasite Interaction" (C. J. Mirocha and I. Uritani, eds.), pp. 80–99. Am. Phytopathol. Soc., St. Paul, Minnesota.

Dvoretskaya, Y. I., Pyrina, I. L., and Feoktisova, O. I. (1959). The physiological nature of tomato resistance to leaf mould. *Biokhim. Plodov Ovoshchei* **5**, 165; cited in B. A. Rubin and Y. V. Artsikhovskaya, "Biochemistry and Physiology of Plant Immunity." Pergamon, Oxford, 1963.

Elgersma, D. W. (1970). Length and diameter of xylem vessels at factors in resistance of elms to *Ceratocystis ulmi*. *Neth. J. Plant Pathol.* **76**, 179–182.

Elnaghy, M. A., and Heitefuss, R. (1976). Permeability changes and production of antifungal compounds in *Phaseolus vulgaris* infected with *Uromyces phaseoli*. II. Role of phytoalexins. *Physiol. Plant Pathol.* **8**, 269–277.

Farrell, G. M., Preece, T. F., and Wren, M. J. (1969). Effects of infection by *Phytophthora infestans* (Mont.) de Bary on stomata of potato leaves. *Ann. Appl. Biol.* **63**, 265–275.

Frič, F. (1975). Translocation of [14]C-labelled assimilates in barley plants infected with powdery mildew (Erysiphe graminis f. sp. hordei Marchal). Phytopathol. Z. 84, 88–95.

Garrett, S. D. (1970). "Pathogenic Root Infecting Fungi." Cambridge Univ. Press, London and New York.

Gassner, G., and Goeze, G. (1936). Einige Versuche über die physiologische Leistungsfähigkeit rostinfizierter Getreidblätter. Phytopathol. Z. 9, 371–386.

Gaumann, E. (1958). The mechanisms of fusaric acid injury. Phytopathology 48, 670–686.

Gerwitz, D. L., and Durbin, R. D. (1965). The influence of rust on the distribution of P[32] in the bean plant. Phytopathology 55, 57–64.

Ghabrial, S. A., and Pirone, T. P. (1967). Physiology of tobacco etch virus-induced wilt of tabasco peppers. Virology 31, 154–162.

Glinka, Z. (1971). The effect of epidermal cell water potential on stomatal response to illumination of leaf discs of Vicia faba. Physiol. Plant. 24, 476–479.

Gondo, M. (1953). Further studies on the transpiration of mosaic diseased tobacco plant. Bull. Fac. Agric., Kagoshima Univ. 2, 71–74.

Gracen, V. E., Grogan, C. O., and Forster, M. J. (1972). Permeability changes induced by Helminthosporium maydis, race T, toxin. Can. J. Bot. 50, 2167–2170.

Graf-Marin, A. (1934). Studies on powdery mildew of cereals. N.Y., Agric. Exp. Stn., Ithaca, Mem. 157, 1–48.

Graniti, A. (1964). The role of toxins in the pathogenesis of infections by Fusicoccum amygdali Del. on almond and peach. In "Host-Parasite Relations in Plant Pathology" (Z. Király and G. Ubrizy, eds.), pp. 211–217. Res. Inst. Plant Prot., Budapest.

Greenall, A. F. (1956). What is the effect of rusts on cereal forage crops? Commonw. Phytopathol. News 2, 60–62.

Gregory, G. F. (1971). Correlation of isolability of the oak wilt pathogen with leaf wilt and vascular water flow resistance. Phytopathology 61, 1003–1005.

Hale, C. N., and Whitbread, R. (1973). The translocation of [14]C-labelled assimilates by dwarf bean plants infected with Pseudomonas phaseolicola (Burk.). Dows. Ann. Bot. (London) [N.S.] 37, 473–480.

Hall, R., and Busch, L. V. (1971). Verticillium wilt of chrysanthemum: Colonization of leaves in relation to symptom development. Can. J. Bot. 49, 181–185.

Hall, R., Ali, A., and Busch, L. V. (1975). Verticillium wilt of chrysanthemum: development of wilt in relation to leaf diffusive resistance and vascular conductivity. Can. J. Bot. 53, 1200–1205.

Hancock, J. G. (1972). Changes in cell membrane permeability in sunflower hypocotyls infected with Sclerotinia sclerotiorum. Plant Physiol. 49, 358–364.

Harrison, A. L. (1935). The physiology of bean mosaic. N.Y., Agric. Exp. Stn., Geneva, Tech. Bull. 235.

Harrison, J. A. C. (1970). Water deficit in potato plants infected by Verticillium albo-atrum. Ann. Appl. Biol. 66, 225–231.

Harrison, J. A. C. (1971). Transpiration in potato plants infected with Verticillium spp. Ann. Appl. Biol. 68, 159–168.

Hartill, W. F. T. (1961). Effects of yellow rust (Puccinia glumarum (Schm.) Erikss. & Henn) upon the development, yield, maturation and composition of wheat. Ph.D. Thesis, University of Southampton.

Harvey, R. B. (1930). The relative transpiration rate at infection spots on leaves. Phytopathology 20, 359–362.

Helms, J. A., Cobb, F. W., Jr., and Whitney, H. S. (1971). Effect of infection by Verticicladiella wageneri on the physiology of Pinus ponderosa. Phytopathology 61, 920–925.

Hess, W. M., and Strobel, G. A. (1970). Ultrastructure of potato stems infected with *Corynebacterium sepedonicum*. *Phytopathology* **60**, 1428–1431.

Hewitt, H. G., and Ayres, P. G. (1975). Changes in $CO_2$ and water vapour exchange rates in leaves of *Quercus robur* infected by *Microsphaera alphitoides* (powdery mildew). *Physiol. Plant Pathol.* **7**, 127–137.

Holmes, F. O. (1964). Symptomatology of viral diseases in plants. *In* "Plant Virology" (M. K. Corbett and H. D. Sisler, eds.), pp. 17–38. University of Florida Press, Gainesville.

Hoppe, H. H., and Heitefuss, R. (1974a). Permeability and membrane lipid metabolism of *Phaseolus vulgaris* infected with *Uromyces phaseoli*. I. Changes in the efflux of cell constituents. *Physiol. Plant Pathol.* **4**, 5–9.

Hoppe, H. H., and Heitefuss, R. (1974b). Permeability and membrane liped metabolism of *Phaseolus vulgaris* infected with *Uromyces phaseoli*. III. Changes in relative concentration of lipid-bound fatty acids and phospholipase activity. *Physiol. Plant Pathol.* **4**, 25–35.

Huang, J.-S., and Goodman, R. N. (1970). The relationship of phosphatidase activity to the hypersensitive reaction in tobacco induced by bacteria. *Phytopathology* **60**, 1020–1021.

Husain, A., and Kelman, A. (1958). Relation of slime production to mechanism of wilting and pathogenicity of *Pseudomonas solanacearum*. *Phytopathology* **48**, 155–165.

Johnson, T. B., and Strobel, G. A. (1970). The active site on the phytotoxin of *Corynebacterium sepedonicum*. *Plant Physiol.* **45**, 761–764.

Johnston, C. O., and Miller, E. C. (1934). Relation of leaf-rust infection to yield, growth and water economy of two varieties of wheat. *J. Agric. Res.* **49**, 955–981.

Johnston, C. O., and Miller, E. C. (1940). Modification of diurnal transpiration in wheat by infections of *Puccinia triticina*. *J. Agric. Res.* **61**, 427–444.

Jones, P., and Ayres, P. G. (1972). The nutrition of the subcuticular mycelium of *Rhynchosporium secalis* (barley leaf blotch); permeability changes induced in the host. *Physiol. Plant Pathol.* **2**, 383–392.

Jones, P., and Ayres, P. G. (1974). *Rhynchosporium* leaf blotch of barley studied during the subcuticular phase by electron microscopy. *Physiol. Plant Pathol.* **4**, 229–233.

Keen, N. T., and Long, M. (1972). Isolation of a protein-lipopolysaccharide complex from *Verticillium albo-atrum*. *Physiol. Plant Pathol.* **2**, 307–315.

Keen, N. T., and Williams, P. H. (1971). Chemical and biological properties of a lipomucopolysaccharide from *Pseudomonas lachrymans*. *Physiol. Plant Pathol.* **1**, 247–264.

Kende, H. (1971). The cytokinins, *Int. Rev. Cytol.* **31**, 301–338.

Király, Z., El Hammady, M., and Pozsár, B. I. (1967). Increased cytokinin activity of rust-infected bean and broad bean leaves. *Phytopathology* **57**, 93–94.

Kozlowski, T. T., ed. (1968a). "Water Deficits and Plant Growth," Vol. 1. Academic Press, New York.

Kozlowski, T. T., ed. (1968b). "Water Deficits and Plant Growth," Vol. 2. Academic Press, New York.

Kozlowski, T. T., ed. (1972). "Water Deficits and Plant Growth," Vol. 3. Academic Press, New York.

Kozlowski, T. T., ed. (1976). "Water Deficits and Plant Growth," Vol. 4. Academic Press, New York.

Kourssanow, A. L. (1928). De l'influence de l'*Ustilago tritici* sur les fonctions physiologiques du froment. *Rev. Gen. Bot.* **40**, 343–371.

Lado, P., Pennachioni, A., and Caldogno, R. F. (1972). Comparison between some effects of fusicoccin and indole-3-acetic acid on cell enlargement in various plant materials. *Physiol. Plant Pathol.* **2**, 75–85.

Larcher, W. (1963). Zur Frage des Zusammenhanges zwischen Austrocknungsresistenz und Frosthärte bei Immergrün. *Protoplasma* **57**, 569–587.

Lewis, D. H. (1973). Concepts in fungal nutrition. *Biol. Rev. Cambridge Philos. Soc.* **48**, 261–278.

Lieberman, M. (1975). Biosynthesis and regulatory control of ethylene in fruit ripening. *Physiol. Veg.* **13**, 489–499.

Lindsey, D. W., and Gadauskas, R. T. (1975). Effects of maize dwarf mosaic virus on water relations of corn. *Phytopathology* **65**, 434–440.

Lumsden, R. D. (1968). Phosphatidase production by *Sclerotinia sclerotiorum* in culture and in association with infected bean hypocotyls. *Phytopathology* **58**, 1058 (abstr.).

Lumsden, R. D., and Bateman, D. F. (1968). Phosphatide-degrading enzymes associated with pathogenesis in *Phaseolus vulgaris* infected with *Thielaviopsis basicola*. *Phytopathology* **58**, 219–227.

MacHardy, W. E., and Beckman, C. H. (1973). Water relations in American Elm infected with *Ceratocytis ulmi*. *Phytopathology* **63**, 98–103.

MacHardy, W. E., Hall, R., and Busch, L. V. (1974). *Verticillium* wilt of chrysanthemum: Relative water content and protein, R.N.A., and chlorophyll levels in leaves in relation to visible wilt symptoms. *Can. J. Bot.* **52**, 49–54.

MacHardy, W. E., Busch, L. V., and Hall, R. (1976). *Verticillium* wilt of chrysanthemum: quantitative relationship between increased stomatal resistance and local vascular dysfunction preceeding wilt. *Can. J. Bot.* **54**, 1023–1034.

McNabb, H. S., Jr., Heybroek, H. M., and MacDonald, W. L. (1970). Anatomical factors in resistance to Dutch elm disease. *Neth. J. Plant Pathol.* **76**, 196–204.

McWain, P., and Gregory, G. F. (1972). A neutral mannan from *Cerotocystis fagacearum* culture filtrate. *Phytochemistry* **11**, 2609–2612.

Majernik, O. (1965). Water balance changes of barley infected by *Erysiphe graminis* D.C. f. sp. *hordei* Marchal. *Phytopathol. Z.* **53**, 145–153.

Majernik, O. (1971). A physiological study of the effects of $SO_2$ pollution, phenyl mercuric acetate sprays, and parasitic infection on stomatal behaviour and ageing in barley. *Phytopathol. Z.* **72**, 255–268.

Manners, J. G., and Myers, A. (1975). Effect of fungi on higher plant physiology. *Symp. Soc. Exp. Biol.* **29**, 279–298.

Martin, J. T., Stuckey, R. E., Safir, G. R., and Ellingboe, A. H. (1975). Reduction of transpiration from wheat caused by germinating conidia of *Erysiphe graminis* f. sp. *tritici*. *Physiol. Plant Pathol.* **7**, 71–77.

Meidner, H., and Willmer, C. (1975). Mechanics and metabolism of guard cells. *Curr. Adv. Plant Sci.* **17**, 1–15.

Melching, J. B., and Sinclair, W. A. (1975). Hydraulic conductivity of stem internodes related to resistance of American elms to *Ceratocystis ulmi*. *Phytopathology* **65**, 645–647.

Minarćic, P., and Paulech, C. (1975). Influence of powdery mildew on mitotic cell division of apical root meristems of barley. *Phytopathol. Z.* **83**, 341–347.

Misaghi, I., DeVay, J. E., and Kosuge, T. (1972). Changes in cytokinin activity associated with the development of *Verticillium* wilt and water stress in cotton plants. *Physiol. Plant Pathol.* **2**, 187–196.

Montemartini, L. (1911). Note de fisiopathologia vegetale. *Atti Ist. Bot. Univ. Pavia* [2] **9**, 39–97.

Muller, D. (1932). Die Assimilation der Blattrollkranken Kartofellpflanzen. *Planta* **16**, 10–16.

Müller-Thurgau, H. (1895). Die Tätigkeit pilzkranker Blätter. *Jahresber. Vers. Stat. Schule Weedensweill* **4**, 54–57 (cited by Weiss, 1924).

Murphy, H. C. (1935). Effect of crown rust infection on yield and water requirement of oats. *J. Agric. Res.* **50**, 387–411.

Nicolas, G. (1930). Sur la transpiration des plantes parasitées par des champignons. *Rev. Gen. Bot.* **42**, 257–271.

Orlob, G. B., and Arny, D. C. (1961). Some metabolic changes accompanying infection by barley yellow dwarf virus. *Phytopathology* **51**, 768–775.

Parodi, P. P., and Bitzer, M. J. (1969). Determinación simultanea de absorción de agua y transpiración en plantas de trigo (*Triticum aestivum* L.) infectades y libres de polvillo de la hoji (*Puccinia recondita* Rob. ex Desm. f sp. *tritici*). *Agri. Tec. (Santiago)* **29**, 191–194.

Paulech, C., and Haspelova-Horvatovicova, A. (1970). Photosynthesis, plant pigments and transpiration in healthy barley and barley infected by powdery mildew. *Biologia (Bratislava)* **25**, 477–487.

Pegg, G. F., and Cronshaw, D. K. (1976). Ethylene production in tomato plants infected with *Verticillium albo-atrum*. *Physiol. Plant Pathol.* **8**, 279–295.

Pomerleau, R., and Mehran, A. R. (1966). Distribution of spores of *Ceratocystis ulmi* labelled with phosphorus-32 in green shoots and leaves of *Ulmus americana*. *Nat. Can. (Que.)* **93**, 577–582.

Powers, H. R., Jr. (1954). The mechanism of wilting in tobacco plants affected by black shank. *Phytopathology* **44**, 513–521.

Priehradny, S. (1971). Water uptake of barley infected by powdery mildew (*Erysiphe graminis* D.C.). *Biologia (Bratislava)* **26**, 507–516.

Priehradny, S. (1975). Response of fungus pathogen in susceptible and resistant barley varieties. I. Transpiration *Phytopathol. Z.* **83**, 109–118.

Rai, P. V., and Strobel, G. A. (1969). Phytotoxic glycopeptides produced by *Corynebacterium michiganense*. II. Biological properties. *Phytopathology* **59**, 53–57.

Reed, H., and Cooley, J. S. (1913). The transpiration of apple leaves infected with *Gymnosporangium*. *Bot. Gaz. (Chicago)* **55**, 421–430.

Ries, S. M., and Strobel, G. A. (1972). Biological properties and pathological role of a phytotoxic glycopeptide from *Corynebacterium insidiosum*. *Physiol. Plant Pathol.* **2**, 133–142.

Robb, J., Busch, L. V., and Lu, B. C. (1975a). Ultrastructure of wilt syndrome caused by *Verticillium dahliae*. I. In chrysanthemum leaves. *Can. J. Bot.* **53**, 901–913.

Robb, J., Busch, L., Brisson, J. D., and Lu, B. C. (1975b). Ultrastructure of wilt syndrome caused by *Verticillium dahliae*. II. In sunflower leaves. *Can. J. Bot.* **53**, 2725–2734.

Safir, G. R., and Schneider, C. L. (1976). Diffusive resistances of two sugarbeet cultivars in relation to their black root disease reaction. *Phytopathology* **66**, 277–280.

Scheffer, R. P., and Samaddar, K. R. (1970). Host-specific toxins as determinants of pathogenicity. *Recent Adv. Phytochem.* **3**, 123–142.

Schuster, G. (1957). Untersuchungen über die Möglichkeiten zum frühzeitigen Nachweis von Pflanzenkrankheiten mit Hilfe der Anwelkmethode. *Nachr. Bl. Dtsch. Pflanzenschutz Dienst* **11**, 241–246.

Selman, I. W. (1945). Virus infection and water loss in tomato foliage. *J. Pomol. Hortic. Sci.* **21**, 146–154.

Sempio, C., Majernik, O., and Raggi, V. (1966). Water loss and stomatal behaviour of bean (*Phaseolus vulgaris* L.) infected by *Uromyces appendiculatus* (Pers.) Link. *Biologia (Bratislava)* **21**, 99–104.

Sequeira, L. (1973). Hormone metabolism in diseased plants. *Annu. Rev. Plant Physiol.* **24**, 353–380.

Shiraishi, T., Oku, H., Isono, M., and Ouchi, S. (1975). The injurious effects of pisatin on the plasmamembrane of pea. *Plant Cell Physiol.* **16,** 939–942.

Sinclair, W. A., Zahand, J. P., and Melching, J. B. (1972). Anatomical marker for resistance of *Ulmus americana* to *Ceratocystis ulmi. Phytopathology* **62,** 789–790 (abstr.).

Slatyer, R. O. (1967). "Plant Water Relationships." Academic Press, New York.

Strobel, G. A. (1970). A phytotoxic glycopeptide from potato plants infected with *Corynebacterium sepedonicum. J. Biol. Chem.* **245,** 32–38.

Strobel, G. A. (1974). Phytotoxins produced by plant parasites. *Annu. Rev. Plant Physiol.* **25,** 541–566.

Strobel, G. A. (1975). A mechanism of disease resistance in plants. *Sci. Am.* **232,** 80–88.

Strobel, G. A., and Hess, W. M. (1968). Biological activity of a phytotoxic glycopeptide produced by *Corynebacterium sepedonicum. Plant Physiol.* **43,** 1673–1688.

Strobel, G. A., Talmadge, K. W., and Albersheim, P. (1972). Observations on the structure of the phytotoxic glycopeptide of *Corynebacterium sepedonicum. Biochim. Biophys. Acta* **261,** 365–374.

Subramanian, D., and Saraswathi-Devi, L. (1959). Water is deficient. *In* "Plant Pathology" (J. G. Horsfall and A. E. Dimond, eds.), Vol. 1, pp. 313–348. Academic Press, New York.

Sutton, J. C., and Williams, P. H. (1970). Relation of xylem plugging to black rot lesion development in cabbage. *Can. J. Bot.* **48,** 391–401.

Takai, S. (1974). Pathogen and cerato-ulmin production in *Ceratocystis ulmi. Nature (London)* **252,** 124–126.

Talboys, P. W. (1968). Water deficits in vascular disease. *In* "Water Deficits and Plant Growth" (T. T. Kozlowski, ed.), Vol. 2, pp. 255–311. Academic Press, New York.

Talboys, P. W. (1972). Resistance to vascular wilt fungi. *Proc. R. Soc. London, Ser. B.* **181,** 319–332.

Teakle, D. S., Smith, P. M., and Steindle, D. R. L. (1975). Ratoon stunting disease of sugarcane; possible correlation of resistance with vascular anatomy. *Phytopathology* **65,** 138–141.

Thatcher, F. S. (1939). Osmotic and permeability relations in the nutrition of fungus parasites. *Am. J. Bot.* **26,** 449–458.

Tinklin, R. (1970). Effects of aspermy virus infection on the water status of tomato leaves. *New Phytol.* **69,** 515–520.

Tseng, T.-C., and Bateman, D. F. (1968). Production of phosphatidases by phytopathogens. *Phytopathology* **58,** 1437–1438.

Turner, N. C. (1972a). K⁺ uptake of guard cells stimulated by fusicoccin. *Nature (London)* **235,** 341–342.

Turner, N. C. (1972b). Stomatal behavior of *Avena sativa* treated with two phytotoxins, victorin and fusicoccin. *Am. J. Bot.* **59,** 133–136.

Turner, N. C., and Graniti, A. (1969). Fusicoccin: A fungal toxin that opens stomata. *Nature (London)* **223,** 1070–1071.

Turner, N. C., and Graniti, A. (1976). Stomatal response of two almond varieties to fusicoccin. *Physiol. Plant Pathol.* **9,** 175–182.

Vaadia, Y. (1976). Plant hormones and water stress. *Philos. Trans. R. Soc. London, Ser. B.* **273,** 513–522.

Van Alfen, N. K., and Turner, N. C. (1975a). Influence of *Ceratocystis ulmi* toxin on water relations of elm *(Ulmus americana). Plant Physiol.* **55,** 312–316.

Van Alfen, N. K., and Turner, N. C. (1975b). Changes in alfalfa stem conductance induced by *Corynebacterium insidiosum* toxin. *Plant Physiol.* **55,** 559–561.

Van der Wal, A. F., and Cowan, M. C. (1974). An ecophysiological approach to crop losses exemplified in the system wheat, leaf rust, and glume blotch. II. Development, growth and transpiration of uninfected plants and plants infected with *Puccinia recondita* f. sp. *triticina* and/or *Septoria nodorum* in a climate chamber experiment. *Neth. J. Plant Pathol.* **80**, 192–214.

Van der Wal, A. F., Smeitink, H., and Maan, G. C. (1975). An ecophysiological approach to crop losses exemplified in the system wheat, leaf rust and glume blotch. III. Effects of soil-water potential on development, growth, transpiration, symptoms and spore production of leaf rust-infected wheat. *Neth. J. Plant Pathol.* **81**, 1–13.

Van Etten, H. D., and Bateman, D. F. (1971). Studies on the mode of action of phaseollin. *Phytopathology* **61**, 915 (abstr.).

Vizárová, G. (1974). Level of free cytokinins in susceptible and resistant cultivars of barley infected by powdery mildew. *Phytopathol. Z.* **79**, 310–314.

Vizárová, G., and Minarčic, P. (1974). The influence of powdery mildew upon the cytokinins and the morphology of barley roots. *Phytopathol. Z.* **81**, 49–55.

Waggoner, P. E., and Dimond, A. E. (1954). Reduction of water flow by mycelium in vessels. *Am. J. Bot.* **41**, 637–640.

Wardlaw, I. F. (1967). The effect of water stress on translocation in relation to photosynthesis and growth. I. Effect during grain development in wheat. *Aust. J. Biol. Sci.* **20**, 25–39.

Wardlaw, I. F. (1969). The effect of water stress on translocation in relation to photosynthesis and growth. II. Effect during leaf development in *Lolium temulentum*. *Aust. J. Biol. Sci.* **22**, 1–6.

Weaver, J. E. (1916). The effects of certain rusts upon the transpiration of their hosts. *Minn. Bot. Stud.* **4**, 379–406.

Weise, M. V., and DeVay, J. E. (1970). Growth regulator changes in cotton associated with defoliating caused by *Verticillium albo-atrum*. *Plant Physiol.* **45**, 304–309.

Weiss, F. (1924). The effect of rust infection upon the water requirement of wheat. *J. Agric. Res.* **27**, 107–118.

Wheeler, H., and Black, H. S. (1963). Effects of *Helminthosporium victoriae* and victorin upon permeability. *Am. J. Bot.* **50**, 686–693.

Wheeler, H., and Hanchey, P. (1968). Permeability phenomena in plant disease. *Annu. Rev. Plant Physiol.* **6**, 331–350.

Whitbread, R., and Bhatti, M. A. R. (1976). The translocation of [14]C-labelled assimilates in dwarf beans infected with *Xanthomonas phaseoli* (E. F. Sm.) Dowson. *Ann. Bot. (London)* [N.S.] **40**, 499–509.

Wilson, C. L. (1973). A lysosomal concept for plant pathology. *Annu. Rev. Phytopathol.* **11**, 247–272.

Wood, R. K. S. (1972). The killing of plant cells by soft rot parasites. *In* "Phytotoxins in Plant Diseases" (R. K. S. Wood, A. Ballio, and A. Graniti, eds.), pp. 272–288. Academic Press, New York.

Wynn, W. K. (1976). Appressorium formation over stomates by bean rust fungus; response to surface contact stimulus. *Phytopathology* **66**, 136–146.

Yarwood, C. E. (1947). Water loss from fungal cultures. *Am. J. Bot.* **34**, 514–520.

# CHAPTER 2

# WATER STRESS AS A PREDISPOSING FACTOR IN PLANT DISEASE

## Donald F. Schoeneweiss

ILLINOIS NATURAL HISTORY SURVEY, URBANA, ILLINOIS

## I. INTRODUCTION

Although much research has been conducted on effects of water deficits on plant growth, comparatively little interest has been shown in the relation of water deficits to plant disease. Water balance in host plants undoubtedly influences initiation and development of disease, but only few experiments have studied this influence in detail.

Considerable information has accumulated in recent years on processes involved in pathogenesis and on host responses to infection, but the role of environmental stresses in modifying or regulating host–pathogen interactions is not well understood. This is not surprising since most research has been conducted with pathogens that attack vigorous host plants and stresses, such as water deficits, exert their greatest effect on diseases caused by pathogens that attack weakened hosts.

Because plant water deficits affect a number of plant processes, even small variations in plant water balance influence host–parasite interactions. In addition, moisture films are required for germination and/or penetration of propagules of nearly all plant pathogenic microorganisms on plant surfaces. These effects of moisture on plant diseases, however, are beyond the scope of this chapter. We are concerned here with the influence of water stress rather than water per se on plant disease. The term water stress implies that an abnormal or unfavorable water status is involved but a clearer definition of the term is needed. For this chapter a terminology adapted from the physical sciences by Levitt (1972) will be used. Levitt defined a biological stress as "any environmental factor capable of inducing a potentially injurious strain in living organisms." In the case of plant disease, the potentially injurious strain is increased disease susceptibility or host predisposition. According to Levitt, strain resulting from stress can be divided into "elastic strain," in which physical or chemical changes are reversible by removal of the stress, and "plastic strain," which involves an irreversible change. Both types of strain may influence plant diseases (Schoeneweiss, 1975a). With these considerations in mind, this chapter will discuss water stress as a predisposing factor in plant disease.

A. The Predisposition Concept

Predisposition refers to host "disposition" (Sorauer, 1874; Hartig, 1882) or "proneness" (Gäumann, 1950) to disease prior to infection. Although the innate capability of a host plant to respond to the presence of a pathogen or its metabolites is determined genetically, its disposition to disease can be shifted by environmental factors toward increased or decreased susceptibility (Barnett, 1959; Yarwood, 1959). Environmental stresses usually increase susceptibility to disease (Schoeneweiss, 1975a), but in the case of viruses and other obligate parasites, water deficit may decrease susceptibility (Yarwood, 1959; Baker and Cook, 1973).

Potential pathogens occasionally are excluded from higher plants by physical barriers (Gäumann, 1950; Akai, 1959) or biochemical ones (Allen, 1959), but in most cases, these organisms enter resistant and susceptible plants with equal frequency (Chester, 1933; Gäumann, 1950; Bollard and Matthews; 1966, Baker and Cook, 1973; Schoeneweiss, 1975a; Wheeler, 1975). Since healthy plants are much more common in nature than diseased ones, most plants are immune to a vast array of potential pathogens that invade them, although immunity may actually be hypersensitivity or a high degree of resistance (Bollard and Matthews, 1966). Therefore the effect of environmental stresses on disease is more often on host plant response than on exclusion of pathogens. When the effect on disease susceptibility occurs prior to establishment of an infection, the host is considered to be predisposed to disease.

The term predisposition was not in common use in plant pathology until recently but the concept of environmental factors affecting disease susceptibility in plants was introduced over 100 years ago (Sorauer, 1874). Historical perspectives on development of the predisposition concept were reviewed by Yarwood (1959), Colhoun (1973), and Schoeneweiss (1975a).

## B. PREDISPOSING FACTORS

According to Yarwood (1959), any change in host plant physiology, regardless of cause, probably has an effect on disease susceptibility. When we consider environmental stresses as predisposing factors, however, we are more concerned with major shifts in disease susceptibility that result in a significant increase, or more rarely a decrease, in disease damage. Of the four major environmental variables—light, water, temperature, and nutrients—variations in light intensity or duration under field conditions are seldom of sufficient magnitude to significantly affect disease predisposition (Schoeneweiss, 1975a). Temperature and the availability of water and nutrients, however, have very pronounced effects, alone or in combination, on susceptibility of plants to disease. It is often difficult to separate the relative importance of one variable without taking into account the other two. This is particularly true of water and temperature. High air temperatures promote increased water loss and desiccation of plant tissues, and low soil temperatures reduce absorption of water by roots, resulting in a similar effect (Levitt, 1972). Freezing of plant tissues may also injure cells by dehydration (Mazur, 1969). In this chapter, no attempt will be made to distinguish between the effects of water deficits and other interacting environmental factors, unless the data reported warrant such a distinction. Colhoun (1973) warned against drawing conclusions from laboratory studies on individual environmental factors, but few experiments have been designed to identify the relative importance of several environmental factors acting simultaneously on predisposition of plants to disease.

## II. WATER DEFICITS ASSOCIATED WITH DISEASE

If an environmental stress is considered to be any factor that causes a potentially injurious strain in plants, water deficit or excess water could induce stress, directly or indirectly. Excess soil moisture due to flooding, poor drainage, or prolonged rainy periods may cause wilting because of reduced soil aeration and consequent reduced water absorption. The effect is usually considered to be an indirect or secondary oxygen-deficit strain (Levitt, 1972) rather than an effect of excess water alone. For example, Stolzy et al. (1965) reported that root growth of citrus seedlings was increased and Phytophthora root rot decreased, regardless of irrigation, if adequate artificial aeration was provided. Prolonged oxygen deficit in the rhizosphere may also lead to accumulation in roots of toxic ions or intermediates of anaerobic respiration (Levitt, 1972).

High air humidity favors development of many foliar diseases and high soil moisture content is often associated with increased root rot damage but the effect is usually on the pathogen rather than on host predisposition. Therefore, excess water alone is seldom a predisposing factor in plant disease and few reports show correlation between excess water and disease susceptibility (Schoeneweiss, 1975a). In addition, problems with excess water in agricultural production on a worldwide basis are rare compared to those arising from water deficits (Kozlowski, 1968a). This chapter will therefore be limited to disease predisposition due to water deficit stress.

Internal plant water deficits usually arise from excess water loss by transpiration, reduced water uptake, or a combination of both (Kramer, 1963). Although the water potentials of all higher plants undergo diurnal and seasonal fluctuations (Kozlowski, 1968a), only prolonged water deficits should be considered as stress factors that affect plant diseases. Prolonged water deficits over several years may have cumulative effects on disease susceptibility of woody perennials, whereas seasonal water deficits may predispose herbaceous plants or woody perennials to disease.

### A. CLIMATIC SHIFTS AND PROLONGED DROUGHT

Annual precipitation is variable in most areas of the world, except where artificial irrigation is practiced. Prolonged drought or several consecutive years of below-normal rainfall have often been associated with disease outbreaks, usually in woody perennial hosts (Hepting, 1963). Diebacks of ash and birch, declines of maple and oak, pole blight of western white pine *(Pinus monticola)*, and dry face and pitch streak of slash pine *(Pinus elliottii)* appeared during and after the drought years of the 1930's in the United States (Ross, 1966; Hepting, 1963). In many cases, microorganisms isolated from affected trees were considered secondary or saprophytic, since inoculation of vigorous hosts seldom produced disease symptoms. It now appears that many of the damaged trees were predisposed by water deficits to attack by relatively nonaggressive pathogens, which caused or contributed to the losses (Crist and Schoeneweiss, 1975; Schoeneweiss, 1975a).

Decline and dieback of white ash *(Fraxinus americana)* and occasionally green ash *(F. pennsylvanica* var. *lanceolata)* in New York State appeared during the drought years of the 1930's and again following periods of low moisture in the 1950's and early 1960's (Ross, 1964, 1966). Silverborg and Ross (1968) hypothesized that ash dieback was induced by periods of low rainfall followed by severe stem and branch cankering by fungi such as *Cytophoma pruinosa* and *Fusicoccum* sp. The fungi were normally present in healthy bark and caused cankers only if trees were predisposed by drought. Tobiessen and Buchsbaum (1976) reported that ash stomata were highly sensitive to water stress and suggested that physiological changes due to reduced $CO_2$ fixation under water stress were involved in predisposing ash to fungal infection.

Another disease associated with climate is pole blight of western white pine. In an interpretation based on analysis of growth rings of trees, Leaphart and Stage (1971) concluded that the long drought of 1916 to 1940, superimposed on sites having low water-holding capacity, triggered a chain of events that ultimately resulted in pole blight. The failure of root regeneration under stress conditions to keep up with death of roots caused by drought and root rot was the main factor in the pole blight syndrome.

According to Secrest *et al.* (1941), a severe drought from 1930 to 1937 was the primary cause of heavy mortality of hemlock *(Tsuga canadensis)* in Wisconsin. Weakened trees were attacked by the root and collar rot fungus, *Armillaria mellea*, and later by boring insects. Since trees that were able to recover formed callus tissue over old *A. mellea* cankers, it appeared that the fungus was a weak pathogen that only attacked trees predisposed by drought. Neely (1968) suggested that an outbreak of bleeding necrosis cankers on sweetgum *(Liquidambar styraciflua)*, caused by *Botryosphaeria ribis (B. dothidea)*, was associated with the dry summers of 1962 to 1964 in Indiana and Illinois. Schoeneweiss (1975a) later confirmed that susceptibility of sweetgum to attack by *B. dothidea* increased dramatically in plants subjected to water deficit stress.

## B. SEASONAL VARIATION AND SHORT-TERM WATER DEFICITS

Many investigators have associated outbreaks of plant diseases with seasonal droughts. For example, Boyce (1961) attributed the appearance of many different tree cankers to the effects of drought periods during the growing season. Denyer (1953) successfully inoculated western hemlock *(Tsuga heterophylla)* with *Cephalosporium* sp. at the beginning of a 90-day summer drought, but inoculations were unsuccessful when precipitation was normal. Bagga and Smalley (1974) reported that any factor which contributed to host moisture stress during the growing season increased susceptibility of aspen *(Populus tremuloides)* to *Hypoxylon* canker.

Water deficits of sufficient magnitude and duration to affect disease susceptibility in plants may also be induced by any factors that prevent or interefere with water uptake, even though soil moisture content may be high. Transplants often undergo massive physiological shock during and after transplanting, due to root injury (Kozlowski, 1968a). Excess transpiration over water absorption occurs in transplanted trees even though soil moisture is near field capacity (Kozlowski, 1967). In a recent study in our laboratory, container-grown European white birch *(Betula alba)* seedlings were removed from containers and root-pruned in a manner similar to that which occurs during field transplanting. Pruned and unpruned plants were immediately repotted, inoculated with the stem canker fungus *B. dothidea*, and placed under high humidity for 16 days. The potting soil was watered to saturation daily and plant water potentials were measured with a pressure chamber just prior to watering each day. Throughout the 16-day period, plant water potentials of unpruned plants ranged from −2 to −4 bars and those of root-pruned plants from −8 to −10 bars. Soil in half of the pruned and unpruned plants was then allowed to dry until plant water potentials ranged from −16 to −18 bars. The fungus colonized only those plants with plant water potentials more negative than −16 bars. These results were in accord with previous studies on predisposition of white birch due to water stress (Crist and Schoeneweiss, 1975) and support the theory that water stress is a main component of transplanting shock.

Plants grown in pots or containers and plants grown in artificial mixes with low water-holding capacity frequently undergo water stress. Such plants often lack vigor and die following transplanting to field soil. Costello and Paul (1975) monitored water loss in the root ball mass of sweetgum plants growing in an artificial container mix after they were transplanted to field soil. Data from tensiometers placed in root balls showed that water loss was greater in the transplanted root balls than in the surrounding field soil or in the root ball mass of plants remaining in containers. They concluded that greater moisture loss from the transplants was due to drainage by the field soil and that container plants may require more irrigation after transplanting than before to avoid water stress. The frequent association of stem cankers and diebacks with transplanted container-grown plants suggests that disease predisposition due to water stress is a common occurrence.

## III. EFFECT OF WATER STRESS ON THE PATHOGEN

The status of water of the external environment and within host tissue may influence growth and reproduction of pathogens, although little is known about the effects of the water status of the host on organisms growing within host tissues. Outbreaks of foliar diseases are commonly associated with damp weather since adequate free water, usually in the form of moisture films, is required for spore germination on plant surfaces (see Yarwood, Chapter 5). Zoospores which constitute the infective stage of many phycomycetes are produced only in soils with water potentials near zero.

The overall influence of water status on pathogens, however, is more complex. Water potential of the growth substrate seldom limits growth of plant pathogens (Griffin, 1969), but water potentials of soils and plants may have indirect effects on pathogen growth and survival. Cook and Papendick (1972) compiled optimal and minimal substrate water potentials for growth of a number of plant pathogenic fungi. All of the fungi were able to grow at water potentials well below the minimum required for growth of most higher plants ($-30$ to $-50$ bars for the fungi compared to 0 to $-15$ bars for higher plants). Fungi that grew best at high water potentials cause severe disease in wet soils and those that grew best at low water potentials are associated with disease in dry soils. Cook and Papendick (1972) suggested that one reason for this association was that antagonistic microorganisms in the soil are unable to compete with pathogens such as *Fusarium roseum* f. sp. *cerealis* ''Culmorum,'' which grow best at water potentials of $-8$ to $-10$ bars or lower. Cook and Christen (1976) found that three cereal root rot fungi required progressively lower water potentials of the growth medium for maximum growth as temperature was increased from $10°$ to $35°C$. Differences in water potential requirement of the three fungi at different temperatures corresponded with their ecological distribution in nature.

Many soil-borne vascular pathogens can grow at substrate water potentials of $-40$ to $-60$ bars (Cook, 1973). Although vascular diseases are commonly associated with wet soils, infection may actually take place under dry conditions, but the distribution of fungal spores in the vascular system depends on the transpiration stream. Hence disease spread and development may increase under wet conditions (Cook, 1973; Papendick and Cook, 1974). Effects of soil moisture on survival and spread of pathogens are discussed further by Griffin in Chapter 6 of this volume.

Bagga and Smalley (1967) reported more growth of *Hypoxylon pruinatum* on 8% than on 1% agar. Since infection of aspen trees by the fungus takes place during periods of moisture stress, they suggested that the effect of moisture stress on the pathogen may be involved in etiology of the disease. In general, however, initial invasion by pathogens is seldom prevented by environmental conditions and the influence of water stress as a predisposing factor in plant disease is on development of established infections (Cook and Papendick, 1972; Papendick and Cook, 1974; Schoeneweiss, 1975a).

## IV. EFFECT OF WATER STRESS ON THE HOST

The concept of predisposition implies that the disposition or proneness of the host to disease prior to infection influences establishment and/or progression of a disease. Therefore, a predisposing factor such as water stress must alter the host in some manner. Because of the complexity and interactions of processes involved in growth and development of higher plants, it is exceedingly difficult to identify those processes that may influence disease susceptibility.

There is an extensive literature on effects of water deficits on plant processes (Vaadia *et al.*, 1961; Henckel, 1964; Kozlowski, 1968a,b, 1972, 1976; Levitt, 1972; Livne and Vaadia, 1972; Naylor, 1972; Hsiao, 1973). The range of sensitivity of various plant processes to water deficits was summarized by Hsiao (Table I). In general, the major effects of water deficits on higher plants include the following: (1) reduced photosynthesis due to stomatal closure, reduced $CO_2$ diffusion, and decreased chloroplast activity (Boyer, 1976); (2) an initial increase followed by a decrease in respiration (Boyer, 1976); (3) increase in abscisic acid and decrease in cytokinins (Hsiao, 1973); (4) an increase in some hydrolytic enzymes (Henckel, 1964), and a reduction or no change in other enzymes (Todd, 1972); (5) a general decrease in nutrient uptake, particularly potassium (Vaadia *et al.*, 1961; Viets, 1972), phosphorus (Vaadia *et al.*, 1961), boron (Viets, 1972), and rubidium (Vaadia *et al.*, 1961), but an increase in nitrogen accumulation (Viets, 1972); (6) reduced protein synthesis and decomposition of proteins and nucleic acids (Vaadia *et al.*, 1961, Henckel, 1964); (7)

TABLE I

GENERALIZED SENSITIVITY TO WATER STRESS OF PLANT
PROCESSES OR PARAMETERS[a,b]

| Process or parameter affected | Sensitivity to stress | | | Remarks |
|---|---|---|---|---|
| | Very sensitive | | Relatively insensitive | |
| | Reduction in tissue $\psi$ required to affect process[c] | | | |
| | 0 bar | 10 bars | 20 bars | |
| Cell growth | | | | |
| Cell wall synthesis | | | | Fast-growing tissue |
| Protein synthesis | | | | Fast-growing tissue |
| Protochlorophyll formation | | | | Etiolated leaves |
| Nitrate reductase level | | | | |
| Abscisic acid accumulation | | | | |
| Cytokinin level | | | | |
| Stomatal opening | | | | Depends on species |
| $CO_2$ Assimilation | | | | Depends on species |
| Respiration | | | | |
| Proline accumulation | | | | |
| Sugar accumulation | | | | |

[a] From Hsiao (1973).

[b] Length of the horizontal lines represents the range of stress levels within which a process becomes first affected. Dashed lines signify deductions based on more tenuous data.

[c] With $\psi$ of well-watered plants under mild evaporative demand as the reference point.

accelerated leaf senescence (Boyer, 1976) and premature increase in the physiological age of the plant (Darbyshire, 1971); and (8) accumulation of proline and sugars (Hsiao, 1973). Determining which of these processes are involved in predisposition of plants to disease is both difficult and speculative. Based on available information on host defense mechanisms, the possible relation of physiological processes to disease will be discussed in Section VIII.

## V. INFLUENCE OF WATER STRESS ON PLANT DISEASES

In the absence of pathogens, water stress alone may induce damage symptoms which are often referred to as "disease" or "physiological disease." Exposure to stress may also predispose or increase the sensitivity of a plant to a subsequent stress, i.e., prolonged drought during the growing season is often followed by increased freezing damage in woody ornamentals. Such effects are beyond the scope of this chapter and the subject matter will be restricted to predisposition of plants to attack by pathogens.

Environmental stresses such as water deficits ultimately cause a reduction in host vigor or vitality (Gäumann, 1950; Yarwood, 1959). The effect of reduced host vigor or susceptibility of tissues to pathogenic attack often varies with the type of pathogen involved (Yarwood, 1959; Schoeneweiss, 1975a).

## A. DISEASES CAUSED BY OBLIGATE PARASITES

Since an obligate parasite is one which only grows and reproduces in living host tissue, or at least has not yet been successfully cultured on artificial media, substrate requirements for growth are more critical for obligate parasites than for saprophytes or facultative parasites. In most cases obligate parasites are considered aggressive pathogens of vigorous plants and any reduction in host vigor due to stress usually reduces the amount of disease damage (Gäumann, 1950; Yarwood, 1959, Baker and Cook, 1973; Schoeneweiss, 1975a). Some exceptions occur, however, and predisposing factors such as wilting increase damage by some viruses and decrease damage by others (Yarwood, 1959). The host factors or processes involved in these phenomena are largely unknown.

## B. DISEASES CAUSED BY FACULTATIVE PARASITES

The most pronounced effects of water stress occur with diseases caused by facultative parasites, which are able to grow saprophytically on nonliving organic matter but can also incite diseases in living plants. The occurrence of a disease in a plant exposed to stress is a common phenomenon, yet this occccurrence is often attributed to a pathogen that spread or was introduced rather recently, rather than to one that was present on or in the host for some time and attacked the host only after the host was weakened. The relative effect of the stress on the disease may depend on a combination of host response and aggressiveness of the pathogen.

### 1. Aggressive Pathogens That Attack Vigorous Hosts

Since invasion of host tissues by both pathogens and nonpathogens is common, aggressive pathogens that cause disease on vigorous or nonstressed plants have great destructive potential and incite many of our most economically important diseases. Environmental stresses such as water deficits generally increase or accelerate development of diseases caused by aggressive pathogens, although, in some cases, water stress

TABLE II

EFFECT OF MOISTURE STRESS ON CANKER DEVELOPMENT ON ASPEN 10 DAYS
AFTER INOCULATION WITH *Hypoxylon pruinatum*[a]

| Watering schedule | Plants cankered (%) | Canker length (mm)[b] |
|---|---|---|
| Every day | 53 | 20 |
| Every third day | 73 | 44 |
| Every fifth day | 93 | 75 |

[a] From Bagga and Smalley (1969).
[b] Canker length LSD (0.05) = 17.

also increases disease incidence. Bagga and Smalley (1969) found that the percentage of cankers formed on aspen inoculated with *Hypoxylon pruinatum* increased from 53% on plants watered every day to 93% on those watered every fifth day (Table II). Canker length also increased from 20 to 75 mm, respectively. Towers and Stambaugh (1967) reported combined effects of temperature and water stress over a 10-month period on infection and mortality of loblolly pine *(Pinus taeda)* seedlings inoculated with *Fomes annosus* (Table III). More seedlings were infected and colonized by *F. annosus* in soil dried to the temporary wilting percentage (TWP) before watering than those in soil maintained at field capacity

TABLE III

INFLUENCE OF SOIL MOISTURE AND TEMPERATURE ON LOBLOLLY PINE SEEDLING
INFECTION AND MORTALITY AND VIABILITY OF *Fomes annosus* IN 20-GM
BEECH INOCULUM SEGMENTS[a]

| | | Treatment[b] | | | |
|---|---|---|---|---|---|
| | | Warm | | Cool | |
| Characteristic | Period (months) | FC | TWP | FC | TWP |
| | | % | % | % | % |
| Seedling mortality | 5 | 0 | 10 | 20 | 40 |
| | 10 | 0 | 30 | 30 | 70 |
| Seedling infection | 5 | 10 | 20 | 80 | 90 |
| | 10 | 20 | 40 | 50 | 100 |
| Inoculum viability | 5 | 20 | 40 | 90 | 90 |
| | 10 | 0 | 50 | 50 | 100 |

[a] From Towers and Stambaugh (1967).
[b] Warm, 15°–38°C; cool, 10°–12°C; FC, field capacity; TWP, temporary wilting percentage.

(FC), regardless of temperature. They also found that the rate of penetration of *F. annosus* in roots of 12-yr-old pines in the field was greatly enhanced by drought. Although cool temperatures also favored the disease, they concluded that pine seedlings were predisposed to root rot by soil moisture stress.

Couch and Bloom (1960b) found that varying the nutrient levels and pH in sand culture had little effect on *Sclerotinia* dollar spot of Kentucky bluegrass *(Poa pratensis)*, except for a reduction in disease under nitrogen deficiency. However, disease severity was significantly greater among plants growing in soils if plants were allowed to extract soil moisture to three-fourths of field capacity or below before rewatering (Fig. 1). The authors suggested that disease proneness of Kentucky bluegrass was accentuated at low soil moisture and could account for epiphytotics of the disease in seasons of low rainfall.

Mortality was high on unirrigated filbert *(Corylus maxima)* trees inoculated with the bacterial blight pathogen *Xanthomonas corylina* (Moore *et al.*, 1974). Results of a 3-yr irrigation study are shown in Fig. 2. Although the pathogen was not isolated from uninoculated trees, it remained viable for long periods in nonsymptomatic tissues of inoculated trees, and Moore *et al.* (1974) warned that periodic irrigation may be essential for the first few years after planting to avoid predisposing trees to bacterial blight.

Other specific examples could be cited of water deficit stress increasing the amount of disease damage by aggressive pathogens. Few studies, however, have been designed to separate the effects of water deficits from those of other environmental variables.

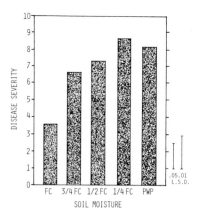

Fig. 1. Effect of soil moisture stresses in the readily available range on the susceptibility of Kentucky bluegrass to *Sclerotinia homeocarpa*. O, no leaf blighting; 10, 100% foliar loss; FC, field capacity; PWP, permanent wilting percent. (From Couch and Bloom, 1960b.)

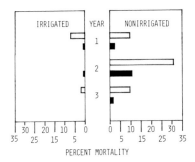

Fig. 2. Yearly mortality of filbert trees as influenced by irrigation and bacterial blight. Each bar represents the percent of trees killed in a given year out of 45 inoculated (☐) trees or 90 noninoculated (■) trees. (From Moore *et al.*, 1974.)

## 2. Nonaggressive Pathogens That Attack Weakened Hosts

Classic cases of disease predisposition occur among diseases caused by relatively nonaggressive pathogens, which may enter host plants but remain latent unless the host has been weakened in some manner, usually by unfavorable environmental conditions. Many stem canker pathogens of woody hosts are typical nonaggressive pathogens. For example, *Botryosphaeria dothidea*, which causes stem cankers on many woody species, does not attack European white birch (Crist and Schoeneweiss, 1975) or sweetgum and red osier dogwood *(Cornus stolonifera)* (Schoeneweiss, 1975a) until host plants are defoliated or exposed to drought or freezing temperatures. Working with a number of host–pathogen combinations, Bier (1959a,b,c 1961a,b,c), Bier and Rowat (1962), Bloomberg (1962), and Landis and Hart (1967) found that disease damage did not appear on inoculated trees until bark moisture content fell below a critical level.

There is considerable variation in the relative aggressiveness of facultative parasites. When European white birch seedlings were inoculated with three canker fungi—*Crypotospora betulae*, *Melanconium betulinum*, and *B. dothidea*—and exposed to drought, canker formation was observed with *M. betulinum* and *B. dothidea* but not with *C. betulae* (Table IV). *Botryosphaeria dothidea* appeared to be the most aggressive of the three and formed the largest cankers. The apparent difference in aggressiveness corresponded to the natural occurrence of the cankers; *B. dothidea* is the most common canker fungus on stressed trees and *C. betulae* is found only on dead or severely weakened branches. Ghaffar and Erwin (1969) reported that cotton *(Gossypium hirsutum)* plants were predisposed by water deficits to root rot caused by *Macrophomina phaseoli*. Only stressed plants were infected, whether or not nutrients were added to the mycelial inoculum (Table V). Since severe root rot damage did not

TABLE IV

INFLUENCE OF PLANT WATER STRESS ON PERCENT OF CANKER FORMATION AND
DIAMETER OF CANKERS FORMED ON *Betula alba* SEEDLINGS
INOCULATED WITH THREE CANKER FUNGI[a]

| Pathogen | Treatment | Plant water potential (bars)[b] | Canker formation (%)[b] | Canker diameter (mm)[b] |
|---|---|---|---|---|
| *Cryptospora betulae* | Turgid | −2.8 | 0[c] | 0[c] |
|  | Wilted | −17.6 | 0 | 0 |
| *Melanconium betulinum* | Turgid | −2.5 | 0 | 0 |
|  | Wilted | −18.0 | 100 | 17.0 |
| *Botryosphaeria dothidea* | Turgid | −2.3 | 0 | 0 |
|  | Wilted | −17.6 | 100 | 35.75 |

[a] Plants incubated at given water potential for 7 days after inoculation, according to the method of Schoeneweiss (1975b).

[b] Means of 4 reps.

[c] Data recorded 1 month after exposure to water stress.

TABLE V

EFFECT OF WATER STRESS ON THE DEVELOPMENT OF ROOT ROT
OF COTTON CAUSED BY *Macrophomina phaseoli*[a]

| | Plants affected/plants treated | | | |
|---|---|---|---|---|
| | Subjected to water stress | | Watered regularly | |
| | Wilted | Infected[c] | Wilted | Infected[c] |
| *M. phaseoli* inoculum[b] |  |  |  |  |
| in water | 6/8 | 8/8 | 0/8 | 0/8 |
| in 1% sucrose | 5/7 | 7/7 | 0/8 | 0/8 |
| in Czapek's solution | 7/8 | 8/8 | 0/8 | 1/8 |
| 1% Sucrose solution | 0/8 | 0/8 | 0/8 | 0/8 |
| Czapek's solution | 0/8 | 0/8 | 0/8 | 0/8 |
| Noninoculated, water only | 0/8 | 0/8 | 0/8 | 0/8 |
| Plants not watered | 2/2 | 0/2 |  |  |

[a] From Ghaffar and Erwin (1969).

[b] 20 ml/pot of a mycelial suspension of *M. phaseoli*.

[c] Infection was indicated by the presence of microsclerotia on roots and discoloration of both cortical and xylem tissues.

occur at different temperatures unless plants were wilted, water stress appeared to be a more important predisposing factor for cotton root rot than temperature or the availability of nutrients.

In many cases, increased susceptibility of plants to attack by nonaggressive pathogens due to water stress can be considered an elastic or reversible strain (Schoeneweiss, 1975a). If the stress is relieved before girdling or death of plant parts occurs, host defense reactions may recover and arrest disease development. For example, cankers continued to develop on white birch seedlings inoculated with *B. dothidea* as long as plants were exposed to water deficit (Fig. 3), but callus tissue formed within one week and cankers stopped expanding two weeks after turgidity was restored by watering (Crist and Schoeneweiss, 1975). A reduction in canker growth after turgidity was restored was also reported by Bloomberg (1962). Cankers caused by *Cytospora chrysosperma* expanded rapidly on inoculated poplar cuttings removed from water, but canker expansion decreased and finally stopped when stem turgidity was restored (Fig. 4). Similar results were obtained by Bloomberg (1962) with rooted cuttings that were wilted and inoculated and then watered after cankers had formed. These effects were attributed to bark moisture content falling below a critical or threshold percentage for infection.

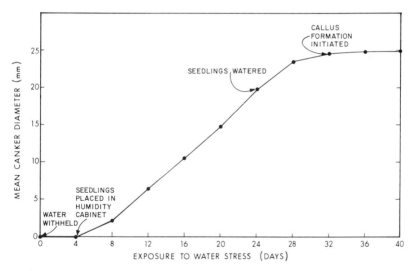

Fig. 3. Relation of length of exposure to water stress to canker diameter in stems of *Betula alba* inoculated with *Botryosphaeria dothidea*. Mean canker diameter is based on measurements at 4-day intervals of maximum diameters of cankers formed on 20 seedlings exposed to water stress levels more negative than −12 bars. (From Crist and Schoeneweiss, 1975.)

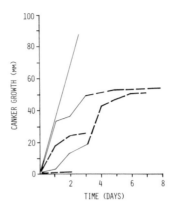

Fig. 4. *Cytospora* canker growth in poplar cuttings under different moisture treatments (———, cuttings dry; - - -, cuttings in water). (From Bloomberg, 1962.)

Apparently, relatively nonaggressive facultative parasites do not cause damage unless the host is predisposed or weakened by exposure to stress sufficient to cause a significant increase in disease susceptibility.

## VI. CONTROL AND MEASUREMENT OF WATER DEFICITS IN DISEASE RESEARCH

Although water deficit, particularly prolonged drought, has been implicated many times as a predisposing factor in plant disease, few studies have been conducted in which water deficits were actually controlled and/or measured. Most associations between water stress and disease are based on field observations of disease damage appearing during or following dry periods. Some progress has been made, however, and the following discussion will attempt to cover the available information on controlled research dealing with water stress predisposition.

### A. Control of Water Deficits

Since disease symptoms usually require days, weeks, or months to develop, most research on the influence of water stress on disease has involved exposure of intact plants to some degree of soil water deficit over an extended period of time. The problems encountered in controlling soil moisture within the readily available range (FC to PWP) were summarized by Couch *et al.* (1967), together with a discussion of methods used to surmount these difficulties in studying plant diseases.

*1. Withholding Irrigation*

By far the most common method of exposing potted or field-grown plants to water stress is to withhold irrigation until plants wilt or until a predetermined soil moisture content is reached. Edmunds (1964) subjected grain sorghum *(Sorghum vulgare)* to water stress by withholding water until available soil moisture (ASM) was below 25%. Watering was resumed 4 to 7 days later and plants were inoculated immediately after watering with the charcoal stalk rot fungus, *Macrophomina phaseoli*. Since severe disease damage appeared in water-stressed plants and disease did not develop in well-watered control plants, they concluded that grain sorghum was predisposed to disease by water stress. Subsequently, Edmunds *et al.* (1964) used controlled irrigation to induce water stress in field plots for screening sorghum lines for resistance to stalk rot. Ghaffar and Erwin (1969) dried potting soil of cotton plants to 10% moisture content (oven dry wt. basis) prior to inoculating the soil with *M. phaseoli*. They reported that typical field symptoms of root rot appeared only on plants that were predisposed to disease by drought (Table V). Potting soil of paper mulberry *(Broussonetia papyrifera* seedlings was dried to TWP prior to inoculation with the stem canker fungus *Fusarium solani* by Schreiber and Dochinger (1967). Wilting predisposed the plants to infection by the pathogen and cankers continued to form even if watering was resumed after inoculation.

A method was developed by Crist and Schoeneweiss (1975) and later refined by Schoeneweiss (1975b) for maintaining relatively constant plant water potentials over a 7- to 9-day incubation period. Potting soil of intact, container-grown tree seedlings was allowed to dry until plant water potentials reached predetermined values. The plants were then placed in a humidity cabinet and held under high humidity, constant temperature, and darkness or reduced photoperiod. Plant water potentials became stable after 48 hr and remained fairly stable throughout the incubation period (Fig. 5). When plants were inoculated after 48 hr in the cabinet and then incubated for 7 days, Crist and Schoeneweiss (1975) correlated plant water deficits with changes in susceptibility of white birch to attack by *B. dothidea* (see Fig. 8); Schoeneweiss made similar correlations with the same pathogen on sweetgum and red osier dogwood (Fig. 6). In each case, stem cankers did not develop unless plants were predisposed to disease by water stress.

Duniway (1977) withheld water from potted safflower *(Carthamus tinctorius)* plants until leaf water potentials decreased to $-13$ and $-15$ bars before the soil was inoculated with zoospores of *Phytophthora cryptogea*. Decreases in root fresh weight caused by disease in both a susceptible and

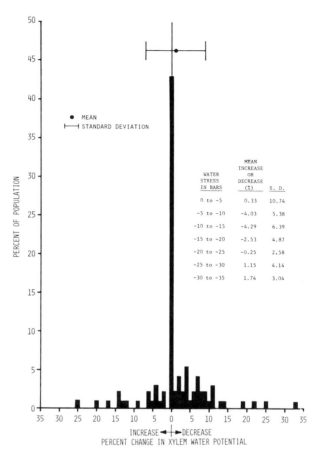

Fig. 5. Percentage of population relative to percentage change in xylem water potential over a 7-day incubation period, under complete darkness at 21°C and 97 ± 2% relative humidity. Water potentials based on pressure bomb measurements of 3 leaves from each of 59 *Cornus stolonifera* and 31 *Liquidambar styraciflua* seedlings, ranging in water potential from −1 to −33 bars. Figures in the table show relative variability at various levels of water stress. (From Schoeneweiss, 1975b.)

a resistant cultivar were significantly greater in plants subjected to water stress than in those maintained in well-watered soil (Table VI). Water stress after inoculation had much less effect on root disease than water stress before inoculation. Since Zimmer and Urie (1967) had reported previously that prolonging the time between irrigations increased the severity of the disease under field conditions, Duniway (1976) concluded that water stress was a significant predisposing factor in root rot of safflower caused by *P. cryptogea*.

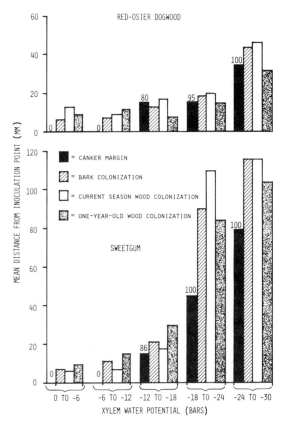

Fig. 6. Relation between xylem water potential, size of stem cankers, and extent of bark and wood colonization by *Botryosphaeria dothidea* in two woody hosts. Plants were wilted to various levels, inoculated, and incubated at relatively constant water potentials for 1 wk, then watered. Measurements were made 1 month after exposure to water stress and the data presented are means of five to ten replicates in each category of water potentials. Numbers indicate percentage of canker formation at inoculation points. Reproduced with permission from ''Predisposition, Stress and Plant Disease,'' by D. F. Schoeneweiss, *Annu. Rev. Phytopathol.* **13.** Copyright © 1975 by Annual Reviews Inc. All rights reserved.

## 2. Watering Regimes

Of the techniques developed to study the effect of soil moisture deficits in the readily available range on disease, varying the frequency of irrigation is probably the most widely used (Couch *et al.*, 1967). This approach consists of growing plants in a uniform amount of soil and allowing plants to dry the soil to a predetermined soil moisture content, then rewatering the soil to FC (Fig. 7). Couch and Bloom (1960b) used this technique to study the effect of soil moisture stress on susceptibility of

TABLE VI
INFLUENCE OF PREINOCULATION WATER STRESS ON THE SEVERITY OF ROOT ROT
CAUSED BY *Phytophthora cryptogea* IN SAFFLOWER[a]

| Experiment | Cultivar | Leaf water potential before inoculation (bars) | No. zoospores used as inoculum (no./crock) | Relative top symptoms (0–5) | Root fresh weight (gm/crock) |
|---|---|---|---|---|---|
| 1 | Nebraska 10 (susceptible) | −5.9 | 0 | 0 | 62 A[b] |
|   |   |   | $2 \times 10^5$ | 1.6 | 61 A |
|   |   | −13.2 | 0 | 0 | 41 B |
|   |   |   | $2 \times 10^5$ | 3.8 | 21 C |
| 2 | Nebraska 10 | −4.5 | 0 | 0 | 83 A |
|   |   |   | $2 \times 10^5$ | 3.6 | 64 B |
|   |   | −17.2 | 0 | 0 | 58 B |
|   |   |   | $2 \times 10^5$ | 4.4 | 28 C |
| 3 | Biggs (resistant) | −4.0 | 0 | 0 | 55 A |
|   |   |   | $4 \times 10^5$ | 0 | 45 AB |
|   |   | −8.8 | 0 | 0 | 41 AB |
|   |   |   | $4 \times 10^5$ | 0.4 | 32 B |
| 4 | Biggs | −4.2 | 0 | 0 | 53 A |
|   |   |   | $5 \times 10^5$ | 0 | 57 A |
|   |   | −12.4 | 0 | 0 | 46 A |
|   |   |   | $5 \times 10^5$ | 1.2 | 27 B |

[a] From Duniway (1977).
[b] Weights within the same experiment that are followed by different letters are significantly different by a Duncan's test at $p = 0.05$.

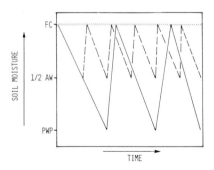

Fig. 7. Illustration of moisture extraction sequence in variation of frequency of irrigation technique for studying soil moisture stresses in the readily available range. Note that in instances of extraction to less than FC the entire soil system is irrigated back to FC after the desired point is reached. FC, field capacity; AW, available water; PWP, permanent wilting percentage. (From Couch *et al.*, 1967.)

Kentucky bluegrass to *Sclerotinia homeocarpa* (see Fig. 1). Moore *et al.* (1963) used turfgrasses to extract soil moisture to FC, one-half of FC, and to PWP before rewatering to FC. Disease proneness of the turfgrasses was greatest in plants grown in soil dried to the PWP before rewatering.

Tobacco *(Nicotiana tabacum)* plants were subjected to three watering regimes by Roten *et al.* (1968) to study the predisposing effect of moisture stress on infection by *Peronospora tabacina*. Plants irrigated at the time of incipient wilting over a 50-day period prior to inoculation were more highly predisposed to disease than those given more (t) water. Infection occurred only if the plants were turgid when inoculated. Therefore, the marked increase in disease associated with water stress pretreatment was considered a valid case of predisposition.

Couch and Bloom (1960a) used a modification of the soil moisture depletion cycle method in a study of development of root knot nematodes. Root systems of intact individual plants were divided into two equal portions and each portion transplanted into one side of a two-part container. One side was kept watered to FC and the other allowed to drop to PWP before rewatering. Low soil moisture reduced root knot damage by inhibiting migration of the nematodes. Although this technique has some merit for studying root diseases, it has had only limited usage. The use of watering regimes in disease studies has been further reviewed by Couch *et al.* (1967), Cook and Papendick (1972), and Schoeneweiss (1975a).

*3. Ceramic Plate Technique*

Miller and Burke (1974) devised a technique for creating water stress, using a porous ceramic plate under suction as one side of a soil chamber. By regulating the amount of vacuum applied to the plate, they were able to maintain constant matric potentials of $-200$ millibars (mb) and $-800$ mb in the root medium over a 4-wk period. Root rot injury caused by *Fusarium solani* f. sp. *phaseoli*, was greatest in infested soil maintained at $-800$ mb soil water potential.

Interpretation of predisposing effects of water deficits on disease, based on alternate wetting and drying of soil, may be subject to criticism since such regular cycles are different than water depletion patterns under field conditions (Schoeneweiss, 1975b). Water stress in nature is more often associated with prolonged drought, where soil water potentials remain low or decrease for a long time and exert more constant stress on plants than wide fluctuations between FC and some lower value. Therefore, the ceramic plate technique (Miller and Burke, 1974) and the humidity cabinet technique (Schoeneweiss, 1975b), in which water stress levels can be held relatively constant, should provide more valid information on water stress predisposition than other methods reported.

## B. Measurement of Water Deficits

Methods for measuring water deficits of plants and soils have been thoroughly reviewed elsewhere (Barrs, 1968; Rawlins, 1976). Most are either too complex or the instrumentation too sensitive to be of practical value in studies on plant disease relationships. Therefore only those methods which have been widely used in disease research or have particular merit for such studies will be discussed here.

### 1. Soil Water Deficits

Until recently, the most common methods of measuring water stress in plant disease studies have been based on some estimate of soil moisture percentage or soil water potential. Often such terms as TWP, available water (AW), PWP, or some fraction thereof were used to characterize degrees of water stress and were accepted as constants (Couch et al., 1967). However, as Slatyer (1957) points out, values such as PWP may be highly variable, depending on the plant species involved; plants often continue to absorb soil moisture well below this value. Determination of soil water potential based on electrical conductance, osmotic pressure, extraction by pressure membranes, and psychrometer measurements have also been used in disease research to estimate water stress. Although each method may find application in certain studies, many of these measure only certain components of water stress rather than total soil moisture stress (TSMS) (Slatyer, 1957; Couch et al., 1967).

Regardless of the confusion over various methods of estimating soil water deficits, all have the same drawback regarding studies on disease predisposition. Unless TSMS reaches equilibrium with plant water stress, estimates based on soil water determinations may not reflect water status at the internal host–pathogen interface where the effect of water stress on disease susceptibility is being exerted (Schoeneweiss, 1975b). According to Kramer (1963), the only way to know whether a plant is being subjected to water stress is by measuring the water deficit in the plant itself.

### 2. Plant Water Deficits

a. Bark Moisture Content. Bier (1959a,b,c, 1961a,b,c, 1964), a strong proponent of water stress predisposition, reported close correlation between development of tree cankers caused by native facultative parasites and moisture content or relative turgidity of living bark. In most cases, when the relative turgidity of bark was expressed as a percentage of the amount of water required to saturate the bark, relative turgidities of 80% or higher inhibited canker development. Since many published reports on tree diseases indicated the natural host factors, such as callus formation and production of tannin and suberin, operate most efficiently with an adequate supply of water, Bier (1964) considered relative bark turgidity an indication of host vigor and proposed that bark turgor of 1-yr-old dormant stems be used as a clinical index of canker vulnerability. Bloomberg (1962) and Landis and Hart (1967) found similar correlations between bark turgidity and canker susceptibility. Filer (1967), however, was not able to correlate bark turgor and rate of infection of cottonwood *(Populus deltoides)* by 3 canker fungi. Bertrand *et al.* (1976) compared the bark turgor method with the pressure chamber method (discussed below) for determining the water status of French prune *(Prunus domestica* "French") trees. Although measurements over a period of weeks with both methods showed a similar trend in plant water status, pressure chamber values were considered to be a far more sensitive measure of water stress than bark turgor. The relation of bark turgor to susceptibility of tissues other than bark to attack by pathogens has not been established (Schoeneweiss, 1975b).

*b. Psychrometer Measurements.* Thermocouple psychrometers have been used to measure leaf (Barrs, 1968) and soil water potentials (Cook, 1973). They have even been implanted into tree branches to measure stem water potential (Wiebe *et al.,* 1970). Duniway (1977) used thermocouple psychrometers to monitor leaf water potential in safflower seedlings in a study on water stress predisposition to root rot. Although thermocouple psychrometers provide an accurate estimate of plant water potential, their extreme sensitivity to temperature limits their use in field studies.

*c. Pressure Bomb Measurements.* The pressure bomb described by Scholander *et al.* (1965) has certain advantages for measuring plant water deficits. The unit is portable and can be taken to the field. Since the pressure bomb requires only a few seconds or minutes to get a reading, which is stable from the beginning, the water status measured is close to that in the intact plant (Tyree and Hammel, 1972). In most cases, pressure bomb readings compare favorably with those made by thermocouple psychrometers (Boyer, 1966; Kaufmann, 1968). Since the pressure bomb measures nonosmotic water potential, it must be calibrated with a psychrometer for each species tested to obtain total leaf water potential (Boyer, 1966; Kaufmann, 1968). However, the technique provides accurate, reproducible estimates of xylem water potential (Schoeneweiss, 1975b) and gives the best field estimates of plant water potential yet devised (Hodges and Lorio, 1971). The pressure bomb technique can easily be standardized to eliminate much of the method error originally encountered (Waring and Cleary, 1967; Schoeneweiss, 1975b).

The pressure bomb has been used for several studies on disease predisposition (Schoeneweiss, 1975a). Cook (1973) and Papendick and Cook (1974) combined pressure bomb and thermocouple psychrometer readings in studies on the influence of low plant and soil water potentials on diseases caused by soil-borne fungi. Crist and Schoeneweiss (1975) and Schoeneweiss (1975a) correlated levels of plant water potential measured with the pressure bomb with changes in susceptibility of tree seedlings to attack by canker fungi. Bertrand *et al.* (1976) studied the influence of moisture stress in the fall on development of *Cytospora* of prune trees during the dormant season. They reported good correlation between pressure bomb readings of plant water potential and canker development.

## VII. DEGREE OF WATER STRESS AS A PREDISPOSING FACTOR

Nearly all plants undergo diurnal and seasonal fluctuations in soil and plant water potentials, but only severe water deficits or prolonged droughts act as predisposing disease factors.

### A. THE CRITICAL WATER DEFICIT CONCEPT

If decreasing plant water potential causes a corresponding increase in susceptibility to disease, a steady increase in disease damage as water potential falls below maximum turgidity should be expected. Obviously such is not the case (Figs. 6 and 8). Where obligate parasites or aggressive facultative parasites are involved, a mild water stress of 2 or 3 bars may have some influence on the total amount of disease damage but the effect may be on growth rate of the host rather than on change in disease susceptibility. Since disease prediposition implies a change in susceptibility of host tissues to attack by a pathogen, it is often difficult to interpret the effects of stress where the pathogen is able to attack turgid or vigorous plants.

In the case of nonaggressive facultative parasites that are unable to attack vigorous plants, stress predisposition is obvious: nonstressed plants remain healthy and stressed plants become diseased (Fig. 9). In nearly 10 years of research on the influence of environmental stresses on diseases caused by nonaggressive stem canker fungi, we have not yet recorded a case of disease development on nonstressed, inoculated plants. Disease symptoms have appeared only in plants that were subjected to some critical degree of freezing, defoliation, or water stress (Schoeneweiss, 1975a).

Crist and Schoeneweiss (1975) and Schoeneweiss (1975a) reported

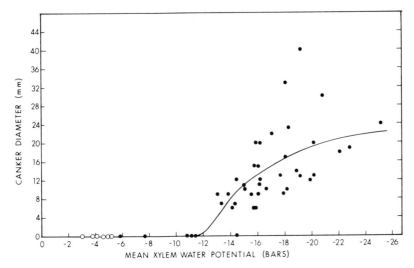

Fig. 8. Relation of xylem water potential to canker diameter in stems of *Betula alba* seedlings inoculated with *Botryosphaeria dothidea*. Water potentials are based on pressure bomb measurements of three shoots on each plant at 4-day intervals during a 16-day period of exposure to water stress. Each data point is the diameter of the canker on an individual plant, 14 days after watering was resumed (○, watered check plants; ●, plants exposed to water stress). (From Crist and Schoeneweiss, 1975.)

that stem cankers caused by *B. diothidea* appeared on several host species when plant water potentials were more negative than −12 bars (Figs. 6 and 8). Bier (1959a,b,c, 1961a,b,c) found that canker susceptibility was associated with a critical bark moisture content of approximately 80 % in many different host–pathogen combinations. Papendick and Cook (1974) reported that Fusarium root rot of wheat appeared when plant water potentials were −33 to −35 bars. They also found that under severe drought conditions a resistant wheat variety was damaged as much by disease as a susceptible variety growing under milder drought conditions. Unfortunately, most studies on water stress predisposition have involved only one or two degrees of water deficit or varied watering regimes. From such data it is seldom possible to make deduction about effects of various degrees of water stress on predisposition. However, the weight of evidence indicates that critical or threshold degrees of water deficit are usually required to alter disease susceptibility.

The use of visual symptoms to assess predisposing effects of water stress can be misleading, since a pathogen may be also to colonize plants exposed to stress some distance beyond the margin of visible lesion (Fig. 6). Criteria used to evaluate the effects of stress on disease susceptibility

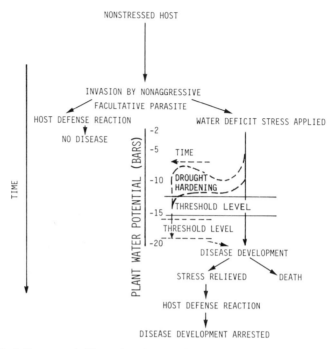

Fig. 9. A diagrammatic illustration of the predisposing effect of water stress on susceptibility of plants to attack by relatively nonaggressive facultative parasites. Hypothetical
change in the predisposing threshold level of plant water potential due to drought hardening
is indicated by broken lines.

should include the relative growth rate and/or distribution of the pathogen
in the host as well as damage symptoms. Laboratory cultures and histological sections are useful in studying the location and extension of
pathogen growth in host tissue.

The concept of critical water deficits in disease predisposition has its
counterpart in the effect of water stress on other plant processes. Hsiao
(1973) defined mild water stress as the lowering of plant water potential by
several bars, moderate stress by more than a few bars but less than $-12$ or
$-15$ bars, and severe stress by more than $-15$ bars. Crafts (1968) considered the plant water potential at which plants enter a stressed condition to
vary with the type of plant: approximately $-5$ bars for flowers and vegetable crops, $-16$ bars for field and forage crops, and $-20$ bars or greater for
desert plants. As plant water potential decreases, more processes are
affected (Table I). In some processes, such as accumulation of abscisic
acid (ABA), an abrupt change occurs in a critical water potential. Hemphill and Tukey (1975) reported a sudden fourfold increase in ABA in

*Euonymus* beginning when plant water potential decreased to near −12 bars. Brix (1962) recorded marked reduction in respiration in tomato *(Lycopersicon esculentum)* and loblolly pine beginning at a leaf water potential of −12 bars. Absorption of P by root cells of maize and soybean was severely inhibited by soil water potentials of −12 and −15 bars, respectively (Dove, 1969). The rates of many other plant processes change rapidly once plant water potential falls below a critical threshhold and often before wilting symptoms appear (Hsiao, 1973).

Many criteria have been used to characterize the degree of plant water stress, and it is literally impossible to convert most of them to a common value such as plant water potential. An interesting point to consider, however, is the relatively large number of physiological processes in plants that show abrupt change as plant water potential drops to −12 to −15 bars, which is the same range as soil water potential considered to be near the PWP for most higher plants. Water stress predisposition also occurs at plant water potentials between −12 and −15 bars with many host-pathogen combinations (Crist and Schoeneweiss, 1975; Schoeneweiss, 1975a; Duniway, 1976). Thus the available information supports the hypothesis that changes in host plant metabolism associated with disease predisposition require a critical or threshhold degree of plant water stress.

## B. Drought Hardening

It is well known that some plants adapt to water deficits or "drought harden" and others do not (Levitt, 1972). Hardening is caused by exposure to a sublethal water stress and results in resistance to an otherwise lethal stress. The mechanisms involved in drought hardening are complex and vary from species to species. In some cases, plants growing under water deficit exhibit xeromorphy or a change in morphology that enables them to tolerate water stress without significant damage (Levitt, 1972). In other plants, hardening is a result of increased capacity to maintain a high ratio of synthetic to hydrolytic reactions when undergoing stress (Henckel, 1964). Even in plants unadapted to drought, exposure to moderate drought may result in a "pseudohardening" or increase in drought avoidance due to deposit of lipids on leaves, causing a reduction in cuticular transpiration (Levitt, 1972). Hardening of plants during growth does not continually increase but falls sharply during the development of reproductive organs (Henckel, 1964). During this period, most physiological or biochemical processes in plants become sensitive to effects of soil and atmospheric drought.

If plant processes involved in drought hardening are also involved in host defense reaction to attack by potential pathogens, the most likely effect of hardening would be to lower the threshold degree of plant water potential that triggers an increase in susceptibility, as illustrated in Fig. 9. Although this lowering of the threshold degree due to hardening probably occurs in nature, no valid research data were found that either confirmed or denied such a conclusion. Our investigations showed that the extent of disease damage in stressed plants decreased when tests were conducted toward the end of the growing season. Whether this decrease was due to drought hardening or some other factor has not been established. Thus far no definite change has been observed in the critical degree or threshold of water stress in disease studies. If drought hardening does in fact increase disease resistance at given degrees of water stress, hardening treatments such as reduced irrigation at critical periods could be of some value in preventing stress-related disease damage.

## VIII. DURATION OF WATER STRESS AS A PREDISPOSING FACTOR

Since severe water deficits require time to develop in plants, it is exceedingly difficult to separate the relative influences of degree and duration of stress. Although there is considerable literature on the degree of water stress as a predisposing factor in plant disease, there is little if any useful information of effects of duration of water stress. Obviously, alteration of most plant processes by sublethal degree of water deficit must involve some minimal period of exposure to the stress. Otherwise diurnal fluctuations in plant water potential, which on a hot dry day may exceed threshold values (Klepper, 1968; Wiebe *et al.*, 1970), would result in continual predisposition to the many potential pathogens that normally gain entrance into plants. The association of disease outbreaks with prolonged dought was discussed in Section II,A, but no literature was found on research in which plant or soil water deficits were monitored throughout a drought period and correlated with the onset of disease.

Disease predisposition may involve a combination of degree and duration of water stress. White birch seedlings exposed to water stress for 20 days did not become susceptible to attack by *B. dothidea* as long as plant water potentials remained above $-12$ bars (Schoeneweiss, 1975a), whereas canker formation was initiated within 4 days (Fig. 8) if plant water potential was more negative than $-12$ bars (Crist and Schoeneweiss, 1975). In a recent test, when red osier dogwood plants were exposed to water stress and inoculated at 3-day intervals with *B. dothidea* during the stress period and after turgidity was restored, canker formation was recorded only on plants with plant water potential more negative than −12 to −15 bars for more than 3 days. Cankers did not form on plants inoculated immediately after turgidity was restored, indicating that host defense reaction recovered rapidly from the elastic strain imposed by water stress. In contrast, Edmunds (1964) reported charcoal stalk rot development in water-stressed grain sorghum inoculated immediately after watering, and Duniway (1977) recorded significant root rot damage caused by *Phytophthora cryptogea* in safflower when plants predisposed by water stress were watered prior to inoculation. Schreiber and Dochinger (1967) also reported that water stress predispostion of paper mulberry to Fusarium canker continued after watering was resumed. In most other studies on predisposing effects of water stress, water was withheld or plants were subjected to watering regimes until significant disease damage appeared (Couch *et al.*, 1967). Therefore valid conclusions concerning the duration of water stress required to induce disease susceptibility at given degrees of stress await more critical studies.

## IX. INFLUENCE OF WATER STRESS ON HOST DEFENSE RESPONSES

A pathogen is by definition capable of causing disease. Whether this capability will be exercised depends on environmental influences and on susceptibility of host cells to attack. When a host is resistant or incompatible, the pathogen is unable to penetrate the host surface, the internal host environment is unsuitable for pathogen growth, the pathogen is attacked or inhibited by the host, or pathogen-secreted toxins are not produced or are prevented from damaging host cells.

Barriers to penetration rarely account for disease resistance, since most pathogens enter resistant and susceptible plants with equal frequency (Tomiyama, 1963). The nutritional suitability of the host as a substrate for pathogen growth could be a resistance factor but it has seldom been demonstrated that deficiency of host nutrients limits growth of pathogens (Bollard and Matthews, 1966; Hare, 1966). In most cases, disease resistance occurs as a result of the dynamic response of living host cells to the presence of a pathogen or its metabolites (Allen, 1959).

Most studies on host defense mechanisms have been conducted with obligate parasites or aggressive facultative parasites that attack vigorous hosts. However, predisposing factors in plant disease usually increase susceptibility in plants that have a high level of resistance when not subjected to stress. The lack of information in the literature on the effects of stresses on host resistance responses makes any conclusion on these effects open to criticism. Therefore, the following discussion of possible relations between known effects of water deficits on plant processes and host defense responses must be considered hypothetical.

## A. Morphological Responses

Morphological or physical barriers to growth of pathogens following their penetration into the host have been implicated in host defense to many plant diseases (Gäumann, 1950; Akai, 1959; Barnett, 1959; Wood, 1967; Baker and Cook, 1973). Production of gums, lignin, suberin, and formation of callus, wound periderm, tyloses, and other barriers to colonization of host tissue are nonspecific host responses to injury or attack by pathogens. These responses are associated with actively growing plants, and exposure to environmental stress can be expected to reduce their effectiveness as disease resistance mechanisms. For example, Butin (1955) attributed resistance of poplars to *Cytospora* canker to the capacity of the host to form wound periderm in the presence of adequate tissue water. In other diseases canker formation often progresses during the dormant season when the host is unable to produce morphological barriers (Bier, 1964). Bertrand *et al.* (1976) reported that development of *Cytospora* canker in prune trees during the dormant season correlated with plant water stress induced by withholding fall irrigation. Drought stress after growth resumed in the spring had no effect on disease development and canker activity decreased in all treatments during the growing season. Plant pathogens may also remain latent but viable in plant tissues after disease development has been arrested by host defense responses, then resume activity when the plant becomes dormant or is exposed to environmental stress (Baker and Cook, 1973).

In attributing disease resistance to morphological barriers, it should be kept in mind that these barriers usually require time to develop and may be the result rather than the cause of host resistance. If pathogen activity is inhibited in some way long enough for the host to form an impenetrable barrier, disease resistance may be considered to be a combination of physiological and morphological responses.

## B. Physiological Responses

When a potential pathogen is in intimate contact with living host cells and physical barriers are not present to prevent a pathogen from attacking the host, either the host substrate is inadequate for pathogen growth or the pathogen or its metabolic products are being inhibited or inactivated in an incompatible relationship. The influence of water stress as a predisposing factor, therefore, is to alter the host substrate or the physiological response of the host to the pathogen. Since predisposition is usually characterized by an increase in susceptibility to nonspecific pathogens, the effect probably is exerted on some basic process or processes sensitive to water stress.

Much evidence indicates that facultative fungal parasites produce toxic substances or enzymes that kill host cells in advance of the fungus, which then grows saprophytically in dead or disorganized tissue (Bollard and Matthews, 1966; Hare, 1966). According to Albersheim *et al.* (1969), virulent plant pathogens are able to produce polysaccharide-degrading enzymes that can attack every component of the primary cell wall of higher plants. They hypothesized that avirulent pathogens were unable to produce one or more of these enzymes and that enzyme production by the pathogen was controlled by the carbohydrate composition of host cell walls. It was subsequently shown, however, that the polysaccharide components of cell walls of higher plants were rather similar and could not account for varietal specificity in plant disease (Albersheim and Anderson-Prouty, 1975). Recent research indicates that endopolygalacturonase enzymes are the first wall-degrading enzymes produced by pathogens and that, associated with their cell walls, all dicotyledons have proteins capable of inhibiting the action of these enzymes. Therefore Albersheim and Anderson-Prouty (1975) suggested that a pathogen cannot attack a dicotyledonous host plant unless conditions within the host are conducive to secretion by the pathogen of a sufficient amount of endopolygalacturonase to overcome the action of the inhibitor in the cell walls of the host. Additional investigations showed that such an enzyme inhibition system is nonspecific and could be involved in general or field resistance of dicotyledons to infection. Since no such inhibitors have been found in monocotyledons, it was suggested that another degradative enzyme may be involved in breakdown of the primary cell wall in monocotyledons.

Production and accumulation of phytoalexins, which are toxic or inhibitory to microorganisms, have been given much credit for resistance of host plants to infection (Kúc, 1972; Ingham, 1972; Wheeler, 1975). Albersheim and Anderson-Prouty (1975) suggested that certain carbohydrates in the cell walls of fungi and bacteria act as elicitors which interact or are "recognized" by specific protein receptors in host cells, triggering the production of phytoalexins by the host. More recent research, summarized by Ebel *et al.* (1976), appeared to support the elicitor hypothesis of phytoalexin production. They reported that a glucan elicitor from cell walls of *Phytophthora megasperma* var. *sojae (Pms)*, a fungal pathogen of soybean, stimulated production of phenylalanine ammonia-lyase (pal) in such diverse species as soybean *(Glycine max)*, parsley *(Petroselinium crispum)*, and sycamore *(Platanus occidentalis)*. Since pal caused accumulation of phytoalexin in soybeans, Ebel *et al.* (1976) hypothesized that glucan elicitors in pathogen cell walls may stimulate responses in host cells that are involved in general or field resistance to pathogens and nonpathogens that enter host plants. In a companion article (Ayers *et al.*, 1976), the same group of researchers summarized results of research indicating that mannin-containing protein in the cell walls of yeasts, which distinguish different strains and mating types, are quite similar to mannin proteins found in cell walls of the soybean pathogen *Pms*. They suggested that these mannin-containing proteins could be factors responsible for race specificity of pathogens. Unfortunately, nearly all research on phytoalexins has been conducted with either nonpathogens or aggressive, host-specific pathogens that are able to attack vigorous hosts. Production of phytoalexins in nonstressed hosts in response to the presence of nonaggressive pathogens and little or no production in stressed hosts has not been reported. However, the absence of such reports may be due more to a lack of research effort than to absence of the response.

Incompatible interactions between host cells and nonpathogens or avirulent races of pathogens are generally manifested in a hypersensitive response which involves rapid death of host cells, often before penetration of the host cell wall has occurred (Stoessl *et al.*, 1976). The hypersensitive reponse may or may not be followed by production of phytoalexins. However, even hypersensitive death of host cells may be preceeded by inhibition of pathogen growth in an incompatible host-pathogen interaction (Király *et al.*, 1972, Stoessl *et al.*, 1976). Király *et al.* (1972) were able to induce hypersensitive cell necrosis in potato *(Solanum tuberosum)* and bean *(Phaseolus vulgaris)* infected with compatible races of *Phytophthora* when pathogen growth was inhibited by treating the host with chloramphenicol or streptomycin. They also reported cell necrosis in wheat *(Triticum vulgare)* infected with *Puccinia graminis* and treated with chemotherapeutants blasticidin S or $Ni(NO_3)_2$, and in *Uromyces phaseoli* infected pinto beans *(P. vulgaris)* in which pathogen growth was inhibited by heat treatment or by application of $Ni(NO_3)_2$. Since neither fungal infection nor treatment with fungal inhibitors alone produced the same response, they concluded that the hypersensitive reaction was a consequence of an unknown defense reaction rather than the cause of host resistance. Király *et al.* (1972) further suggested that release of endotoxins from pathogen cells damaged by the unknown defense reaction induced necrosis and phytoalexin production in host cells. According to Stoessl *et al.* (1976), however, electron microscope studies have shown that death of host cells in hypersensitive reactions clearly precedes the death of pathogen cells. Thus, although toxin production by living cells of a pathogen may be involved in the hypersensitive reaction, damage to cells of the pathogen by host defense reaction is not a prerequisite for the hypersensitive response. In regard to disease predisposition due to water stress, the pathogens that attack stressed plants are usually facultative parasites and hypersensitive cell necrosis should favor rather than repress disease development.

Pathogens that attack plants predisposed to disease by water stress either grow poorly or are inhibited in nonstressed plants. If accumulation of phytoalexins or other inhibitors is responsible for general or field resistance to disease, as suggested by Ebel *et al.* (1976), growth of nonaggressive pathogens that enter nonstressed hosts must be slow enough for inhibitors to accumulate in sufficient quantity to completely inhibit further growth of the pathogens. Conversely, pathogen growth in a stressed plant must be rapid enough to precede accumulation of inhibitors. Thus the effect of water stress as a predisposing factor could be to increase pathogen growth or to reduce the rate of host defense reaction. Since sugars accumulate in plants under water stress (Hsiao, 1973), an increase in the carbohydrate pool available to the pathogen may result in increased pathogen growth. Reduced protein synthesis and decomposition of proteins and nucleic acids in plants under water stress (Vaadia *et al.*, 1961; Henckel, 1964) may reduce the ability of the host plant to respond to the presence of a pathogen and hence reduce production and accumulation of compounds that inhibit pathogen growth.

Although information is lacking on the effects of water deficits on physiological processes responsible for resistance of nonstressed plants to attack by pathogens, carbohydrate metabolism of host plants subjected to water stress appears to be involved. In addition to the increase in total sugars in plants under water stress, reducing sugar such as glucose are converted to nonreducing sugars (Hsiao, 1973). High concentrations of glucose repress synthesis by pathogens of several polysaccharide-degrading enzymes (Albersheim *et al.*, 1969) and conversion to sucrose or other sugars under water stress may derepress enzyme synthesis, allowing the pathogen to degrade host cell walls.

Other than carbohydrate metabolism, plant processes sensitive to water stress that may be involved in disease predisposition include reduced energy transport due to reduction in P absorption and phosphorylation (Vaadia *et al.*, 1961). An increase in the physiological age of the plant induced by water stress (Darbyshire, 1971) also reflects loss of vigor and reduction in energy transport. If cell wall degradation is the initial event that determines success or failure of an infection, the capacity of the host to repair cell walls rapidly would be a critical factor in disease resistance.

Secretion of materials from the cytoplasm through the plasmalemma and into the cell wall by the Golgi apparatus and the endoplasmic reticulum, especially the synthesis of polysaccharides or polysaccharide–protein complexes, requires large amounts of energy (Wheeler, 1975). Thus reduction in energy transport may reduce the host's ability to repair cell walls and resist pathogenic attack.

Amounts of regulatory compounds such as abscisic acid and cytokinin, which appear to change markedly at water stress levels similar to those associated with disease predisposition, may also have an effect on host responses, although not enough evidence is available to draw conclusions from such associations. In general, much in known about the effects of water deficits on plant processes but little information is available on specific processes involved in host defense responses to infection. Therefore much more research is needed on the physiology of host predisposition to disease.

## X. CONCLUDING REMARKS

Water deficits have many adverse effects on plant growth and development. One of these effects is an increase in susceptibility to attack by plant pathogens. If susceptibility of the host is altered before an infection is established, the plant is said to be predisposed. Since predisposition implies a change in host disposition or proneness to disease, water stress must affect processes in the host that enable it to defend itself against attack by a pathogen. Until we understand how the host is able to resist this attack, it is nearly impossible to determine how water stress causes the plant to become predisposed. Unfortunately, our understanding of this central consideration is incomplete. Much valuable information has accumulated on effects of water deficits on plant processes and considerable research has been conducted on the infection process and host responses to infection. Nearly all of this work, however, has involved host–pathogen combinations in which the success of infection was determined by genetic factors acting under conditions favorable for the host. The expression of these factors when the host is exposed to environmental stress such as water deficit may be quite different. The fact that many plants, which successfully resist infection when not stressed, become predisposed by exposure to water stress is strong evidence that host defense responses are affected.

Pathogens that cause increased damage in plants predisposed by stress usually have little difficulty in gaining entrance into host tissues. Once inside the plant, they are in intimate contact with host cells and often can survive and even colonize a nonstressed host to a limited extent without causing visible damage. If the host is exposed to water stress until plant water potential drops below a predisposing critical value and remains there for some minimal period of time, the balance shifts in favor of the pathogen and disease development occurs.

Methods and techniques have been developed to expose plants to controlled water deficit stress and to identify and characterize host responses to infection. Only a lack of research effort remains to be overcome to unlock the riddle of predisposition and provide information needed to understand how stress causes plants to become vulnerable to pathogenic attack.

## REFERENCES

Albersheim, P., Jones, T. M., and English, P. D. (1969). Biochemistry of the cell wall in relation to infective processes. *Annu. Rev. Phytopathol.* **7**, 171–194.

Albersheim, P., and Anderson-Prouty, A. J. (1975). Carbohydrates, proteins, cell surfaces, and the biochemistry of pathogenesis. *Annu. Rev. Plant Physiol.* **26**, 31–52.

Akai, S. (1959). Histology of defense in plants. *In* "Plant Pathology" (J. G. Horsfall and A. E. Dimond, eds.), Vol. 1, p. 391. Academic Press, New York.

Allen, P. J. (1959). Physiology and biochemistry of defense. *In* "Plant Pathology" (J. G. Horsfall and A. E. Dimond, eds.), Vol. 1, p. 435. Academic Press, New York.

Ayers, A. R., Valent, B., Ebel, J., and Albersheim, P. (1976). Host–pathogen interactions. XI. Composition and structure of wall-released elicitor fractions. *Plant Physiol.* **57**, 766–774.

Bagga, D. K., and Smalley, E. B. (1967). Water stress in relation to initiation and development of *Hypoxylon* canker of aspen. *Phytopathology* **57**, 802 (abstr.).

Bagga, D. K., and Smalley, E. B. (1969). Factors affecting canker development on *Populus tremuloides* artificially inoculated with *Hypoxylon pruinatum*. *Can. J. Bot.* **47**, 907–914.

Bagga, D. K., and Smalley, E. B. (1974). The development of *Hypoxylon* canker of *Populus tremuloides:* Role of interacting environmental factors. *Phytopathology* **64**, 658–662.

Baker, K. F., and Cook, R. J. (1973). "Biological Control of Plant Pathogens." Freeman, San Francisco, California.

Barnett, H. C. (1959). Plant disease resistance. *Annu. Rev. Microbiol.* **13**, 191–209.

Barrs, H. D. (1968). Determination of water deficits in plant tissues. In "Water Deficits and Plant Growth" (T. T. Kozlowski, ed.), Vol. 1, p. 235. Academic Press, New York.

Bertrand, P. F., English, H., Uriu, K., and Schick, F. J. (1976). Late season water deficits and development of *Cytospora* canker in French prune. *Phytopathology* **66**, 1318–1320.

Bier, J. E. (1959a). The relation of bark moisture to the development of canker diseases caused by native facultative parasites. I. *Cryptodiaporthe* canker on willow. *Can. J. Bot.* **37**, 229–238.

Bier, J. E. (1959b). The relation of bark moisture to the development of canker diseases caused by native facultative parasites. II. *Fusarium* canker on black cottonwood. *Can. J. Bot.* **37**, 781–788.

Bier, J. E. (1959c). The relation of bark moisture to the development of canker diseases caused by native facultative parasites. III. *Cephalosporium* canker on western hemlock. *Can. J. Bot.* **37**, 1140–1142.

Bier, J. E. (1961a). The relation of bark moisture to the development of canker diseases caused by native facultative parasites. IV. Pathogenicity studies of *Cryptodiaporthe salicella* (Fr.) Petrak, and *Fusarium lateritium* Nees. on *Populus trichocarpa* Torry and Gray, *P.* 'Robusta', *P. tremuloides* Michx., and *Salix* sp. *Can. J. Bot.* **39**, 139–144.

Bier, J. E. (1961b). The relation of bark moisture to the development of canker diseases caused by native facultative parasites. V. Rooting behavior and disease vulnerability in cuttings of *Populus trichocarpa* Torrey and Gray, and *P.* 'Robusta.' *Can. J. Bot.* **39**, 145–154.

Bier, J. E. (1961c). The relation of bark moisture to the development of canker diseases caused by native facultative parasites. VI. Pathogenicity studies of *Hypoxylon pruinatum* (Klotzch) Cke., and *Septoria musiva* Pk. on species of *Acer, Populus,* and *Salix. Can. J. Bot.* **39**, 1555–1561.

Bier, J. E. (1964). The relation of some bark factors to canker susceptibility. *Phytopathology* **54**, 250–253.

Bier, J. E., and Rowat, H. (1962). The relation of bark moisture to the development of canker diseases caused by native facultative parasites. VIII. Ascospore infection of *Hypoxylon pruinatum* (Klotzch) Cke. *Can. J. Bot.* **40**, 897–901.

Bloomberg, W. J. (1962). *Cytospora* canker of poplars: Factors influencing the development of the disease. *Can. J. Bot.* **40**, 1271–1280.

Bollard, E. G., and Matthews, R. E. F. (1966). The physiology of parasitic disease. *In* "Plant Physiology" (F. C. Steward, ed.), Vol. 4B, p. 417. Academic Press, New York.

Boyce, J. S. (1961). "Forest Pathology." McGraw-Hill, New York.

Boyer, J. S. (1966). Leaf water potentials measured with a pressure chamber. *Plant Physiol.* **42**, 133–137.

Boyer, J. S. (1976). Water deficits and photosynthesis. *In* "Water Deficits and Plant Growth" (T. T. Kozlowski, ed.), Vol. 4, p. 153. Academic Press, New York.

Brix, H. (1962). The effect of water stress on the rates of photosynthesis and respiration in tomato plants and loblolly pine seedlings. *Physiol. Plant.* **15**, 10–20.

Butin, H. (1955). Über den Einfluss des Wassergehaltes der Peppel auf irhe Resistenz gegenüber *Cytospora chrysosperma* (Pers.) Fr. *Phytopathol. Z.* **24**, 245–264.

Chester, K. S. (1933). The problem of acquired physiological immunity in plants. *Q. Rev. Biol.* **8**, 129–154 and 275–324.

Colhoun, J. (1973). Effects of environmental factors on plant disease. *Annu. Rev. Phytopathol.* **11**, 343–364.

Cook, R. J. (1973). Influence of low plant and soil water potentials on diseases caused by soilborne fungi. *Phytopathology* **63**, 451–458.

Cook, R. J., and Christen, A. A. (1976). Growth of cereal root rot fungi as affected by temperature-water potential interactions. *Phytopathology* **66**, 193–197.

Cook, R. J., and Papendick, R. I. (1972). Influence of water potential of soils and plants on root diseases. *Annu. Rev. Phytopathol.* **10**, 349–374.

Costello, L., and Paul, J. L. (1975). Moisture relations in transplanted container plants. *HortScience* **10**, 371–372.

Couch, H. B., and Bloom, J. R. (1960a). Influence of siol moisture stresses on the development of the root knot nematode. *Phytopathology* **50**, 319–321.

Couch, H. B., and Bloom, J. R. (1960b). Influence of environment on diseases of turfgrasses. II. Effect of nutrition, pH, and soil moisture on *Sclerotinia* dollar spot. *Phytopathology* **50**, 761–763.

Couch, H. B., Purdy, L. H., and Henderson, D. W. (1967). Application of soil moisture principles to the study of plant disease. *Va. Polytech. Inst. State Univ., Dep. Plant. Pathol., Bull.* No. 4.

Crafts, A. S. (1968). Water deficits and physiological processes. *In* "Water Deficits and Plant Growth" (T. T. Kozlowski, ed.), Vol. 2, p. 85. Academic Press, New York.

Crist, C. R., and Schoeneweiss, D. F. (1975). The influence of controlled stresses on susceptibility of European white birch stems to attack by *Botryosphaeria dothidea*. *Phytopathology* **65**, 369–373.

Darbyshire, B. (1971). The effect of water stress on indoleacetic acid oxidase in pea plants. *Plant Physiol.* **47**, 65–67.

Denyer, W. B. G. (1953). *Cephalosporium* canker of western hemlock. *Can. J. Bot.* **31**, 361–366.

Dove, L. D. (1969). Phosphate absorption by air-stressed root systems. *Planta* **86**, 1–9.

Duniway, J. M. (1977). Predisposing effect of water stress on the severity of Phytophthora root rot in safflower. *Phytopathology* **67**, 884–889.

Ebel, J., Ayers, A. R., and Albersheim, P. (1976). Host–pathogen interactions. XII. Response of suspension-cultured soybean cells to the elicitor isolated from *Phytophthora megasperma* var. *sojae*, a fungal pathogen of soybeans. *Plant Physiol.* **57**, 775–779.

Edmunds, L. K. (1964). Combined relation of plant maturity, temperature, and soil moisture to charcoal stalk rot development in grain sorghum. *Phytopathology* **54**, 514–517.

Edmunds, L. K., Voigt, R. L., and Carasso, F. M. (1964). Use of Arizona climate to induce charcoal rot in grain sorghum. *Plant Dis. Rep.* **48**, 300–302.

Filer, T. H., Jr. (1967). Pathogenicity of *Cytospora*, *Phomopsis*, and *Hypomyces* on *Populus deltoides*. *Phytopathology* **57**, 978–980.

Gäumann, E. (1950). "Principles of Plant Infection." Hafner, New York.

Ghaffar, A., and Erwin, D. C. (1969). Effect of soil water stress on root rot of cotton caused by *Macrophomina phasioli*. *Phytopathology* **59**, 795–797.

Griffin, D. M. (1969). Soil water in the ecology of fungi. *Annu. Rev. Phytopathol.* **7**, 284–310.

Hare, R. C. (1966). Physiology of resistance to fungal diseases in plants. *Bot. Rev.* **32**, 95–137.

Hartig, R. (1882). "Lehrbuch der Baumkrankheiten" [H. M. Ward, ed., Engl. transl. by W. Somerville). Macmillan, New York, 1894].

Hemphill, D. D., Jr., and Tukey, H. B., Jr. (1975). Effect of intermittent mist on abscisic acid levels in *Euonymus alatus* Sieb.: Leaching vs. moisture stress. *HortScience* **10**, 369–370.

Henckel, P. A. (1964). Physiology of plants under drought. *Annu. Rev. Plant Physiol.* **15**, 363–386.

Hepting, G. H. (1963). Climate and forest diseases. *Annu. Rev. Phytopathol.* **1**, 31–50.

Hodges, J. D., and Lorio, P. L., Jr. (1971). Comparison of field techniques for measuring moisture stress in large loblolly pines. *For. Sci.* **17**, 220–223.

Hsiao, T. C. (1973). Plant responses to water stress. *Annu. Rev. Plant Physiol.* **24**, 519–570.

Ingham, J. L. (1972). Phytoalexins and other natural products as factors in plant disease resistance. *Bot. Rev.* **38**, 343–424.

Kaufmann, M. R. (1968). Evaluation of the pressure chamber technique for estimating plant water potential of forest tree species. *For. Sci.* **14**, 369–374.

Király, Z., Barna, B., and Ersek, T. (1972). Hypersensitivity as a consequence, not the cause, of plant resistance to infection. *Nature (London)* **239**, 456–458.

Klepper, B. (1968). Diurnal pattern of water potential in woody plants. *Plant Physiol.* **43**, 1931.

Kozlowski, T. T. (1967). Diurnal variations in stem diameters of small trees. *Bot. Gaz. (Chicago)* **128**, 60.

Kozlowski, T. T. (1968). Introduction. *In* "Water Deficits and Plant Growth" (T. T. Kozlowski, ed.), Vol. I, p. 1. Academic Press, New York.

Kozlowski, T. T., ed. (1968a). "Water Deficits and Plant Growth," Vol. I. Academic Press, New York.

Kozlowski, T. T., ed. (1968b). "Water Deficits and Plant Growth," Vol. 2. Academic Press, New York.

Kozlowski, T. T., ed. (1972). "Water Deficits and Plant Growth," Vol. 3. Academic Press, New York.

Kozlowski, T. T., ed. (1976). "Water Deficits and Plant Growth," Vol. 4. Academic Press, New York.

Kramer, P. J. (1963). Water stress and plant growth. *Agron. J.* **55**, 31–35.

Kuć, J. (1972). Phytoalexins. *Annu. Rev. Phytopathol.* **10**, 207–232.

Landis, W. R., and Hart, J. H. (1967). Cankers of ornamental crabapples associated with *Physalospora obtusa* and other microorganisms. *Plant Dis. Rep.* **51**, 230–234.

Leaphart, C. D., and Stage, A. R. (1971). Climate: A factor in the origin of the pole blight disease of *Pinus monticola* Dougl. *Ecology* **52**, 229–239.

Levitt, J. (1972). "Responses of Plants to Environmental Stresses." Academic Press, New York.

Livne, A., and Vaadia, Y. (1972). Water deficits and hormone relations. *In* "Water Deficits and Plant Growth" (T. T. Kozlowski, ed.), Vol. 3, p. 241. Academic Press, New York.

Mazur, P. (1969). Freezing injury in plants. *Annu. Rev. Plant Physiol.* **20**, 419–448.

Miller, D. E., and Burke, D. W. (1974). Influence of soil bulk density and water potential on *Fusarium* root rot of beans. *Phytopathology* **64**, 526–529.

Moore, L. D., Couch, H. B., and Bloom, J. R. (1963). Influence of environment on diseases of turfgrasses. III. Effect of nutrition, pH, soil temperature, air temperature, and soil moisture on Pythium blight of Highland bentgrass. *Phytopathology* **53**, 53–57.

Moore, L. W., Lagerstedt, H. B., and Hartmann, N. (1974). Stress predisposes young filbert trees to bacterial blight. *Phytopathology* **64**, 1537–1540.

Naylor, A. W. (1972). Water deficits and nitrogen metabolism. *In* "Water Deficits and Plant Growth" (T. T. Kozlowski, ed.), Vol. 3, p. 241. Academic Press, New York.

Neely, D. (1968). Bleeding necrosis of sweetgum in Illinois and Indiana. *Plant Dis. Rep.* **52**, 223–225.

Papendick, R. J., and Cook, R. J. (1974). Plant water stress and development of *Fusarium* foot rot of wheat subjected to different cultural practices. *Phytopathology* **64**, 358–363.

Rawlins, S. L. (1976). Measurement of water content and the state of water in soils. *In* "Water Deficits and Plant Growth" (T. T. Kozlowski, ed.), Vol. 4, pp. 1–55. Academic Press, New York.

Ross, E. W. (1964). Cankers associated with ash dieback. *Phytopathology* **54**, 272–275.

Ross, E. W. (1966). Ash dieback: Etiological and developmental studies. *N. Y. State Coll. For Syracuse Univ., Tech. Publ.* **88**.

Roten, J., Cohen, Y., and Spiegel, S. (1968). Effect of soil moisture on the predisposition of tobacco to *Peronospora tabacina*. *Plant Dis. Rep.* **52**, 310–313.

Schoeneweiss, D. F. (1975a). Predisposition, stress, and plant disease. *Annu. Rev. Phytopathol.* **13**, 193–211.

Schoeneweiss, D. F. (1975b). A method for controlling plant water potentials for studies on the influence of water stress on disease susceptibility. *Can. J. Bot.* **53**, 647–652.

Scholander, P. F., Hammel, H. T., Bradstreet, E. D., and Hemmingsen, E. A. (1965). Sap pressure in vascular plants. *Science* **148**, 339–346.

Schreiber, L. P., and Dochinger, L. S. (1967). *Fusarium* canker on paper mulberry *(Broussonetia papyrifera)*. *Plant Dis. Rep.* **51**, 531–532.

Secrest, H. C., MacAloney, H. J., and Lorenz, R. C. (1941). Causes of the decadence of hemlock at the Menominee Indian Reservation, Wisconsin. *J. For.* **39**, 3–12.

Silverborg, S. B., and Ross, E. W. (1968). Ash dieback disease development in New York State. *Plant Dis. Rep.* **52**, 105–107.

Slatyer, R. O. (1957). The significance of the permanent wilting percentage in studies of plant and soil water relations. *Bot. Rev.* **23**, 585–636.

Sorauer, P. (1874). "Handbuch der Pflanzenkrankheiten." Weigandt, Berlin.

Stoessl, A., Stothers, J. B., and Ward, E. W. B. (1976). Sesquiterpenoid stress compounds of the Solanaceae. *Phytochemistry* **15**, 855–872.

Stolzy, L. H., Letey, J., Klotz, L. J., and Labanauskas, C. K. (1965). Water and aeration as factors in root decay of *Citrus sinensis. Phytopathology* **55**, 270–275.

Tobiessen, P., and Buchsbaum, S. (1976). Ash dieback and drought. *Can. J. Bot.* **54**, 543–545.

Todd, G. W. (1972). Water deficits and enzymatic activity. *In* "Water Deficits and Plant Growth" (T. T. Kozlowski, ed.), Vol. 3, p. 177. Academic Press, New York.

Tomiyama, K. (1963). Physiology and biochemistry of disease resistance in plants. *Annu. Rev. Phytopathol.* **1**, 295–324.

Towers, B., and Stambaugh, W. J. (1967). The influence of induced soil moisture stress upon *Fomes annosus* root rot of loblolly pine. *Phytopathology* **58**, 269–272.

Tyree, M. T., and Hammel, H. T. (1972). The measurement of the turgor pressure and the water relations of plants by the pressure bomb technique. *J. Exp. Bot.* **23**, 267–282.

Vaadia, Y., Raney, F. C., and Hagan, R. M. (1961). Plant water deficits and physiological processes. *Annu. Rev. Plant Physiol.* **21**, 265–292.

Viets, F. G., Jr. (1972). Water deficits and nutrient availability . *In* "Water Deficits and Plant Growth" (T. T. Kolowski, ed.), Vol. 3, p. 217. Academic Press, New York.

Waring, R. H., and Cleary, B. D. (1967). Plant moisture stress: Evaluation by pressure bomb. *Science* **155**, 1248–1254.

Wheeler, H. (1975). "Plant Pathogenesis." Springer-Verlag, Berlin and New York.

Wiebe, H. H., and Brown, R. W., Daniel, T. W., and Campbell, E. (1970). Water potential measurements in trees. *BioScience* **20**, 225–226.

Wood, R. K. S. (1967). "Physiological Plant Pathology." Blackwell, London.

Yarwood, C. E. (1959). Predisposition. *In* "Plant Pathology" (J. G. Horsfall, A. E. Dimond, ed.), Vol. 1, p. 521. Academic Press, New York.

Zimmer, D. E., and Urie, A. L. (1967). Influence of irrigation and soil infestation with strains of *Phytophthora drechsleri* on root rot resistance of safflower. *Phytopathology* **57**, 1056–1059.

CHAPTER 3

# ABIOTIC DISEASES INDUCED BY UNFAVORABLE WATER RELATIONS

## R. D. Durbin

PLANT DISEASE RESISTANCE RESEARCH UNIT, ARS, USDA, AND DEPARTMENT OF PLANT
PATHOLOGY, UNIVERSITY OF WISCONSIN, MADISON, WISCONSIN

## I. INTRODUCTION

In nature, plants are periodically subjected to both water deficits and excesses. When man has partial control of water use, as for example by irrigating plants, long periods of water stress usually do not occur. However, internal water deficits in plants can develop rapidly so that even irrigated plants are likely to undergo some degree of water stress. Also, short periods of water excess will still occur. Under natural conditions these water imbalances clearly constitute a major limiting factor for plant growth and productivity. In some cases they reach such a magnitude that they cause abiotic diseases. Even so, for the most part, their effects, however important, remain unnoticed.

Plants are particularly sensitive to water imbalance at specific times in their life cycle: (a) germination, (b) flowering, and (c) fruit set and development. The effects of water imbalance can extend beyond the stress period and be outwardly manifested in portions of the plant at a distance from the original site of stress, the root. Ultimately, tissues that are metabolically active and import large amounts of organic and inorganic solutes are those that most often become diseased. Thus, it is not surprising that many abiotic diseases involve leaves and fruits.

Like most diseases, abiotic diseases are controllable. However, the necessary expenditure of funds, even if practical to consider, may not be justified solely on the basis of disease control. But when the additional values that accompany disease control are considered, the costs may be warranted. For example, proper moisture levels can improve soil tilth, augment the rate of nitrogen fixation and mineralization (Witkamp, 1969), and increase yields beyond those ascribable to the disease. Irrigation systems, if present, also offer a means for applying fertilizer. If properly manged, such systems will produce plants that will live longer and be better able to resist facultative parasites; they will also be more winter hardy.

Many influences determine whether and when an abiotic disease will become manifest. Edaphic and climatic influences, farming practices, and the plant all contribute. Because the situation is so complex and dynamic, it is very difficult to separate the various influences and determine the role they play and their magnitude. Exacerbating factors can easily obscure the basic interactions. Nevertheless, some progress has been made in understanding these complex situations. Further advancement will depend on development of quantitative techniques, the ingenuity of researchers, and the ability to promote team approaches to studying the individual diseases.

This chapter will discuss how periods of water imbalance affect the root environment, the interaction of this environment with the plant, and how the resulting physiological dysfunctions lead to abiotic diseases.

## II. DISEASES INDUCED BY WATER EXCESSES

The effects of excessive soil moisture on plants are exceedingly complex and result from both direct and indirect actions. Initially, the roots are responding to the rapidly changing gaseous conditions around them—oxygen is being depleted and $CO_2$ is increasing. These changes affect not only the plant but also the surrounding soil chemistry, structure, and microbiota. Alterations in these components initially affect the plant, each other, and again, the plant. Within the plant, physiological processes, perhaps initially only a few, are altered (Boyer, 1973). Each alteration by itself soon induced alterations in related processes until, at an increasingly rapid pace, all are ultimately perturbed.

The severity and duration of these perturbations determine the amount of plant damage. The species involved, their ability to resume rapid growth following periods of water excess, and the time of year are also important considerations determining the extent of plant damage. Plants subjected to relatively short periods of waterlogging will soon regain their normal appearance. However, their physiological processes may be disrupted for much longer periods, resulting in a disproportionately large reduction in yield (van't Woudt and Hagan, 1957). For instance, anaerobiosis for a single day can reduce the yield of tobacco *(Nicotiana tabacum)* and the foliar dry weight increment of tomato *(Lycopersicon esculentum)* by as much as one-third (Fulton and Erickson, 1964), whereas, under similar conditions, growth of soybean *(Glycine max)* is not inhibited. Likewise, avocado *(Persea americana)*, cherry *(Prunus* spp.*)*, citrus *(Citrus* spp.*)*, fig *(Ficus carica)*, olive *(Olea europaea)*, and peach *(Prunus persica)* are quite sensitive to flooding, whereas apple *(Malus* spp.*)*, grape *(Vitis* spp.*)*, and persimmon *(Diospyros virginiana)* are more resistant, and rough lemon *(Citrus limon)* is quite tolerant (Parker and Rounds, 1944; Rowe and Beardsell, 1973).

The extent of plant damage is also related to the conditions of growth prior to the time of waterlogging. Roots formed under conditions of high soil-water potential are structurally and physiologically adapted to such conditions so that a sudden excess, such as a heavy rain, has a much smaller effect on them, and in turn on the plant *in toto*, than on roots formed under low soil-water potentials. The higher the soil-water potential under which a root forms, the greater is the tendency for (a) increased root diameter; (b) fewer root hairs; (c) more branched and shorter roots; (d) larger and more numerous intercellular spaces; (e) shorter cells with increased wall suberization and lignification; and (f) reduced root dry weight increment which contributes to a decrease in the root:shoot ratio (Baker and Cook, 1974; Burström, 1965).

## A. SYMPTOMS

As the stress period lengthens, disease symptoms become more severe and numerous, until eventually death of the plant results. Some of these symptoms are expressed uniformly over the entire plant whereas others are more restricted to specific portions; e.g., older leaves, young leaves and stem apex, or fruit. They include leaf epinasty followed by temporary wilting (Fig. 1), chlorosis and premature defoliation commencing with the older leaves, stem hypertrophy and reduction in elongation, loss of geotropism, adventitious root formation on the stem near the water surface (Fig. 2), abscission of flowers and fruits, fruit blemishes and pre-

Fig. 1. Petiole epinasty of tomato due to excessive soil moisture.

mature ripening, and death of roots starting with the younger portions (Kramer, 1951; Burrows and Carr, 1969; van't Woudt and Hagan, 1957).

The exact physiological dysfunctions responsible for these symptoms are unknown (Rowe and Beardsell, 1973). However, a number of the symptoms, e.g., epinasty, adventitious root formation, and stem hypertrophy, are indicative of imbalances in plant growth substances. Furthermore, such imbalances have been shown to occur in waterlogged plants (Kramer, 1969). Indeed, ethylene accumulation has been proposed to account for most, but not all, of the symptoms exhibited by flooded plants (Abeles, 1973; Kawase, 1976). A reduction in translocation of gibberellins and cytokinins from the root also occurs, presumably resulting from a lowered production by the root (Burrows and Carr, 1969; Reid and Crozier, 1971). This, in turn, would influence hormone-dependent processes in the stem and leaves, and lead to a reduction in stem growth (gibberellin effect) and chlorophyll maintenance (cytokinin effect). Reid and Crozier (1971) reported that gibberellic acid ($GA_3$) applications overcame the inhibition of stem elongation during the first week after waterlogging. Later, when adventitious roots started to form, the amount of extractable GA in the shoot increased and exogenous applications were

Fig. 2. Adventitious roots and hypertrophied lenticels on a stem of *Populus deltoides* formed in response to flooding. (Courtesy, J. S. Pereira and T. T. Kozlowski.)

ineffectual. This indicates that GA levels could be important in inhibiting stem elongation during the first week, but subsequently other factors become more important. Finally, auxin concentrations in the stem are increased. Phillips (1964) found over three times as much auxin in shoots from flooded sunflower *(Helianthus annuus)* plants as in shoots from control plants. Whether this reflects an effect on synthesis, transport, or activity of auxin is unclear. He suggested that the increase in auxin was sufficient to account for root formation and suppression of stem elongation. It appears that this increase could be related to the plant's ability to withstand water excesses. Cell lignification and suberization, adventitious root development, and production of large air spaces in roots have all been cited as auxin responses that allow the plant to survive unfavorable conditions (Grable, 1966; Kramer, 1951). In light of the interactions among plant growth substances, as for example the effect of auxin in stimulating ethylene production, it seems likely that these substances interact. Another probable example of hormone imbalance during waterlogging involves the formation of dark green intumescences on bean *(Phaseolus vulgaris)* stems. These formations burst, if high moisture continues, causing a disease referred to as "edema" (Zaumeyer and Thomas, 1957).

Symptoms of abiotic diseases can be easily and quickly obscured if they provide conditions conducive to secondary invaders (see Yarwood, Chapter 5). For example, fruit from waterlogged apple and pear *(Pyrus communis)* trees may develop soft areas (van't Woudt and Hagan, 1957). These are readily damaged, and subsequently invaded by fungi, bacteria, and fruit-feeding insects. Similar situations exist underground.

## B. FACTORS INFLUENCING DISEASE DEVELOPMENT

### 1. Oxygen Content

As previously mentioned, one of the primary effects of excessive soil moisture is on the oxygen content of the soil. In arable soils oxygen content may vary widely, even without the presence of excessive water. In some cases, because of an adverse soil structure or farming practices that limit aeration, the oxygen level may be below 1% (Pearson, 1966; Yelenosky, 1963). In other situations the oxygen content may not be much below that of air (Grable, 1966). Because these data represent measurements from relatively large soil masses, it seems probable that oxygen concentrations immediately around and in the rhizosphere are lower. In addition, it is known that anaerobic microenvironments exist even in well-aerated soils (Greenwood, 1969; Stolzy *et al.*, 1975). Whatever the circumstances, as soil moisture increases, the rate of oxygen diffusion decreases. This becomes particularly acute as the soil moisture increases above about 30% of moisture-holding capacity. In part, this is because oxygen diffuses some $10^4$ times more slowly in the liquid phase than in the gaseous phase of the soil. In addition, roots, as well as many soil microbes, being essentially aerobic, will deplete the available oxygen within a matter of hours (van't Woudt and Hagan, 1957).

The partial oxygen pressure at which plant metabolism is limited varies among species (Carr, 1961). If not completely submerged, soybeans are quite tolerant of low oxygen tension. *Salix* spp. are even more tolerant, whereas cotton *(Gossypium hirsutum)*, snapdragon *(Antirrhinum majus)*, tobacco, and tomato are examples of quite intolerant species. Sensitivity also varies among cultivars and rootstocks (Rowe and Beardsell, 1973). Water used to flood cranberries *(Vaccinium macrocarpon)* in winter can also become oxygen deficient because of inadequate aeration and oxygen depletion by algae. When this happens the leaves turn brown and the terminal shoots die. The following spring, the leaves abscise and the flower may fail to set fruit (Anderson, 1956). Other diseases attributed to oxygen deficiency are listed by Bergman (1959).

Within the plant fermentative processes are initiated as a direct result of oxygen deficiency. Fulton and Erickson (1964) found that toxic amounts of ethanol were produced in overwatered tomatoes, with the amount increasing sharply below oxygen diffusion rates of $20 \times 10^{-8}$ gm $cm^{-2}$ $min^{-1}$. They also suggested that, in addition to ethanol toxicity, the diversion of metabolism from aerobic to fermentative pathways could have serious consequences to the energy charge of the cell. Mature potato *(Solanum tuberosum)* tubers when covered by a film of water will become anaerobic within $2\frac{1}{2}$ hr at 21°C (Burton and Wigginton, 1970), and begin to produce ethanol. Since the enzymes necessary for this pathway are essentially ubiquitous in plants, it seems probable that ethanol formation is a general, but not necessarily universal, response to lowered oxygen levels.

## 2. Plant Anatomy

Some of the variations in sensitivity of plants to flooding are determined by plant structure. The roots of some species obtain oxygen by diffusion from the atomsphere through continuous pathways composed variously of intercellular spaces, lysigenous spaces, and specialized cells within the plant (plant aeration) rather than through the soil (soil aeration). These spaces form by several processes. The most common type, called schizogenous, results from a separation at the point of contact between the middle lamella of a new cell and wall of its mother cell (Sifton, 1957). In some plant families, schizogenous spaces form specialized secretory ducts. Another type of intercellular space, called lysigenous, arises by dissolution of entire cells. These are well-known anatomical adaptations of bog and aquatic plants and rice *(Oryza sativa)*, but they have been relatively little studied in other plants, even though large air spaces are common in many vascular plants. Recent studies strongly suggest that such structural features are involved in supplying a significant proportion of the oxygen requirement of mesophytic roots, especially in young plants (Chirkova, 1968; Greenwood and Goodman, 1971; Luxmoore *et al.*, 1970).

## 3. CO₂ Content

The $CO_2$ content of soils under waterlogged conditions can reach levels that are directly toxic to plants (Enoch and Dasberg, 1971). Normally, the $CO_2$ concentration in soils of the temperate zone varies from 0.1 to 1.5% in uncropped soil to about 3% in cropped soils (Russell and Appleyard, 1915). Where large amounts of organic matter are present, especially in tropical areas, $CO_2$ concentration may reach 10–15%, and after heavy rains, values as high as 20% have been recorded. Similarly high $CO_2$ concentrations occur in waterlogged soils. At such high $CO_2$ concentrations, the capacity of roots to absorb water and, to a lesser extent, ions, is sharply reduced (Chang and Loomis, 1945). Under severe conditions these reductions alone can cause shoots to wilt very rapidly, as in tobacco "drown" (Fig. 3) (Glinka and Reinhold, 1962; Lucas, 1975).

### 4. Soil Solution pH

Interactions of $CO_2$ with the soil also may have a major indirect effect on root function (Grable, 1966). Above pH 5.0, the $CO_2$ reacting with water produces hydrogen and bicarbonate and, at higher pHs, carbonate ions. In calcareous soils a further interaction yields calcium and additional bicarbonate ions. In such soils small increases in $CO_2$ cause a proportionately large pH drop, a 10% $CO_2$ level decreases the pH by approximately 2 units. In other soil types, however, waterlogging may increase soil pH.

### 5. Nutrient Availability

Relatively little is known about the direct effect of pH changes in the soil solution on plants. The most important effect appears to be indirect, and results from changes in the availability of ions to the root (van Cleemput and Patrick, 1974; Wallihan *et al.*, 1961). The increased availability of Al below pH 6.0 constitutes a major crop hazard, especially in the high rainfall regions which contain some 40% of the world's arable soils. Manganese also becomes more available as soil pH decreases, whereas Ca, Fe, K, Mg, and P become less available (Truog, 1946). This latter group of elements commonly is deficient in acid soils when they are waterlogged. Fruit crops are especially susceptible to iron chlorosis under waterlogged conditions. Vegetables and conifers are also readily affected. Numerous

Fig. 3. Tobacco drowning. The wilted leaf blades partially supported by the arched midribs are characteristic of the disease. (Courtesy, G. B. Lucas.)

other nutrient disorders, depending upon the soil, climate, and species involved, can also be triggered by excessive soil water.

Compounding these ion imbalances are the soluble salts brought into solution because of the excess water (Reeve and Fireman, 1967). They may originate from natural sources or as a result of previous fertilizer applications. The roots of many plants are injured by the salts, resulting in "chlorosis" and reduced plant growth. In addition, the salts are translocated to the leaves where they accumulate. Tolerant species, such as carnation *(Dianthus caryophyllus)* and stock *(Matthiola incana)* will show little effect other than a slow decline in vigor and a yellowing of the older leaves. On the other hand, in sensitive species, localized toxic levels can be reached, especially around the hydathodes. Death of the leaf margin and sometimes of the entire leaf ensues. This disease, called "leaf scorch," can be found on a variety of vegetables and ornamental plants (Ivanoff, 1944; Schoonover and Sciaroni, 1957). The necrotic tissues provide an entry point for secondary invaders such as species of *Botrytis, Rhizopus,* and *Erwinia* (Baker et al., 1954).

### 6. Redox Potential

The availability of solutes to roots is significantly influenced by the decrease in the oxidation-reduction or redox potential usually associated with waterlogging (Grable, 1966; Turner and Patrick, 1969; Patrick and Mahapatra, 1968). Well-aerated soils have redox potentials ranging from $+400$ to $+700$ mV whereas most waterlogged soils have values from $+200$ to $-300$ mV. $Eh$ is not, however, a good indicator of the state of soil aeration. The rate of decrease after waterlogging is dependent upon a host of complex factors, and is accelerated by oxidizable organic matter. Under reducing conditions denitrification is first accelerated (van Cleemput and Patrick, 1974) and toxic levels of nitrite may accumulate. At lower redox values the oxidation state of sulfate and carbonate declines. There is also an increase in availability of many elements, including Bo, Co, Fe, Mn, Mo, Ni, P, Pb, S, and Zn (Ng and Bloomfield, 1962; Hodgson, 1963). In some cases their concentrations may reach levels toxic to the plant.

### 7. Soil Microorganisms

Besides the predisposing effects of water imbalance on plant pathogens, microorganisms not normally considered to be pathogens often act in concert with unfavorable water conditions to cause disease. Such facultative pathogens generally are benefited by conditions unfavorable for the host. These microorganisms often colonize the plant, remaining quiescent and causing little injury until the environment favors their growth over that of the host. Once this happens plants can be injured very rapidly,

especially in waterlogged soils. Within a matter of hours the soil microflora changes dramatically, particularly in the rhizosphere. Fungi and actinomycetes almost disappear while many largely anaerobic bacterial
species now flourish (Baker and Cook, 1974; Cook and Papendick, 1970).
Fermentations are initiated by these organisms, producing an array of
compounds, e.g., hydrogen, ammonia, methane, hydrogen sulfide, $CO_2$,
and ethylene, in concentrations toxic to the root.

*8. High Temperature*

When excess soil moisture is accompanied by high air temperature,
many more plant species are affected and disease symptoms develop very
rapidly. This enhancing effect is undoubtedly more complex than simply
one of increasing respiration and transpiration rates under conditions of
decreasing oxygen solubility (Cooper, 1973). For example, Lucas (1975)
postulated that in tobacco "drowning" high temperatures favor soilborne
anaerobic microorganisms which together with the dying roots produce
toxins. These substances are absorbed by the plant where they induce
"gumming." The resulting mechanical occlusion of the water-conducting
system, along with death of roots and the inhibitory effect of lowered
oxygen tension on water absorption, all contribute to sudden cessation of
water transport within the plant at a time when transpiration demands are
high. The leaf blades quickly become completely flaccid and the plants are
said to "flop."

## III. DISEASES INDUCED BY WATER DEFICITS

It is now generally accepted that as the soil moisture declines from
field capacity to the permanent wilting point, there is a concomitant, but
not direct, decrease in plant growth (Kramer, 1969). The level at which
the resulting water stress induces an abiotic disease is not a sharp one, just
as there is no definite level at which water becomes unavailable to the
plant. However, even a small decrease in water potential over time will
reduce yield. Obviously, a "disease" that reduces yield without exhibit-
ing specific symptoms is difficult to identify as such unless suitable controls are available for comparison. Consider identifying a transitory water
stress during seed germination which would have the effect of reducing
emergence or lengthening the germination period. For instance, Doneen
and MacGillivray (1943) found that the rate of germination of vegetable
seeds in Yolo fine sandy loam was progressively lower as the percentage
of soil moisture was reduced. Additionally, percentage germination varied
with the degree of soil moisture availability: germination of celery seed
*(Apium graveolens)* required a continuously high soil-water potential and it

would not germinate below 11% soil moisture, a value well above the wilting percentage of 8.6% for the soil used; germination of lima bean *(Phaseolus limensis)*, beet *(Beta vulgaris)*, lettuce *(Lactuca sativa)* and pea *(Pisum sativum)* was affected below 11% soil moisture; while germination of sweet corn *(Zea mays)*, crucifers, and watermelon *(Citrullus vulgaris)* was insensitive over the moisture range used, 9–18%.

Unlike an excess of water, which generally appears suddenly, water deficits may develop relatively slowly over several days to months. The more time a plant has to adapt, the less it is injured. Often, though, an unthrifty plant results that is more easily damaged by other stresses, diseases, or insects, and these, rather than the water stress, are blamed for death of the plant. The benign effect of yield reduction will still occur. Robins and Domingo (1953) showed that depleting soil moisture to the incipient wilting point for only 1–2 days during the tasseling-silking period of corn reduced grain yield by 22%.

A. SYMPTOMS

*1. Leaves*

The most characteristic early symptoms of water stress are leaf cupping and foliar wilt. Such symptoms are very striking on affected solanaceous species. Another prominent example is the "needle droop" exhibited by various *Pinus* species (Fig. 4) (Bergdahl and French, 1976). As the stress period continues, leaf tissues become desiccated, yellow, and necrotic, and eventually die. Commonly, the necrosis is evidenced at the leaf tip, as in "terminal bleach" of cereals (Treshow, 1957), along the margins, and more rarely as irregular interveinal lesions. Tobacco leaves, for example, develop numerous, large reddish-brown interveinal spots bordered by a chlorotic zone, a condition called "drought spot." Later, the leaf margins curve down and die. This phase has been called "rim fire" (Lucas, 1975). During the winter conifer needles may become desiccated because of the plant's inability to withdraw water from the soil (Treshow, 1970). The needles turn reddish brown and in severe cases, death of buds and twigs follows. This disease has been called "needle burn" or "red belt." Affected trees are degraded or are unsalable as Christmas trees, and their growth during the following year is drastically inhibited.

At a later stage of stress, the leaves, especially of woody plants, may abscise. The stem and fruit, if present, are then exposed and may sunscald. The skin of the fruit turns excessively brown, and often a corky layer forms just beneath the affected areas. The loss of leaves weakens the

Fig. 4. Red pine, *Pinus resinosa*, showing abnormal drooping of the previous year's needles. Note the resultant reduction in current year's growth. (Courtesy, D. W. French.)

plant to the extent that yield is reduced, and if the stress becomes chronic, the plant will decline each year until it dies. The loss of foliage also exposes the fruit to excessive high temperatures which will contribute to development of other diseases (Gates, 1968) such as "water core" of apple (Anderson, 1956).

### 2. Flowers

The number and size of flowers on water-stressed plants are less than normal. Their "keeping quality" is reduced, and they commonly abscise. The stem, as in the case of Easter lily *(Lilium longiflorum)*, may be longer so that the unthrifty, open appearance of the inflorescence is accentuated (Stuart *et al.*, 1952). Colors are less vivid and odors are weaker.

### 3. Fruit

Developing fruit are especially sensitive to water stress, partially because unless they are quite young they cannot compete successfully with leaves for water when it becomes limiting (Shaw and Laing, 1966). Water stress induces disease in fruit, with the type of disease expression depending on the stage of fruit development and the species involved. If the fruit is immature, it often abscises as in "June drop" of citrus (Fawcett and Lee, 1926). Later in the season, it usually remains attached but is undersized and in extreme cases, shriveled. Abscission can occur during the advanced stages of maturation as well. Besides these obvious symptoms, fruit quality may be reduced by poor color, aroma, or lack of keeping ability.

Various types of external and internal necrotic lesions also commonly develop on fruit. The stylar end is most frequently involved and may assume a different color from the rest of the fruit, and become sunken. In addition, the surrounding tissues may break down and a darkish lesion associated with gum deposits develops. This is the case in "endoxerosis" or "internal decline" of lemon (Fawcett and Lee, 1926). As much as half of the tissue in a lemon fruit may be involved and gum formation is common. Usually when so much of the fruit tissue is affected, the fruit eventually abscises. Small, isolated necrotic lesions may also appear on other portions of fruit. In "bitter pit" of apple, the symptoms do not appear until the fruit is more than half grown. The lesions start as sunken, water-soaked spots, underneath which spongy, brown masses of necrotic cells form. Later the epidermis above the spots dies, and the affected areas become more sunken and darker in color (Anderson, 1956).

## B. COMPLEX ETIOLOGIES

A number of abiotic diseases have complex etiologies, about which there is little agreement beyond the fact that water deficits contribute to disease development. In some diseases water stress appears to disrupt solute translocation, producing a localized deficiency, particularly of the fruit. Fundamentally then, as in the case of nutritional diseases induced by water excess, the solute is responsible for the disease symptoms. Calcium deficiency represents the most common and best-studied abiotic disease of this type. It includes such diseases as "lettuce and crucifer tipburn," "bitter pit" of apple, "blackheart" of celery, and "blossom end rot" of tomato (Fig. 5). These diseases most probably are induced because of the relatively large Ca requirement at the sink (fruit or leaf margin), and the immobility of Ca once it has entered another part of the plant. Thus, a continual supply of Ca must be provided from the root via the transpiration stream. In tipburn of lettuce the symptoms result from a brownish dieback of the tissue along the leaf margins. Any of the leaves in leaf lettuce can be affected. In head-type cultivars the symptoms appear more often on interior leaves than on exterior leaves. Cultivars vary

**R. D. Durbin**

Fig. 5. Tomato blossom end rot showing darkened, sunken lesions at the blossom scar. (Courtesy, T. Barksdale.)

widely in their susceptibility. Another example, blossom end rot of tomato, is common and can be quite destructive. This disease also occurs, but less frequently, on pepper. The symptoms begin as water-soaked lesions at the blossom end of the ripening fruit. The lesion enlarges, becoming sunken and blackish in color (Walker, 1952).

## IV. DISEASES INDUCED BY ALTERNATING STRESSES

A third, really a composite, effect of adverse water relations results from alternating water deficits and excesses. Since there is a greater disparity between these two conditions than between either condition and normalcy, the abiotic diseases that result are more severe and involve a wider range of species. These diseases can include those previously mentioned, and can occur simultaneously in that symptoms from prior periods of water excess and deficiency may be present on a single plant at the same time.

SYMPTOMS

The most common abiotic disease associated with alternating water deficits and excesses is "fruit cracking." Brought about by excessive moisture following a dry period, the internal tissues of the fruit absorb water and expand faster than they can be accommodated by the outer tissues and epidermis. Hence the fruit cracks, usually beginning at the stem end. This condition is prevalent on apple, grape, stone fruit, strawberry *(Fragaria chiloensis)*, and tomato. The cracks can be quite deep and they serve as an ideal infection court for fruit-rotting pathogens. Similar environmental changes are conducive to the development of celery blackheart (Walker, 1952). The primary symptom of this disease consists of necrosis of the young leaflets. This is rapidly followed by death and blackening of the entire heart.

When water excesses early in the growing season adversely affect root development, plants may be more sensitive to drought conditions at a later time. In other cases, water stress is enhanced by diseases that impede water movement into and within the plant, such as root rots, nematode diseases, and vascular pathogens. Another example is the increased availability of Fe, sometimes to toxic levels, that occurs as wet soil dries. In a similar situation Bo availability decreases.

## REFERENCES

Abeles, F. B. (1973). "Ethylene in Plant Biology." Academic Press, New York.
Anderson, H. W. (1956). "Diseases of Fruit Crops." McGraw-Hill, New York.
Baker, K. F., and Cook, R. J. (1974). "Biological Control of Plant Pathogens." Freeman, San Francisco, California.
Baker, K. F., Matkin, O. A., and Davis, L. H. (1954). Interaction of salinity injury, leaf age, fungicide application, climate, and *Botrytis cinerea* in a disease complex of column stock. *Phytopathology* **44**, 39–42.
Bergdahl, D. R., and French, D. W. (1976). Needle droop: An abiotic disease of plantation red pine. *Plant Dis. Rep.* **60**, 472–476.
Bergman, H. F. (1959). Oxygen deficiency as a cause of disease in plants. *Bot. Rev.* **25**, 417–485.
Boyer, J. S. (1973). Response of metabolism to low water potentials in plants. *Phytopathology* **63**, 466–472.
Burrows, W. J., and Carr, D. J. (1969). Effects of flooding the root system of sunflower plants on the cytokinin content in the xylem sap. *Physiol. Plant.* **22**, 1105–1112.
Burström, H. G. (1965). The physiology of plant roots. *In* "Ecology of Soil-Borne Plant Pathogens" (K. F. Baker *et al.*, eds,), pp. 154–169. Univ. of California Press, Berkeley.
Burton, W. G., and Wigginton, M. J. (1970). The effect of a film of water upon the oxygen status of a potato tuber. *Potato Res.* **13**, 180–186.
Carr, D. J. (1961). Chemical influences of the environment. *Encycl. Plant Physiol.* **16**, 737–794.
Chang, H. T., and Loomis, W. E. (1945). Effect of carbon dioxide on absorption of water and nutrients by roots. *Plant Physiol.* **20**, 221–232.
Chirkova, T. V. (1968). Oxygen supply to roots of certain woody plants kept under anaerobic conditions. *Fiziol. Rast.* **15**, 565–568.
Cook, R. J., and Papendick, R. I. (1970). Effect of soil water on microbial growth, antagonism, and nutrient availability in relation to soil-borne fungal diseases of plants. *In*

"Root Diseases and Soil-Borne Pathogens" (K. F. Baker, R. V. Bega, and P. E. Nelson, eds.), pp. 81–88. Univ. of California Press, Berkeley.

Cooper, A. J. (1973). "Root Temperature and Plant Growth." Commonw. Agric. Bur., Slough, England.

Doneen, L. D., and MacGillivray, J. H. (1943). Germination (emergence) of vegetable seed as affected by different soil moisture conditions. *Plant Physiol.* **18**, 524–529.

Enoch, H., and Dasberg, S. (1971). The occurrence of high carbon dioxide concentrations in soil air. *Geoderma* **6**, 17–21.

Fawcett, H. S., and Lee, H. A. (1926). "Citrus Diseases and Their Control." McGraw-Hill, New York.

Fulton, J. M., and Erickson, A. E. (1964). Relation between soil aeration and ethyl alcohol accumulation in xylem exudate of tomatoes. *Soil Sci. Soc. Am., Proc.* **28**, 610–614.

Gates, D. M. (1968). Energy exchange between organisms. *Aust. J. Sci.* **31**, 67–74.

Glinka, Z., and Reinhold, L. (1962). Rapid changes in permeability of cell membranes to water brought about by carbon dioxide and oxygen. *Plant Physiol.* **37**, 481–486.

Grable, A. R. (1966). Soil aeration and plant growth. *Adv. Agron.* **18**, 57–106.

Greenwood, D. J. (1969). Effect of oxygen distribution in the soil on plant growth. *In* "Root Growth" (W. J. Whittington, ed.), pp. 202–223. Plenum, New York.

Greenwood, D. J., and Goodman, D. (1971). Studies on the supply of oxygen to the roots of mustard seedlings (*Sinapis alba* L.). *New Phytol.* **70**, 85–96.

Hodgson, J. F. (1963). Chemistry of the micronutrient elements in soils. *Adv. Agron.* **15**, 119–159.

Ivanoff, S. (1944). Guttation-salt injury on leaves of cantaloupe, pepper and onion. *Phytopathology* **34**, 436–437.

Kawase, M. (1976). Ethylene accumulation in flooded plants. *Physiol. Plant.* **36**, 236–241.

Kramer, P. J. (1951). Causes of injury to plants resulting from flooding of the soil. *Plant Physiol.* **26**, 722–736.

Kramer, P. J. (1969). "Plant and Soil Water Relationships. A Modern Synthesis." McGraw-Hill, New York.

Lucas, G. B. (1975). "Diseases of Tobacco," pp. 583–584. Biol. Consult. Assoc., Raleigh, North Carolina.

Luxmoore, R. J., Stolzy, L. H., and Letey, J. (1970). Oxygen diffusion in the soil–plant system. I. A model. *Agron. J.* **62**, 371–332.

Ng, S. K., and Bloomfield, C. (1962). The effect of flooding and aeration on the mobility of certain trace elements in soils. *Plant Soil* **16**, 108–135.

Parker, E. F., and Rounds, M. B. (1944). Avocado tree decline in relation to soil moisture and drainage in certain California soils. *Proc. Am. Soc. Hortic. Sci.* **44**, 71–79.

Patrick, W. H., Jr., and Mahapatra, I. C. (1968). Transformation and availability to rice of nitrogen and phosphorus in waterlogged soils. *Adv. Agron.* **20**, 323–359.

Pearson, R. W. (1966). Soil environment and root development. *In* "Plant Environment and Efficient Water Use" (W. H. Pierre *et al.*, eds.), pp. 95–126. Am. Soc. Agron., Madison, Wisconsin.

Phillips, I. D. J. (1964). Root-shoot hormone relations II. Changes in endogenous auxin concentration produced by flooding of the root system in *Helianthus annuus. Ann. Bot. (London)* [*N. S.*] **28**, 37–45.

Reeve, R. C., and Fireman, M. (1967). Salt problems in relation to irrigation. *Agron. Monogr.* **11**, 988–1008.

Reid, D. M., and Crozier, A. (1971). Effects of waterlogging on the gibberellin content and growth of tomato plants. *J. Exp. Bot.* **22**, 39–48.

Robins, J. S., and Domingo, C. E. (1953). Some effects of severe soil moisture deficits at specific growth stages of corn. *Agron. J.* **45**, 618–621.

Rowe, R. N., and Beardsell, D. V. (1973). Waterlogging of fruit trees. *Hortic. Abstr.* **43**, 534–548.

Russell, E. J., and Appleyard, A. (1915). The atmosphere of the soil: Its composition and the causes of variation. *J. Agric. Sci.* **7**, 1–48.

Schoonover, W. R., and Sciaroni, R. H. (1957). The salinity problem in nurseries. *Calif., Agric. Exp. Stn., Bull.* **23**, 52–67.

Shaw, R. H., and Laing, D. R. (1966). Moisture stress and plant response. *In* "Plant Environment and Efficient Water Use" (W. H. Pierre *et al.*, eds.), pp. 73–94. Am. Soc. Agron., Madison, Wisconsin.

Sifton, H. B. (1957). Air space in plants. II. *Bot. Rev.* **23**, 303–312.

Stolzy, L. H., Zentmyer, G. A., and Roulier, M. H. (1975). Dynamics and measurement of oxygen diffusion and concentration in the root zone and other microsites. *In* "Biology and Control of Soil-Borne Plant Pathogens" (G. W. Bruehl, ed.), pp. 50–54. Am. Phytopathol. Soc., St. Paul, Minnesota.

Stuart, N. W., Skou, W., and Kiplinger, D. C. (1952). Further studies on causes and control of leaf scorch in Croft Easter lily. *Proc. Am. Soc. Hortic. Sci.* **60**, 434–438.

Treshow, M. (1957). Terminal bleach of cereals. *Plant Dis. Rep.* **41**, 118–119.

Treshow, M. (1970). "Environment and Plant Response." McGraw-Hill, New York.

Truog, E. (1946). Soil reaction influence on availability of plant nutrients. *Soil Sci. Soc. Am., Proc.* **11**, 305–308.

Turner, F. T., and Patrick, W. H., Jr. (1969). Chemical changes in waterlogged soils as a result of oxygen depletion. *Proc. Int. Congr. Soil Sci., 9th, 1968* Vol. 4, pp. 53–65.

van Cleemput, O., and Patrick, W. H., Jr. (1974). Nitrate and nitrite reduction in flooded gamma-irradiated soil under controlled pH and redox potential conditions. *Soil Biol. & Biochem.* **6**, 85–88.

van't Woudt, B. D., and Hagan, R. M. (1957). III. Crop responses at excessively high soil moisture levels. *Agron. Monogr.* **7**, 514–578.

Walker, J. C. (1952). "Diseases of Vegetable Crops." McGraw-Hill, New York.

Wallihan, E. F., Garber, M. J., Sharpless, R. G., and Printy, W. L. (1961). Effect of soil oxygen deficit on iron nutrition of orange seedlings. *Plant Physiol.* **36**, 425–428.

Witkamp, M. (1969). Environmental effects on microbial turnover of some mineral elements. Part I. Abiotic factors. *Soil Biol. & Biochem.* **1**, 167–176.

Yelenosky, G. (1963). Tolerance of trees to deficiencies of soil aeration. *Proc. Int. Shade Tree Conf., 39th, 1963* pp. 16–25.

Zaumeyer, W. J., and Thomas, H. R. (1957). A monographic study of bean diseases and methods for their control. *U. S., Dep. Agric., Tech. Bull.* **868**, 135.

CHAPTER 4

# WATER AND SPORE LIBERATION

## C. T. Ingold

BIRKBECK COLLEGE, UNIVERSITY OF LONDON, LONDON ENGLAND

## I. INTRODUCTION

In the reproductive activity of a fungus parasitizing a plant, it is usually possible to distinguish a number of episodes relating to the spores, namely, production, liberation, dispersal, deposition on the host, and, finally, infection. Water may be an important factor in any or all of these five episodes, but our concern is mainly with liberation and dispersal with a backward glance occasionally at production and sometimes a forward one to deposition under conditions that may lead to infection.

Although, for purposes of discussion, it is convenient to consider these episodes in isolation, one may pass imperceptibly into the next and in a particular instance it may be difficult to decide exactly where the water factor is exerting its influence. For example, if discharged spores are being trapped above the active perithecia of an ascomycete on a diseased leaf under different water regimes, the rate of trapping may reflect the influence of water on ripening of the asci (i.e., on spore production), or on bursting of the asci (spore liberation), or more likely on both processes.

Again it must be emphasized that care must be taken in interpretation of field studies in which the rate of spore liberation and the relative humidity of the ambient air are followed simultaneously with a view to elucidating the water relations of liberation. For most plant pathogens there is a circadian rhythm in the concentration of the spores in the air around diseased plants, probably related to a rhythm in spore release. The spores of some species have a maximum concentration in the air on either side of noon, while others concentrate around midnight or near dawn. This involves a close correlation, positive or negative, with the relative humidity of the air. However, this does not necessarily imply that humidity is the determining factor, since, in addition, both temperature and light have circadian rhythm under natural conditions. In particular it should be said that some phytopathologists have assumed that humidity is the master factor without accounting for the possible involvement of light.

Students of spore liberation have usually recognized two contrasting types: one in which the spores are violently discharged and the other in which release is a passive process on the part of the fungus.

There are two types of violent discharge. First, those in which turgid living cells are involved, and therefore high humidity is essential; and, second, those in which active spore release is triggered by sudden drying.

Many spores are, however, set free passively in the sense that external kinetic energy is involved. This energy may come from moving air, passing animals, falling rain, or drifting mist. Clearly, water is often an important factor. In wet conditions there is, particularly, the process of splash dispersal, while in dry conditions there is the reduction of some spores to a powdery state so that they can readily be blown or shaken free.

## II. ACTIVE SPORE LIBERATION

### A. Spore Liberation Involving Turgid Cells

The spores of many fungi pathogenic to plants are violently discharged by some mechanism dependent on the turgidity of living cells. In this category belong ascospores, basidiospores, aeciospores of rusts, and a few conidia such as those of *Sclerospora* spp. (downy mildew of maize) and *Pyricularia oryzae* Cav. (rice blast).

### 1. Ascomycetes

In many Ascomycetes which parasitize plants and in which ascospores play a highly significant role in the epidemiology of disease, spore release occurs only after wetting of the perithecia which allows ripe asci to expand and burst. Normally, ascospores appear in the air in the immediate vicinity of the diseased plants shortly after the onset of rain, a certain minimum fall being needed if significant spore discharge is to occur.

This situation is well illustrated by the work of Hirst *et al.* (1955) on the concentration of ascospores of *Venturia inaequalis* (Cooke) Wint. (apple scab) in the air of apple orchards. Perithecia develop on dead fallen leaves and mature in late spring. At that time of year there is a close correlation between periods of rain and high ascospore content of the orchard air. Dew has little influence on discharge, but a rainfall of as little as 0.2 mm induces abundant release of ascospores. A similar picture emerged in a study by McOnie (1964) of *Guignardia citricarpa* Kiely, which causes black spot on leaves of citrus trees in South Africa. Another fungus that has been the subject of considerable study in relation to the water factor is *Eutypa armeniacae* Hansf. and Carter, a pathogen of apricot trees in many parts of the world. Moller and Carter (1965) studied the concentration of ascospores in the air of apricot orchards. After several dry days, a minimum of 1–2 mm of rain was needed to initiate significant ascospore release. As might be expected, there was a lag between the onset of rain and the appearance of abundant ascospores in the ambient air (Fig. 1).

It should be remarked that relating the concentration of spores in the air to local rainfall is valid only if there is reasonable certainty that the ascospores are of local origin. This point is illustrated by the work of Ramos *et al.* (1975) on *Eutypa armeniacae* in California. In an apricot orchard at Tracey in the San Joaquin Valley 50 km east of any known source of inoculum of the fungus, high concentrations of ascospores were recorded in the orchard air in mid-November, but there was little or no correlation with local rainfall. High concentrations in the Tracey air depended on rainfall many kilometers to the west and on winds of appropriate direction and velocity to provide for transport of the spores.

In most plant diseases caused by Ascomycetes, wetting by rain or by sprinkler irrigation seems to be necessary to initiate spore discharge and normally this precipitation must exceed a certain minimum. The question arises whether, in a period free from rain, dew may produce sufficient wetting to allow ascospore release. This state of affairs has been reported for *Mycosphaerella pinodes* (Berk. and Blox.) Vestergr., but it seems to be rare. *M. pinodes* is a parasite of peas and the perithecia are produced on old pea straw. Carter (1963) showed that dew alone is enough to bring about considerable discharge of ascospores from straw that is air-dry prior to dew formation each night (Fig. 2).

Discharge may not continue indefinitely from wetted ascocarps. In-

C. T. Ingold

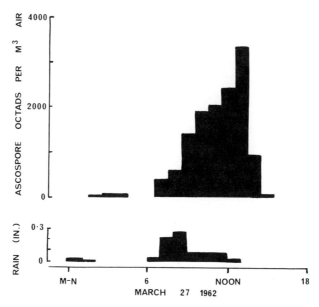

Fig. 1. Hourly estimates (Hirst trap technique) of *Eutypa armeniacae* ascospore octads and hourly rainfall. Spore trap surrounded by a ring (radius about 1 m) of infected apricot branches. (Modified from Moller and Carter, 1965.)

deed, a regime involving alternate wet and dry periods may be necessary for maximum spore discharge. This is the case in *Ophiobolus graminis* (Sacc.) Sacc., the organism causing take-all of cereals. After wetting wheat straw-bearing perithecia, the rate of ascospore release rises to a maximum in about half an hour and then rapidly declines so that, a few hours later, hardly any spores are liberated (Fig. 3). A period of dryness is then needed before further wetting results in significant discharge.

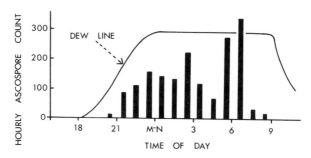

Fig. 2. *Mycosphaerella pinodes.* Hourly ascospore counts (Hirst spore trap technique) in air flowing over infected pea straw. The dew line is derived from a surface wetness recorder. Dew started to form at 18 hr. (After Carter, 1963.)

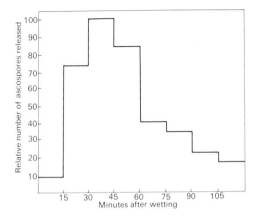

Fig. 3. *Ophiobolus graminis*. Rate of release of ascospores (spores trapped per 15 min as percentage of peak rate) in wind tunnel plotted against time. (After Gregory and Stedman, 1958.)

In some Ascomycetes, although wetting is needed to prepare the fungus for spore release, actual discharge is mainly associated with the early stages of subsequent drying. This was found, for example, for the apple canker fungus *Nectria galligena* Bres. by Lortie and Kuntz (1963), and Rowe and Beute (1975) obtained similar results with *Calonectria crotalariae* (Loos) Bell and Sobers, a fungus causing disease in peanuts (Fig. 4). The phenomenon has received considerable study in the saprophytic pyrenomycete *Sordaria fimicola* Ces. and de Not. (Austin, 1968).

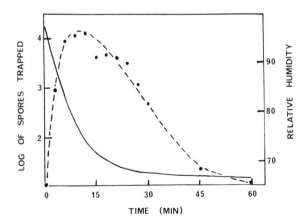

Fig. 4. *Calonectria crotalariae*. Rate of trapping of ascospores discharged from perithecia in relation to reduction in the humidity of the ambient air. (After Rowe and Beute, 1975.)

When a nearly saturated airstream bathing the exposed parts of the perithecia was suddenly changed to one of much reduced humidity, there was a dramatic, but temporary, increase in the rate of spore liberation. Hadley (1968) recorded much the same effect in discharge of ascospores from the perithecial stromata of ergot, *Claviceps purpurea* (Fr.) Tul.

In *Nectria*, *Calonectria*, *Claviceps*, and *Sordaria* the perithecial cavity is surrounded by soft tissues and it is probable, therefore, that incipient drying puts pressure on the contents of the perithecium, thereby forcing asci more rapidly up the neck canal to the outside and thus leading, for a time, to an increased rate of ascospore discharge.

There are a few Ascomycetes in which surface wetting is not necessary for spore discharge. In these the ascigerous stage is closely and intimately associated with the sappy living tissues of the host. A well-known example is *Epichloe typhina* (Pers. ex Fr.) Tul., a fungus causing the "choke" of grasses. Provided the water supply in the soil is sufficient to maintain turgidity of the host plant, the asci of the perithecia, embedded in the stroma tightly wrapped around the grass stalk, can freely discharge their ascospores even when continuously bathed in air of low relative humidity (Ingold, 1948). Essentially the same situation obtains in *Taphrina deformans* (Berk.) Tul., the fungus causing peach leaf curl, where the asci form an extensive hymenium closely adhering to the living epidermal cells of the host (Yarwood, 1941).

Before leaving the question of spore liberation in Ascomycetes in relation to the water factor, it should be noted that violent spore discharge does not always occur. In some species asci deliquesce while still within the perithecium and the mass of ascospores simply oozes out at the ostiole. This is a regular feature, for example, in species of the large genus *Ceratocystis*. Further, in one and the same fungus both passive oozing and violent discharge may occur, the dominant process depending on the water factor. For example, in *Calonectria crotalariae*, during very wet weather, the ascospores tend to ooze out and any dispersal is by rain splash, while under slightly drier, but still damp, conditions violent discharge occurs and the ascospores become airborne (Rowe and Beute, 1975).

## 2. Basidiomycetes

Since ascospore discharge involves bursting of turgid cells, the reason for its dependence on water is obvious. However, so far as basidiospores are concerned, the reason for the need for water to sustain discharge is not as clear, since there is no firm agreement yet on how the basidium functions as a spore gun (Ingold, 1971). However, there can be little doubt that, whatever may be the mechanism, it can operate only under conditions of relative humidity at or around 100%. On the other hand, in contrast to asci, direct wetting with water prevents basidiospore discharge from a hymenium of basidia.

The most precise information about the humidity factor in basidiospore discharge comes from a study of fungi that are not pathogenic to plants. Thus work on *Schizophyllum commune* Fr. has shown that an immediate and drastic reduction in the rate of spore liberation follows a change from damp to relatively dry ambient air (Zoberi, 1964).

From the point of view of the plant pathologist, the liberation of basidiospores from basidia derived from germination of teliospores of rusts is of special importance. The evidence suggests that discharge occurs only under conditions of high relative humidity. Carter and Banyer (1964), for example, studied the concentration of basidiospores of *Puccinia malvacearum* Mont. in the air near rusted *Malva* sp. in Australia (Fig. 5). The numbers of spores were negligible during the daytime hours, but spores were abundant at night, with a peak concentration around midnight. Probably this is determined by the high humidity at this time, but the light factor might also be involved. Exactly the same picture emerges from a study by Snow and Froelich (1968) of dispersal of basidiospores in *Cronartium fusiforme* Hedgc. and Hunt ex Cumm., the fusiforme rust of southern pines.

### 3. Discharge Involving Change of Shape in Turgid Cells

Another type of violent spore release dependent on high humidity involves sudden rounding-off of turgid spores which, up to the moment of

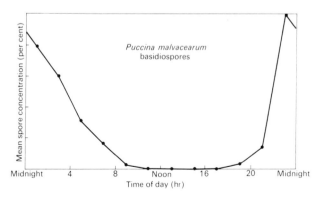

Fig. 5. *Puccinia malvacearum*. Concentration of basidiospores (as percentage of peak geometric mean) throughout the day averaged over a period of 21 days in air near rusted *Malva* in Australia. (After Carter and Banyer, 1964.)

discharge, are prevented from assuming their stable condition of minimum surface by adhesion to other spores or to conidiophores. For the plant pathologist the most familiar example is in Uredinales where, under appropriate conditions of moisture, aeciospores spring out either singly or in groups from aecia. In these, which usually occur on the underside of the rusted leaf, the spores are tightly packed in growing files produced from the basal cells. At the exposed surface of the aeciospore mass, turgid spores are constrained to polyhedral forms but have a tendency to round off, being prevented from so doing only by adhesion to neighboring spores. The strains created are, in due course, relieved violently and the aeciospores spring out to a distance of a centimeter or more.

For this type of spore release to occur, relative humidity must be high so as to maintain the turgidity of the spores. In this connection it is interesting to consider the work of Kramer *et al.* (1968), who studied liberation of aeciospores in *Puccinia andropogonis* Schw. on *Zanthoxylum* and *Uromyces psoraleae* Peck on *Psoralea*. On rainless days there was a marked periodicity of release, with discharge mainly at night presumably related to the high humidity at that time (Fig. 6). However, the nocturnal pattern was not invariable, and on wet days the peak of discharge sometimes occurred in the daytime.

Although this kind of discharge may be the rule for aeciospores, there are a few rusts in which this does not occur and in some cases dry conditions may even be necessary if spores are to escape. This will be considered later in relation to *Gymnosporangium*.

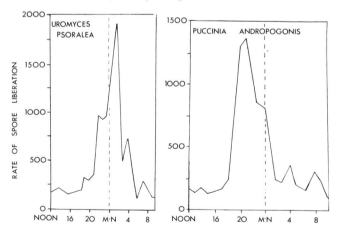

Fig. 6. Periodicity of aeciospore release on days when no rainfall occurred. Left: average number of spores from thirty-one 24-hr collections from *Uromyces psoraleae*. Right: average from twenty-two 24-hr collections from *Puccinia andropogonis*. (After Kramer *et al.*, 1968.)

A rounding-off mechanism requiring saturated conditions for its operation is responsible for the violent discharge of conidia (sporangia) of a downy mildew of maize, *Sclerospora philippinensis* Weston (Weston, 1923). A fresh crop of conidiosphores is produced daily in the hours after midnight when the diseased leaves have a film of dew through which the fertile tips of the conidiophores project. At the narrow junction between each fine branch (sterigma) of the conidiophore and its conidium there is a cross-wall where the two are in flat contact. Both conidium and sterigma are in a turgid state and strains exist where the two adhere. Eventually the strains are suddenly relieved and the conidium springs off. For this to occur the saturated air of a dewy morning is necessary, in striking contrast, as we shall see later, to the low relative humidity needed for liberation of conidia in some other downy mildews.

## 4. Discharge of Conidia in Pyricularia

*Pyricularia oryzae* Cav., the agent causing blast of rice, needs a relatively high humidity for spore release. The affected rice leaves bear branched conidiophores, each with several conidia. It appears that these are set free actively, although the distance of discharge is only a fraction of a millimeter. It has been suggested that the bursting of the stalk cell of the conidium may be the mechanism concerned with liberation (Ingold, 1964) but electrostatic forces may be involved (Leach, 1976). In addition to active discharge, conidia are, no doubt, also splash-dispersed.

## B. Spore Discharge Triggered by Drying

Violent spore liberation is sometimes triggered by rapid change in the ambient air from high to low relative humidity.

## 1. Discharge Due to Hygroscopic Movements

Violent hygroscopic movements may be responsible for discharge without water rupture occurring. The only example of this kind of mechanism among plant pathogens would seem to be the sudden twirling of drying conidiophores in certain species of *Peronospora* and *Pseudoperonospora* (and possibly in some other genera of Peronosporaceae) resulting in the shaking free of the finely attached conidia. This behavior has been described by Pinckard (1942) for *Peronospora tabacina* Adam. In this parasite a new crop of conidiophores is produced on the diseased tobacco leaves overnight. Drying by the sun's rays at dawn causes the conidiophore to lose water rapidly. Its cylindrical stalk collapses to a ribbon as water is lost and twists spirally, thereby shaking off the conidia. Later in the day, around sunset, the conidiophore performs the reverse movements by absorbing water as the relative humidity of the air quickly increases, but by that time it has usually shed all its spores so that no further discharge results. The effect of this hygroscopic discharge is an enrichment of the air spora in the early hours. This is well illustrated in a study by Royle (1967) of the air spora in a hop field attacked by *Pseudoperonospora humuli* (Miyabe and Takah.) G. W. Wilson (Fig. 7). It should, however, be noted that spores can be jerked off their conidiophores in other ways than be hygroscopic twirling. The conidiophores occur on the undersurfaces of the hop leaves. When a diseased leaf, held in the natural position, was struck by a large falling drop of water, Hirst and Stedman (1963) saw a small black eddying cloud of spores issuing from below the point of impact.

*2. Discharge Triggered by Water Rupture Resulting from Drying*

Due largely to the work of Meredith (e.g., 1961, 1962, 1963) there is now a body of information concerning another mechanism of violent dis-

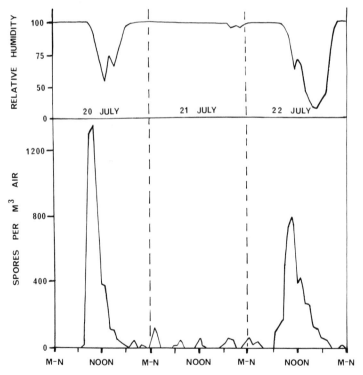

Fig. 7. Changes in concentration of airborne conidia of *Pseudoperonospora humuli* in the air of a hop field in late July at hourly intervals. (After Royle, 1967.)

charge of conidia brought about by lowering humidity in the surrounding air. This mechanism, which involves breaking of tensile water in a cell of the conidiophore distorted by drying, is well known in ferns and leafy liverworts (Ingold, 1939).

*Deightoniella torulosa* (Syd.) M. B. Ellis, which is responsible for a fruit spot of banana, clarifies this principle. A single septate conidium crowns the apex of a simple conidiophore, the top cell of which has an unequally thickened wall. The wall is relatively thick except in the uppermost regions including the zone of contact with the base of the conidium. This unequal thickening of the wall is essential for operation of the mechanism. Under dry conditions, water evaporates from this cell and, because of the unequal wall thickening, the resulting decrease in volume involves a change in shape with the attached conidium sucked downward. This distorted cell of the conidiophore is straining to return to its former shape, but is prevented from doing so by cohesion of the water molecules and by their adhesion to the cell wall. Indeed, the water in the cell is ever so slightly stretched. Eventually the strain becomes so great that the stretched (tensile) water breaks. The cell returns immediately to approximately its former shape and a gas phase (equal in volume to the difference between the cell in its original and in its distorted condition) is now visible. This sudden and violent change in shape of the terminal cell of the conidiophore jerks the conidium into the air.

Consistent with this mechanism of discharge in *Deightoniella*, a study of spores in the air near a banana plantation showed an abrupt peak of concentration between 6 and 8 A.M. (Sreeramulu, 1962) (Fig. 8). This is

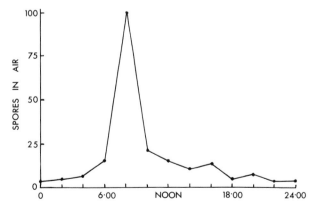

Fig. 8. *Deightoniella torulosa*. Spores in the air near a banana plantation as percentage of peak mean. Average values at 2-hr intervals based on data for 53 days. (After Sreeramulu, 1962.)

closely correlated with a sharp fall in relative humidity from about 95 to around 60%.

Meredith (1962) has suggested that a water-rupture mechanism operates in a considerable range of conidial fungi including *Drechslera turica* (Pass.) Subram. and Jain which parasitizes *Sorghum* and *Zea*. However, Kenneth (1964) was not able to confirm the discharge of conidia in *D. turica*. In this species the large conidium is finely attached (in contrast to the broad attachment in *Deightoniella torulosa*) and no doubt can be blown off from diseased plants by wind or shaken free by leaf flutter induced by wind or rain.

Using *D. turica*, Leach (1975) studied liberation of conidia from infected maize leaves confined in a small experimental chamber through which air of controlled humidity was passed at known speed. The conidia released into the airstream were subsequently impacted on a greased slide where they could be counted. When air humidity was rapidly reduced, there was a great, but temporary, increase in spore release which would be expected if the water-rupture mechanism were in operation. However, on subsequently switching from this dry airstream to one of high relative humidity, there was another peak, though a minor one, in the rate of deposition of conidia on the greased slide (Fig. 9).

In a recent paper dealing with *D. turica*, but with implications for a

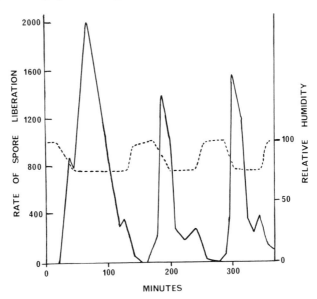

Fig. 9. *Drechslera turica*. Influence of changes in relative humidity (dashed line) on the rate of spore release (solid line) under experimental conditions. (Modified from Leach, 1975.)

wide range of dry-spore conidial fungi, Leach (1976) suggested that static electric charges on the conidiophores, or on the diseased leaf generally, may play a significant part in violently repelling conidia of like charge into the air. Further, it is suggested that the buildup of the charges is related to changes in humidity, particularly if these are drastic and rapid. Should this type of discharge be substantiated, it would represent a further but indirect way in which water relations could influence spore liberation.

## III. PASSIVE SPORE LIBERATION

### A. SPLASH DISPERSAL

The pioneer work on splash dispersal of plant pathogens was published some 50 years ago by Faulwetter (1917a,b) who was concerned with the role of falling water drops in the local spread of a bacterial disease, namely, angular leaf spot of cotton caused by *Xanthomonas malvacearum*. A few years later Weston (1923) noted that drops dripping from the upper foliage of diseased maize as well as heavy rain could play a role in dispersal of conidia of downy mildew of maize (*Sclerospora* spp.).

The fundamental principles of splash dispersal were demonstrated by Gregory and his co-workers (1959). Water drops 2–5 mm in diameter were caused to fall from a height of several meters on to a horizontal film of an aqueous suspension of *Fusarium* conidia. From the point of impact of each drop, a spatter of reflected droplets resulted, each composed of an intimate mixture of water of the original falling drop and of the conidium-containing film. For example, a single drop 5 mm in diameter (the maximum size that is stable) falling from a height of 7.4 meters on to a film of aqueous spore suspension 0.1 mm deep produced about 5000 reflected droplets varying in size from 5 to 2400 $\mu$m. These droplets were thrown to a horizontal distance of up to 100 cm and a large proportion contained conidia.

Gregory and his co-workers also studied splash liberation from a natural target, namely a twig of *Acer* studded with the erumpent conidial stromata of *Nectria cinnabarina* (Tode ex Fr.) Fr. Under dry conditions the conidia are firmly cemented to the stromatal surface and cannot be removed by wind. However, under wet conditions the surface of each stroma becomes a slimy layer teeming with detached conidia. With the wet twig inclined at an angle of 45°, a large falling drop produced over 2000 droplets after impact. Most of these were less than 100 $\mu$m in diameter and all carried spores.

In nature the drops involved in splash dispersal are those of heavy showers and others dripping from foliage wetted by rain or dew. Generally a raindrop is likely to be spore-free, but a "drip-drop" from an infected shoot may gather spores of the pathogen before falling. On impact its contained spores may then be splash dispersed.

The effective distance of splash dispersal rarely exceeds a meter. However, from a source of inoculum, spores may be splashed to such a distance and these in turn be carried still further by other drops. Thus in a series of hops some spores may be distributed by rain splash perhaps to distances of several meters, but the dilution at each stage is so great that such secondary and tertiary splashing is not likely to be very significant. Splash dispersal is fundamentally short range. The general situation is summarized in Fig. 10.

Although splash dispersal on its own has a short range, the combination of splash with wind may spread spores over considerable distances. For example, with *Phytophthora palmivora* Butler infecting papaya fruits,

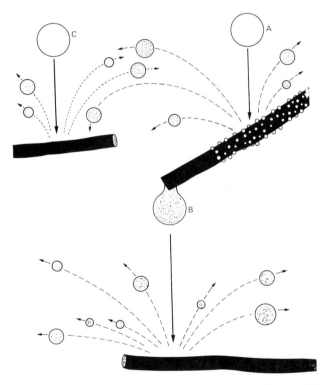

Fig. 10. Diagram of splash dispersal from an inclined twig bearing conidial stromata of *Nectria*. Incident paths of drops shown by solid lines; reflected ones by dashed lines. (A) Spore-free drop; (B) drop accumulated at end of twig and rich in spores; (C) spore-free drop involved in "secondary" dispersal.

epidemics occur in wet weather with the conidia (sporangia) dispersed by wind-driven rain (Hunter and Kunimoto, 1974). Unlike the conidia of *P. infestans* (Mont.) deBary, those of *P. palmivora* are not released under the influence of dry winds.

A further point should be made. Some of the droplets resulting from the impact of a large drop on a source of inoculum are small enough (5–20 $\mu$m) to stay suspended in the air for a considerable time, and, if these contain spores, a contribution will be made to the drifting air spora. Most of the fungal spores that the aerobiologist finds in this spora get there either through violent discharge or are dry-spore types shaken or blown into the air. Nevertheless, some slime-spore fungi, normally splash dispersed, such as species of *Phoma* and *Fusarium*, frequently make a small contribution to the air spora. It is likely that rain splash is largely responsible for their presence.

Particularly in relation to conidial fungi, mycologists have made the biological distinction between dry-spore and slime-spore types (Mason, 1937). In the latter the conidia are associated with slime and cannot be liberated either directly by wind or by mechanical agitation. Further, under dry conditions the conidia are cemented by the dehydrated slime and normally are not in a suitable state for take-off into the air. Thus dispersal of slime-spore types is either by insects or by rain splash. In recent years there has been a growing awareness of the importance of splashing rain in the local spread of plant diseases. Well-established examples are the conidial stage of the apple-canker fungus (*Nectria galligena* Bres.), *Gloeosporium* spp. causing bitter rot of apples, and *Colletotrichum lindemuthianum* (Sacc. and Magn.) Bri. and Cav. responsible for anthracnose in beans *(Phaseolus)* and many other plants.

Splash dispersal of slime-spore pathogens is very important in some orchard crops, first within an individual tree and then, in combination with wind-driven rain, in spreading infection from tree to tree. So far as dispersal within a single tree is concerned, the problem has been studied in apple orchards (Corke, 1966) in relation to bitter rot, and in connection with the coffee berry disease in Kenya caused by *Colletotrichum coffearum* Noack. This fungus attacks both the twigs and the berries. Infection on the upper parts of a coffee bush is spread downward by splash and also by flow of spore-containing water over the surface of the branches. Wind-driven rain is responsible for dispersal from bush to bush (Waller, 1972).

Heavy rain, while often essential for spread of slime-spore fungi, may also be highly significant in relation to dry-spore types by producing stem vibration and leaf flutter of the host that can be responsible for spore liberation on a large scale. A striking example is illustrated in a study by Jarvis (1962) of the spore content of the air in a raspberry plantation with

the berries heavily infected with the grey mold, *Botrytis cinerea* Pers. ex
Fr. In this fungus the conidia are loosened from attachment to the conidiophores by hygroscopic movements of the latter and are thus in a state
to become airborne. In fine settled weather, with high humidity at night
coupled with calm conditions, few conidia are found in the air surrounding
the diseased plants, but with the dry windy weather of daytime, as the
result of thermal turbulence, the concentration of spores in the air greatly
increases. However, this pattern is altered when rain occurs. Thus a
heavy shower at night may lead quickly to high concentrations of conidia
in the air. Again the effect of rain and of overhead irrigation on liberation
of urediospores have been studied by Carter *et al.* (1970) in relation to the
rust *Transchelia discolor* (Fckl.) Tranz. and Litv. which attacks plum and
apricot trees. The concentration of spores in the orchard air rises steeply
immediately after rain or following the turning-on of the sprinkler system
(Fig. 11).

Hirst and Stedman (1963) have analyzed the liberation of dry spores
by large rain drops. Two processes seem to be involved. Many spores are
shaken free, but others are blown away by local high-pressure waves
produced by the drops just ahead of impact. For brevity they refer to
these two processes as "tap and puff." Normally they act together and
their effects cannot be distinguished separately.

Although heavy rain may be a highly effective agent in liberation of
spores in such dry-spore fungi as *Botrytis,* the urediospores of rusts and the

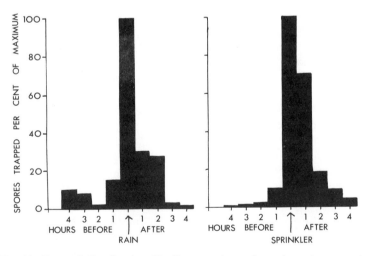

Fig. 11. *Tranzschelia discolor.* Urediospores trapped per hour (as percentage of
maximum) in relation to the onset of rain and of sprinkler irrigation. (After Carter *et al.*,
1970).

brand spores of smuts, it may militate against effective dispersal, for such rain tends to comb spores out of the air and deposit them in the immediate vicinity of the source. The extreme case is the tropical downpour. For example, in plantations of the Queensland nut *(Macadamia)*, prolonged heavy rain is not conducive to development of grey mold *(Botrytis cinerea)*. Not only does the rain remove nearly all spores from the flowering parts, but these spores are almost immediately scrubbed out of the air by the raindrops and brought to the ground (Hunter *et al.*, 1972).

## B. Mist Pickup

Diameters of stable falling drops cannot exceed 5 mm and in effective splash dispersal it is the large drops (3–5 mm diam.) that are involved. However, at the other end of the scale, the minute drops of drifting mist may play some part in spore liberation and dispersal. The possible mechanism of mist pickup is illustrated in Fig. 12. This type of dispersal was first suggested by Glynne (1953) in connection with conidia of *Cercosporella herpotrichoides* Fron, causing eye spot of wheat. Later Davies (1959) showed that a vigorous spray of water droplets from an atomizer could detach spores from certain molds, but this situation bears little resemblance to what might occur in nature. Conditions more nearly natural were realized in a careful study by Dowding (1969). He used mist, with droplets of defined size ranging from 2 to 60 $\mu$m in diameter, moving at from 0.48 to 10.3 ms$^{-1}$ over colonies of *Ceratocystis piceae* (Munch) Bakshi, which is responsible for blue stain in felled pine trunks. After

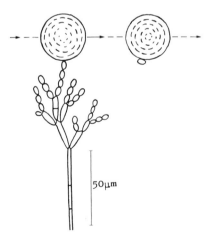

50 μm

Fig. 12. Diagram of how a drifting mist droplet might pick up spores from the conidiophore of *Cladosporium*.

passing over the fungus, the droplets in the mist were impacted on nutrient agar in petri dishes stacked in an Andersen sampler (Ingold, 1971). Colonies of *C. piceae*, derived from the slimy conidial stage, developed in the dishes. Even at the lowest speed used (0.48 ms$^{-1}$) spores were picked up by the mist, but there was no liberation into dry airstreams of equal velocity.

### C. Blow-Off of Dry Spores

Low humidity may be an important factor in liberation in certain plant pathogens where the spores are not actively discharged. This factor operates largely by reducing exposed masses of spores to a powdery state in which they can easily be blown or shaken into the air. This is the situation for the urediospore stage of rusts, the brand spores of smuts, and conidia of powdery mildews and of molds such as *Botrytis cinerea*. Where dry conditions favor blow-off or shake-off the aerobiologist usually finds that the spore concentration in the air is at a maximum around midday. At this time not only are the spore masses likely to be most powdery, but also in fine weather it is the period of maximum gustiness.

Attention was drawn earlier to the need of damp conditions for violent discharge of aeciospores of most rusts with the result that spore release occurs mainly at night when humidity is high. However, in certain species of *Gymnosporangium* the situation differs. In these the aeciospores do not bounce out of the aecium and, in fact, spore release is often a daytime affair. This condition is known in *G. juniperi-virginianae* Schw., the cedar rust with the aecial stage on apple leaves. The emergent peridium around the aecium is dissected into ribbons of tissue. The peridial cells are more heavily thickened on the inner than on the outer side, the reverse of the condition in most rusts. The result is that when the air is damp the peridial ribbons cover the mouth of the aecium, preventing the spores from falling out, but bend back and allow them to escape when drying occurs. The effect of this is that liberation of aeciospores tends to take place in the middle of the day when relative humidity is likely to be low (Pady *et al.*, 1968).

### IV. DRY AND WET AIR SPORA

Suspended in the air is a population of spores, pollen grains, and some other living cells collectively known as the air spora (Gregory, 1973). So far as the fungal element is concerned, its composition at any place and time depends on the balance between liberation and deposition of spores, as well as on the eddy-diffusion situation that influences mixing with neighboring air masses. Both in the processes of liberation and of deposition, water relations play a highly significant role. First, as discussed in detail above, the water factor is likely to affect the types and numbers of spores added to the air spora. Second, water relations, especially the incidence of rain, may profoundly influence the spora by scrubbing spores out of the air. In this connection it should be pointed out that the scrubbing effect is not felt equally by spores of all sizes. Large spores are more readily combed out by rain than are small ones. This may be illustrated by reference to a graph given by Chamberlain (1967) describing the efficency of fair-sized raindrops (2 mm diam.) in collecting particles of different sizes. By reference to Fig. 13 it may be seen that few spores less than 5 μm in diameter (such as the conidia of *Penicillium* and the brand spores of *Ustilago* spp.) are likely to be picked up by the drops, while a majority of those above 10 μm (e.g., urediospores of Uredinales and conidia of Peronosporales and Erysiphales) will be combed out by rain if it persists.

Early in the extensive Rothamsted studies of the microbiology of the air (Hirst, 1953) it was pointed out that there is a striking difference between dry- and wet-air spora. In dry weather the summer spora of country air, particularly in the daytime, is dominated by conidia of such fungi as *Cladosporium, Alternaria, Epicoccum, Erysiphe,* and the urediospores of rusts. The immediate effect of a rainstorm may be an increase of spores of these fungi in the air, but some hours later after the rain, the characteristic wet-air spora develops. By this time many of the normal elements of the dry-air spora have been greatly reduced in numbers by the scrubbing action of the rain, and at the same time there is a striking increase of elongated, and mostly septate, spores. These are mainly as-

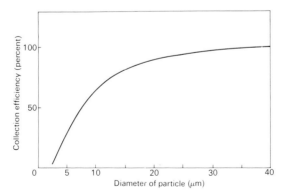

Fig. 13. Efficiency of collection of particles of increasing diameter by falling raindrops 2 mm in diameter. (After Chamberlain, 1967.)

cospores derived from the abundant perithecia, occurring on dead vegetation, which must be wetted if violent spore discharge is to occur. Liberation does not end the problems of a spore in relation to water. During dispersal there are further hazards. Dispersal is of no avail if at the end of the journey the spore is no longer viable. One of the factors that may affect this viability is drying, although it must be remembered that exposure to ultraviolet light may also be important. Fungal spores appear to vary in their capacity to survive desiccation during transport. The nature of the problem may be illustrated by reference to rust fungi. Thus urediospores, with their fairly thick walls, can be borne for hundreds of miles through air without losing their ability to cause infection. On the other hand, it appears that the thin-walled basidiospores soon lose their viability when suspended in air, and this may be the result of desiccation. The same is probably true of the conidia (sporangia) of Peronosporales. There is, however, rather little critical evidence relating to the effect of drying on air-borne spores.

It should be noted that where delicate spores are involved, and which are liable to reduced viability as a result of desiccation, the timing of liberation may be important to the pathogen so that the spores have a good chance of completing their journey in air that is damp. In this connection the discharge of the delicate basidiospores of rusts around midnight or in the early hours of morning would seem to be of biological advantage. There is a further problem in relation to water and spore dispersal. With the exception of certain members of the Erysiphales, nearly all the fungi that attack leaves or herbaceous stems can succeed only if the spores germinate in droplets of water on the host. These infection drops occur mostly as dew during the night and early morning, and also as the result of rain. A successful pathogen must be in a position to seize the transient opportunities for infection. This is particularly true of fungi with spores that retain their viability for only a brief period.

To return to the point made at the beginning of this chapter, the water relations of spore liberation cannot be considered in a meaningful way as an isolated problem in the life cycle of a pathogen. It is necessary to consider, as well, the water relations of spore production, the water problems during dispersal itself, and those involved in establishment on its host.

REFERENCES

Austin, B. (1968). Effects of airspeed and humidity changes on spore discharge in *Sordaria fimicola*. *Ann. Bot. (London)* [N. S.] **32**, 251–260.
Carter, M. V. (1963). *Mycosphaerella pinodes*. II. The phenology of ascospore release. *Aust. J. Biol. Sci.* **16**, 800–817.

Carter, M. V., and Banyer, R. J. (1964). Periodicity of basidiospore release in *Puccinia malvacearum. Aust. J. Biol. Sci.* **17**, 800–802.

Carter, M. V., Moller, W. J., and Pady, S. M. (1970). Factors affecting production and dispersal in *Tranzschelia discolor. Aust. J. Agric. Res.* **21**, 905–914.

Chamberlain, A. C. (1967). Deposition of particles to natural surfaces. In "Airborne Microbes," *Proceedings of the seventeenth symposium of the Society for General Microbiology"* Cambridge. **17**, 138–164.

Corke, A. T. K. (1966). The role of rainwater in the movement of *Gloeosporium* spores on apple trees. *In* "The Fungus Spore" (M. F. Madelin, ed.), pp. 143–150. Butterworth, London.

Davies, R. R. (1959). Detachment of conidia by cloud droplets. *Nature (London)* **183**, 1695.

Dowding, P. (1969). The dispersal and survival of spores of fungi causing bluestain in pine. *Trans. Br. Mycol. Soc.* **52**, 125–137.

Faulwetter, R. F. (1917a). Dissemination of the angular leafspot of cotton. *J. Agric. Res.* **8**, 457–475.

Faulwetter, R. F. (1917b). Wind-blown rain, a factor in disease dissemination. *J. Agric. Res.* **10**, 639–648.

Glynne, M. D. (1953). Production of spores by *Cercosporella herpotrichoides. Trans. Br. Mycol. Soc.* **36**, 46–51.

Gregory, P. H. (1973). "The Microbiology of the Atmosphere," 2nd ed. Leonard Hill, London.

Gregory, P. H., and Stedman, O. J. (1958). Spore dispersal in *Ophiobolus graminis* and other fungi of cereal foot-rots. *Trans. Br. Mycol. Soc.* **41**, 449–456.

Gregory, P. H., Guthrie, E. J., and Bunce, M. E. (1959). Experiments on splash dispersal of fungus spores. *J. Gen. Microbiol.* **20**, 328–354.

Hadley, G. (1968). Development of stromata in *Claviceps purpurea. Trans. Br. Mycol. Soc.* **51**, 763–769.

Hirst, J. M. (1953). Changes in atmospheric spore content: Diurnal periodicity and the effects of weather. *Trans. Br. Mycol. Soc.* **36**, 375–393.

Hirst, J. M., and Stedman, O. J. (1963). Dry liberation of fungus spores by raindrops. *J. Gen. Microbiol.* **33**, 335–344.

Hirst, J. M., Storey, I. F., Ward, W. C., and Wilcox, H. J. (1955). The origin of apple scab epidemics in the Wisbech area in 1953 and 1954. *Plant Pathol.* **4**, 91–96.

Hunter, J. E., and Kunimoto, R. K. (1974). Dispersal of *Phytophthora palmivora* sporangia by wind-blown rain. *Phytopathology* **64**, 202–206.

Hunter, J. E., Rohrbach, K. G., and Kunimoto, R. K. (1972). Epidemiology of *Botrytis* blight of *Macadamia* racemes. *Phytopathology* **62**, 316–319.

Ingold, C. T. (1939). "Spore Discharge in Land Plants." Oxford Univ. Press (Clarendon), London and New York.

Ingold, C. T. (1948). The water-relations of spore discharge in *Epichloe. Trans. Br. Mycol. Soc.* **31**, 277–280.

Ingold, C. T. (1964). Possible spore discharge mechanism in *Pyricularia. Trans. Br. Mycol. Soc.* **47**, 573–575.

Ingold, C. T. (1971). "Fungal Spores: Their Liberation and Dispersal." Oxford Univ. Press (Clarendon), London and New York.

Jarvis, W. R. (1962). The dispersal of spores of *Botrytis cinerea* Fr. *Trans. Br. Mycol. Soc.* **45**, 549–559.

Kenneth, R. (1964). Conidial release in some Helminthosporia. *Nature (London)* **202**, 1025–1026.

Kramer, C. L., Pady, S. M., Clary, R., and Haard, R. (1968). Diurnal periodicity in aeciospore release of certain rusts. *Trans. Br. Mycol. Soc.* **51**, 679–687.

Leach, C. M. (1975). Influence of relative humidity and red-infrared radiation on violent spore release by *Drechslera turica* and other fungi. *Phytopathology* **65,** 1303–1312.

Leach, C. M. (1976). An electrostatic theory to explain violent spore liberation by *Drechslera turica* and other fungi. *Mycologia* **68,** 63–86.

Lortie, M., and Kuntz, J. E. (1963). Ascospore discharge and conidium release by *Nectria galligena* Bres. under field and laboratory conditions. *Can. J. Bot.* **41,** 1203–1210.

McOnie, K. C. (1964). Orchard development and discharge of ascospores of *Guignardia citricarpa* and the onset of infection in relation to the control of citrus black spot. *Phytopathology* **54,** 1448–1453.

Mason, E. W. (1937). "Annotated Account of Fungi Received at the Imperial Mycological Institute," List 2, Fasc. 3 (Gen. Pt.). I.M.I., Kew, Surrey.

Meredith, D. S. (1961). Spore discharge in *Deightoniella torulosa* (Syd.) Ellis. *Ann. Bot. (London)* [N. S. ] **25,** 271–278.

Meredith, D. S. (1962). Spore discharge in *Cordana musae* (Zimm.) Höhnel and *Zygosporium oscheoides* Mont. *Ann. Bot. (London)* [N. S. ] **27,** 233–241.

Meredith, D. S. (1963). Violent spore release in some Fungi Imperfecti. *Ann. Bot. (London)* [N. S. ] **27,** 39–47.

Moller, W. J., and Carter, M. V. (1965). Production and dispersal of ascospores in *Eutypa armeniacae*. *Aust. J. Biol. Sci.* **18,** 67–80.

Pady, S. M., Kramer, C. L., and Clary, R. (1968). Periodicity in aeciospore release in *Gymnosporangium juniperi-virginianae*. *Phytopathology* **58,** 329–331.

Pinckard, J. A. (1942). The mechanism of spore dispersal in *Peronospora tabacina* and certain other downy mildew fungi. *Phytopathology* **32,** 505–511.

Ramos, D. E., Moller, W. J., and English, H. (1975). Production and dispersal of ascospores of *Eutypa armeniacae* in California. *Phytopathology* **65,** 1364–1371.

Rowe, R. C., and Beute, M. K. (1975). Ascospore formation and discharge by *Calonectria crotalariae*. *Phytopathology* **65,** 393–398.

Royle, D. J. (1967). Diurnal and seasonal fluctuations in spore concentrations of airborne pathogens. *Wye Coll., Dep. Hop. Res., Annu. Rep.* for 1966 pp. 49–56.

Snow, G. A., and Froelich, R. C. (1968). Daily and seasonal dispersal of *Cronartium fusiforme*. *Phytopathology* **58,** 1532–1536.

Sreeramulu, T. (1962). Some observations on *Deightoniella* fruit- and leaf-spot disease of banana. *Curr. Sci.* **31,** 258–259.

Waller, J. M. (1972). Water-borne spore dispersal in coffee berry disease and its relation to control. *Ann. Appl. Biol.* **71,** 1–18.

Weston, W. H. (1923). Production and dispersal of conidia in the Philippine Sclerosporas of maize. *J. Agric. Res.* **23,** 239–278.

Yarwood, C. E. (1941). Diurnal cycle of ascus maturation of *Taphrina deformans*. *Am. J. Bot.* **28,** 355–357.

Zoberi, M. H. (1964). Effect of temperature and humidity on ballistospore discharge. *Trans. Br. Mycol. Soc.* **47,** 109–114.

CHAPTER 5

# WATER AND THE INFECTION PROCESS

## C. E. Yarwood

DEPARTMENT OF PLANT PATHOLOGY, UNIVERSITY OF CALIFORNIA, BERKELEY,
CALIFORNIA

## I. INTRODUCTION

Knowledge of the relation of moisture to the infection process began before the nature of disease was understood. When late blight of potatoes caused the Irish Famine in 1845–1846, it was recognized that the disease was associated with cloudy, rainy, foggy, and cool weather, while the experts were still arguing whether the disease was caused by the will of God, the weather itself, or by the fungus *Phytophthora infestans*, which was constantly associated with the disease (Large, 1940). DeBary (1863), who was instrumental in establishing the germ theory of disease, was also one of the first to show the necessity of free water for the infection of potatoes by *Phytophthora*. For many years after this, and to a certain extent even today, beliefs about the infection process have been and are dominated by the idea that "moisture generally augments the production of spores, and is of course essential to their germination" (Duggar, 1909). Unfortunately, this early knowledge of the necessity of water for the infection process came to be applied to pathogens in general, whereas we now know there are important exceptions and extensive variations.

The subject of water and the infection process has been a major aspect of plant pathology studies for many years, and has been briefly reviewed by Boughey (1949), Foister (1935, 1946), Yarwood (1956, 1959b), Cochrane (1958), Waggoner (1965), Wallin (1967), Rotem and Palti (1969), Colhoun (1973), and likely many others.

The infection process is defined here as the invasion of the host by the pathogen. Whetzel *et al.* (1925) define infection as the act of producing a diseased condition in the tissues of the suscept. With fungus diseases, this includes as a minimum, spore germination and host penetration. As a maximum, and in Whetzel's terminology, it includes inoculation, incubation, infection, and sporulation. Incubation period is used here in the sense of Whetzel *et al.* (1925) and Gaumann and Brierley (1950) as the time from inoculation until the pathogen establishes nutritive relations with the host, though Stakman and Harrar (1957) define incubation period as the time from inoculation until manifestation of symptoms, and this definition is in common use. In the terminology of Riker (1936), infection includes transfer, entrance, multiplication, and exit. While fungi are responsible for most parasitic diseases (and most diseases are parasitic), this treatment will include bacterial, fungus, and virus diseases. Diseases caused by higher plants, nematodes, mycoplasms, and infections of roots will generally be omitted because there is less literature on the relation of moisture to these agents or situations and infections of roots will be treated elsewhere in this volume. Although this is primarily a review of the literature, I will draw on my unpublished observations in several places.

## II. MOISTURE AND SPORE GERMINATION

Fungus spores, like seeds, are typically reproductive structures of low water content, which function for survival and reproduction of the species. Storage of spores under cool dry conditions, like storage of seeds under similar conditions, has become a major method of long-time preservation of fungi (Bromfield, 1967; Mazur, 1968; Rotem, 1968).

With most fungus diseases, spore germination is a necessary part of incubation and a necessary preliminary to infection. Because detailed observations of spore germination on leaves are difficult and because regulation and measurement of humidity on the surfaces of plant parts is difficult (Longree, 1939; Stoll, 1941; Delp, 1954), much of the information on water requirements for spore germination has come from trials on glass or other inert and transparent substrates. It is logical to expect, but by no means certain, that the water requirement for germination on leaves and glass will be similar. This uncertainty is perhaps most conspicuous with powdery mildews; Yarwood (1936a), Peries (1962), Burchill (1966), Mence and Hildebrandt (1966), and Jhooty (1971) believe the host exerts a stimulus on germination, although Longree (1939) and Zaracovitis (1966) feel otherwise. There is also disagreement about the relative humidity at leaf surfaces. Longree (1939) and Frampton and Longree (1941) apparently believe that the humidity at a leaf surface in a dry environment approaches 100%, whereas Thut (1938), Yarwood and Hazen (1944), and Rogers (1959), on the basis of different methods of measurement, are convinced it is much lower. On glass the problem of regulating humidity is much less than on leaf surfaces but still exists. In spite of difficulties, much information (and opinions) exist concerning the water requirements for spore germination.

Representative data, but only a fraction of the information available, on the humidity requirements for spore germination are given in Table I. These are arranged as miscellaneous fungi, downy mildews, powdery mildews, and rusts. Among the miscellaneous fungi, the lowest humidity requirement recorded (76%) is for *Aspergillus niger*. For *Monilinia fructicola* and *Venturia inaequalis* there is apparent disagreement between investigators in that two of them found germination at 95 or 99% relative humidity (RH) and two found germination only in the presence of free water. This discrepancy could be due to inadequate control of humidity, since at RH approaching 100% a slight lowering of temperature could cause condensation of water. For the downy mildews both investigators found free moisture necessary for germination, though Yarwood (1939b) believed that more water was necessary for *Pseudoperonospora* with swarm spores than for *Peronospora*, which does not form swarm spores. Of the 11 records for rusts, 5 indicated that free moisture is necessary and 6 that germination can occur in a humid atmosphere without free water. Of the 35 records of miscellaneous fungi, downy mildews, and rusts, there is no record of germination at less than 76% RH; 21 indicate that 100% RH is necessary, and 16 indicate that free water is necessary. Schein (1964) seems to doubt if any plant pathogenic fungi, other than powdery mildews, can germinate without free water; I agree that if there are

TABLE I

HUMIDITY REQUIREMENTS FOR GERMINATION OF FUNGUS SPORES

| Species | Spore type | Source | Minimum RH |
|---|---|---|---|
| Miscellaneous | | | |
| *Alternaria solani* | Conidia | Doran (1922) | 100 |
| *Aspergillus niger* | Conidia | Bonner (1948) | 76 |
| *Beauveria bassiana* | Conidia | Hart and MacLeod (1955) | 94 |
| *Botrytis cinerea* | Conidia | Snow (1949) | 93 |
| *Colletotrichum lindemuthianum* | Conidia | Chowdhury (1937) | 95 |
| *Cylindrocladium scoparium* | Conidia | Anderson (1918) | 100[a] |
| *Monilinia fructicola* | Conidia | Clayton (1942) | 100[a] |
| *Monilinia fructicola* | Conidia | Doran (1922) | 100[a] |
| *Penicillium glaucum* | Conidia | Lesage (1895) | 82 |
| *Phyllosticta antirrinui* | Conidia | Smiley (1920) | 100[a] |
| *Piricularia oryzae* | Conidia | Suzuki (1975) | 100[a] |
| *Septoria apii* | Conidia | Schneider (1953) | 92 |
| *Ustilago hordei* | Chlamydospores | Clayton (1942) | 95 |
| *Ustilago nuda* | Chlamydospores | Clayton (1942) | 93 |
| *Venturia inaequalis* | Conidia | Clayton (1942) | 99 |
| *Venturia inaequalis* | Conidia | Doran (1922) | 100 |
| *Venturia inaequalis* | Conidia | Keitt and Jones (1926) | 100[a] |
| *Venturia inaequalis* | Ascospores | Clayton (1942) | 99 |
| Downy mildews | | | |
| *Peronospora destructor* | Sporangia | Yarwood (1939b) | 100[a] |
| *Peronospora effusa* | Sporangia | Yarwood (1939b) | 100[a] |
| *Peronospora pygmaea* | Sporangia | Doran (1922) | 100[a] |
| *Pseudoperonospora cubensis* | Sporangia | Yarwood (1939b) | 100[a] |
| *Pseudoperonospora humuli* | Sporangia | Yarwood (1939b) | 100[a] |

144

Rusts

| | | | |
|---|---|---|---|
| *Cronartium camandre* | Aeciospores | Powell (1974) | 100 |
| *Cronartium ribicola* | Basidiospores | Hirt (1942) | 97 |
| *Gymnosporangium claviceps* | Aeciospores | Doran (1922) | 100 |
| *Gymnosporangium juniperi-virgina* | Teliospores | Weimer (1917) | 100[a] |
| *Hemileia vastatrix* | Uredospores | Nutman *et al.* (1963) | 100[a] |
| *Puccinia graminis* | Uredospores | Clayton (1942) | 98 |
| *Puccinia graminis* | Uredospores | Stock (1931) | 100 |
| *Puccinia coronata* | Uredospores | Durrell (1918) | 100 |
| *Puccinia coronata* | Uredospores | Clayton (1942) | 99 |
| *Puccinia helianthi* | Uredospores | Yarwood (1939b) | 100[a] |
| *Uromyces fallens* | Uredospores | Yarwood (1939b) | 100[a] |
| *Uromyces phaseoli* | Uredospores | Yarwood (1939b) | 100[a] |

Powdery mildews

| | | | |
|---|---|---|---|
| *Erysiphe cichoracearum* | Conidia | Schnathorst (1960) | 0.1 |
| *Erysiphe graminis* | Conidia | Arya and Ghemawat (1954) | 93 |
| *Erysiphe graminis* | Conidia | Brodie and Neufeld (1942) | 0 |
| *Erysiphe graminis* | Conidia | Cherewick (1944) | 0 |
| *Erysiphe graminis* | Conidia | Clayton (1942) | 95 |
| *Erysiphe graminis* | Conidia | Grainger (1947) | 90 |
| *Erysiphe graminis* | Conidia | Yarwood (1936b) | 0 |
| *Erysiphe polygoni* | Conidia | Brodie and Neufeld (1942) | 0 |
| *Erysiphe polygoni* | Conidia | Yarwood (1936b) | 0 |
| *Microsphaera alni* | Conidia | Brodie and Neufeld (1942) | 0 |
| *Podosphaera leucotricha* | Conidia | Berwith (1936) | 100 |
| *Sphaerotheca humuli* | Conidia | Longree (1939) | 95 |
| *Uncinula necator* | Conidia | Delp (1954) | 0 |
| *Uncinula necator* | Conidia | Yossifovitch (1923) | 20 |

[a] Germination only in presence of free water.

any, they are rare and need further confirmation. Fungal spores on substrates which absorb atmospheric moisture (agar, textiles, grain, leaves) seem to be able to germinate at lower humidity than on nonabsorbtive substrates like glass (Cochrane, 1958; Tarr, 1972).

With powdery mildews, on the other hand, five independent investigators found that at least four species (*Erysiphe graminis*, *E. polygoni*, *Microsphaera alni* and *Uncinula necator*) can germinate at relative humidities approaching 0, as over concentrated sulfuric acid. With *Erysiphe graminis* the evidence is confusing in that three investigators found germination at 0% RH, while three found no germination below 90% RH. This is one of the many cases where positive evidence is more significant than negative evidence; and three cases of germination at 0% RH are more convincing than three failures to get germination at less than 90% RH. Nevertheless, these failures are disturbing.

In terms of long-time survival, however, powdery mildew conidia are likely less tolerant of low humidity than most fungi, with the likely exception of some basidiospores (Hirt, 1942). *Erysiphe graminis*, *E. polygoni*, and *Sphaerotheca fuliginea* lose their germinability in about 24 hours under ordinary dry laboratory conditions (Yarwood *et al.*, 1954) whereas spores of 34 miscellaneous nonerysiphaceous fungi lived from 21 days to 60 years (Altman and Dittmer, 1962) and tobacco mosaic virus has been viable for 52 years (Johnson and Valleau, 1935) under comparable conditions. I believe that under ordinary conditions powdery mildew conidia must germinate or die within 48 hours, though under low temperatures and medium humidity they have been kept viable for 2 years (Hermansen, 1972). Most other fungus spores, in contrast, are resting organs of infectivity which are best preserved at low water content (Bromfield, 1967; Hawker and Madelin, 1976).

Schnathorst (1960) believes that powdery mildews can be divided into three groups: (1) species tolerant to low humidity, presumably including *Erysiphe polygoni*, *E. graminis*, *Microsphaera alni*, and *Uncinula necator*; (2) species that germinate best at high RH but capable of germinating at low RH (including *Erysiphe cichoracearum* from lettuce); and (3) species that germinate only at high RH and presumably include *Podosphaera leucotricha* and *Sphaerotheca pannosa*. I suggest there is a continuous gradient from those preferring and/or tolerating low humidity, to those less tolerant of low humidity, but that all powdery mildews are more tolerant of low humidity during the germination process than any other recognized group of pathogens. Humidity requirements for spore germination are also expressed as bars of water potential (Griffin, 1972).

## III. WATER CONTENT OF SPORES

Three interpretations have been offered for the germination of Erysiphaceae at low humidities but none has been adequately confirmed. Graf-Marin (1934) and Brodie (1945) indicate that conidia of Erysiphaceae have a high osmotic pressure, and Brodie suggests that this high osmotic pressure ( = low water content) permits the fungus to remove moisture from the air and use this moisture for germination. Yarwood (1950a, 1952a) reports that viable conidia of Erysiphaceae in ambient air have a high water content, that they differ in this respect from most if not all other air-borne spores (see Table II), and that this internal water is adequate for germination. The high water content is confirmed in principle by Corner (1935) and quantitatively by Jhooty and McKeen (1965), Zaracovitis (1974), and Drandarevski (1969). The presence of vacuoles containing water in the conidia of Erysiphaceae (Yarwood, 1952a), but not in most air-borne fungus spores (Ingold, 1954), would seem a confirmation. The presence in these vacuoles of particles with active Brownian motion (Yarwood, 1952a), which may not have been seen in any other fungi, supports the idea that the vacuoles contain water. The rapid movement of particles in the protoplasm of powdery mildews generally (Leveille, 1851), and *Erysiphe cichoracearum* in particular (McKeen *et al.*, 1967), indicates that the protoplasm is rather dilute.

However, Somers and Horsfall (1966) challenge the above; their data, if valid, clearly indicate that my ideas on water content of fungus spores are untenable. Richmond and Somers (1963) and Somers and Horsfall (1966) found that the water content of 10 fungi which I have placed in the miscellaneous group (Table II) averaged 62.5%, which is about equal to the water content of powdery mildews as found by myself, but greater than that of powdery mildews as found by Somers and Horsfall. But the water content for this group, as cited by Richmond and Somers and Somers and Horsfall, is 220% higher than the water content of 14 fungi in this group (excluding *Claviceps purpurea*) as measured by nine other investigators. Even within the same three genera, values by Somers and Horsfall are much higher than those found by four other investigators. For the single species, *Aspergillus niger*, the value of Richmond and Somers is 293% higher than those of Yarwood (1950a) and Zobl (1950), while values of Yarwood and Zobl are about equal to each other. Just as Foster (1949) considered Amos' value of 42% water for conidia of *Aspergillus oryzae* to be abnormally high, so I consider 82% water for conidia of *Alternaria tenuis*, 88% water for *Botrytis fabae*, and 82% water for *Venturia in-*

TABLE II

WATER CONTENT OF FUNGUS SPORES

| Species | Spore type | Source | Water (%) |
|---|---|---|---|
| Miscellaneous | | | |
| *Aethalium septicum* | Conidia | Aso (1900) | 7 |
| *Alternaria tenuis* | Conidia | Richmond and Somers (1963) | 85 |
| *Alternaria tenuis* | Conidia | Somers and Horsfall (1966) | 50 |
| *Aspergillus brevicaule* | Conidia | Zobl (1950) | 10 |
| *Aspergillus fumigatus* | Conidia | Zobl (1950) | 7.4 |
| *Aspergillus niger* | Conidia | Richmond and Somers (1963) | 55 |
| *Aspergillus niger* | Conidia | Yarwood (1950a) | 13 |
| *Aspergillus niger* | Conidia | Zobl (1950) | 15 |
| *Aspergillus oryzae* | Conidia | Aso (1900) | 42 |
| *Aspergillus oryzae* | Conidia | Sumi (1928) | 17 |
| *Aspergillus terreus* | Conidia | Stapleton and Hollaender (1952) | 43 |
| *Botrytis allii* | Conidia | Richmond and Somers (1963) | 30 |
| *Botrytis cinerea* | Conidia | Yarwood (1950a) | 17 |
| *Botrytis fabae* | Conidia | Richmond and Somers (1963) | 75 |
| *Claviceps purpurea*[a] | Conidia | Glaz (1955) | 70–76 |
| *Monilinia fructicola* | Conidia | Yarwood (1950a) | 25 |
| *Mucor* sp. | Sporangiospores | Cramer (1894) | 42 |
| *Neurospora crassa* | Conidia | Richmond and Somers (1963) | 75 |
| *Neurospora tetrasperma* | Ascospores | Sussman (1968)[b] | 5 |
| *Penicillium digitatum* | Conidia | Yarwood (1950a) | 6 |
| *Penicillium expansum* | Conidia | Richmond and Somers (1963) | 65 |
| *Penicillium glaucum* | Conidia | Cramer (1894) | 35 |
| *Penicillium italicum* | Conidia | Richmond and Somers (1963) | 65 |

148

| | | | |
|---|---|---|---|
| *Peronospora destructor* | Sporangia | Yarwood (1950a) | 17 |
| *Rhizopus nigricans* | Sporangiospores | Richmond and Somers (1963) | 48 |
| *Stemphyllium sarcinaeforme* | Conidia | Somers and Horsfall (1966) | 44 |
| *Tilletia levis* | Chlamydospores | Zellner (1911) | 7.9 |
| *Tilletia tritici* | Chlamydospores | Zellner (1911) | 8 |
| *Venturia inaequalis* | Conidia | Richmond and Somers (1963) | 82 |
| Rusts | | | |
| *Cronartium fusiforme* | Aeciospores | Walkinshaw *et al.* (1967) | 10–15 |
| *Melampsora lini* | Uredospores | Turel and Ledingham (1959) | 10–66[d] |
| *Puccinia coronata* | Uredospores | Sharp and Smith (1957) | 8.4–16[c] |
| *Puccinia graminis* | Uredospores | Bromfield (1967) | 15–30 |
| *Puccinia graminis* | Uredospores | Shaw (1964) | 10–15 |
| *Uromyces phaseoli* | Uredospores | Somers and Horsfall (1966) | 18–51[d] |
| *Uromyces phaseoli* | Uredospores | Yarwood (1950a) | 12 |
| Powdery mildews | | | |
| *Erysiphe cichoracearum* | Conidia | Somers and Horsfall (1966) | 18, 31, 69[d] |
| *Erysiphe cichoracearum* | Conidia | Yarwood (1950a) | 52 |
| *Erysiphe graminis* | Conidia | Somers and Horsfall (1966) | 10, 35, 66[d] |
| *Erysiphe graminis* | Conidia | Yarwood (1950a) | 75 |
| *Erysiphe graminis* | Conidia | Zaracovitis (1964) | 65 |
| *Erysiphe polygoni* | Conidia | Jhooty and McKeen (1965) | 67–75 |
| *Erysiphe polygoni* | Conidia | Yarwood (1950a) | 72 |
| *Sphaerotheca macularis* | Conidia | Jhooty and McKeen (1965) | 45–69 |

[a] Spores produced in liquid culture. All other spores were produced in air.

[b] Data taken from Sussman but credited to Lingappa and Sussman.

[c] Data for *Puccinia coronata* and *P. graminis*.

[d] The lower value, or the two lower values are for ambient conditions (dry air); the higher value is for humidified air. Humidified conditions are considered abnormal for powdery mildews and rusts, and data on spores produced under such conditions are excluded from calculations in the text.

*aequalis*, as found by Richmond and Somers, to be not only abnormally high, but also questionable.

With rusts also, values by Somers and Horsfall for water content are much higher than those found by six other investigators.

With powdery mildews, data by Somers and Horsfall are much lower than those of the only three other investigators whose data are available. Data by Jhooty and McKeen (1965), Zaracovitis (1974), and Drandarevski (1969) agree reasonably well with mine.

The third interpretation is that of McKeen (1970), who suggests that respiration of oils could yield some of the water needed for spore germination of the powdery mildews. Yarwood (1952a), Kimbrough (1963), Johnson *et al.* (1976), and McKeen (1970) have all emphasized the high oil, fat, or lipid content of conidia of powdery mildews. These oils or other nonaqueous volatile constituents could be driven off in drying the spores and be misinterpreted as water.

It is apparent from the above that knowledge of the water content of fungus spores is in an unsatisfactory state, mainly because the recent findings by Somers and Horsfall differ so markedly from other investigators. Three factors which may have an important bearing on the controversy are the oil and/or other volatile or respirable contents of spores, the conditions under which the spores are produced, and the method of determination of water content.

The effect of the environment on production of spores and their water content has been emphasized by Turel and Ledingham (1959) and Somers and Horsfall (1966), and both groups agree that under high humidity the water content of spores is much higher than under ambient (dry) conditions. I feel that the water content of spores of the types studied here should be measured with spores produced in dry air, since that is the condition under which they are normally produced and/or disseminated.

## IV. SWELLING OR SHRINKING OF SPORES DURING GERMINATION

I believe that most air-borne fungus spores absorb external water during germination and therefore increase in volume (Table III), though I have found reported increases in volume for only 10 species. The maximum increase may be about twentyfold for *Rhizopus arrhizus* as observed by Ekundayo and Carlisle (1964). The failure to detect increase in volume in *Aspergillus niger, Botrytis* sp., and *Uromyces fallens* is considered more likely a failure to detect the increase, than that no increase occurred. Arthur (1929) indicated that a slight increase in volume of spores during germination of rust spores may be a general phenomenon. In contrast Yarwood

TABLE III

SWELLING OR SHRINKAGE OF SPORES DURING GERMINATION

| Species | Source | Change in volume (%) |
|---|---|---|
| Aspergillus sp. | Armolik and Dickson (1956) | +Severalfold |
| Aspergillus fumigatus | Mandels and Darby (1953) | +65 |
| Aspergillus niger | Mandels and Darby (1953) | 0 |
| Bacillus subtilis | Mandels and Darby (1953) | +950 |
| Botrytis sp. | Mandels and Darby (1953) | 0 |
| Cincinnobolus cesatii | Yarwood (1936b) | +58 |
| Colletotrichum trifolii | Yarwood (1936b) | +74 |
| Erysiphe polygoni | Yarwood (1936b) | −30 |
| Erysiphe polygoni | Brodie and Neufeld (1942) | 0 |
| Erysiphe graminis | Yarwood (1936b) | −9 |
| Fusarium culmorum | Marchant and White (1966) | +35 |
| Monilinia fructicola | Yarwood (1936b) | +146 |
| Penicillium sp. | Armolik and Dickson (1956) | +Severalfold |
| Rhizopus arrhizus | Ekundayo and Carlisle (1964) | +2000 |
| Trichoderma sp. | Barnes and Parker (1946) | +300 |
| Uromyces fallens | Yarwood (1936b) | 0 |
| Sphaerotheca macularis | Jhooty and McKeen (1965) | 0 |

(1936b, 1952a) found that two species of Erysiphaceae shrank during germination. This was confirmed by Nour (1958) through Brodie and Neufeld (1942) and Jhooty and McKeen (1965) failed to confirm this.

## V. MOISTURE AND INFECTION

### A. BACTERIA

Plant pathologists seem to agree that free water is necessary for infections with most bacteria, though experimental data in support of this are limited. Most plant pathogenic bacteria are motile, and this motility can be well expressed only in a water suspension. That this motility aids in infection is indicated by Panopoulos and Schroth (1974), but Diachun et al. (1944) thought otherwise. Clayton (1936) and Johnson (1947) emphasize that water congestion of leaves is necessary for extensive infection of tabacco by Bacterium tabacum and B. ungulatum. Williams and Keen (1967) found a similar relation with Pseudomonas lachrymans in cucumber. Increasing the water content of potato tubers increased the invasiveness of Bacillus carotovorus (Gregg, 1952) and other bacteria (Fernando and Stevenson, 1952; Lapwood, 1957). Water soaking of tissues before or during inoculations has become a standard procedure in inoculations with bacteria (Williams, 1967).

Shaw (1935) found maximum invasiveness of pear shoots or leaves was at 100% RH, and at 99% RH invasiveness was only one-sixth of that at 100% RH. Thomas and Ark (1934) report that the sugar content of pear nectar in laboratory air was 35–45%, in moist chambers only 6–7%, and that invasiveness of *Erwinia amylovora* was inversely proportional to the sugar concentration ( = directly proportional to the water concentration).

Plant pathogenic bacteria invade the intercellular spaces or the xylem vessels of their hosts. Xylem vessels are typically filled with a dilute water solution, and the intercellular spaces are readily filled with water by internal pressure, by driving rain, or as a result of injury.

While water-soaking and moist chamber incubation, separately or combined, favor infection with bacteria, rub inoculation with abrasive (Bohn and Maliot, 1945; Yarwood, 1969b) commonly gives as much infection, without either water-soaking or moist chamber incubation. This rub method of inoculation with bacteria, presumably adapted from inoculation with viruses (Rawlins and Tompkins, 1934; Yarwood, 1957), probably deposits the bacteria in the host cells or in the intercellular spaces in contact with cellular water. Rub inoculation with abrasive is so standard for virus inoculation that transmission of an unknown pathogen by this method may have led to the belief that a bacterial infection was a virus infection (Yarwood *et al.*, 1961).

Surprisingly, quick drying in abrasive inoculation with bacteria, as in inoculation with viruses, and especially with high dosages of crude inoculum, may increase infection (Yarwood, 1969b). This lends further support to the idea, also applicable to virus infections, that part of the increase in infection due to quick drying, may be due to the removal of toxic metabolites which tend to inhibit the infection process, since it is unlikely that removal of water per se increases infection with bacteria.

## B. FUNGI

Fungi with motile spores, like motile bacteria, are believed to need free water for germination and penetration. These include all the Plasmodiophorales and Chitridiales, and most of the Peronosporales. Among the Peronosporales are many species with spores which germinate only by swarm spores, some which germinate only by a germ tube, and some which can germinate either by swarm spores or a germ tube. Forms which germinate by swarm spores, such as *Pseudoperonospora cubensis* and *Pseudoperonospora humuli*, appear to require more water for infection than forms which germinate by a germ tube, such as *Peronospora destruc-*

*tor* or *Peronospora effusa* (Yarwood, 1939b). However, I cannot accept without further qualification Tarr's (1972) statement that "Sporangia and oospores of some Phycomycetes produce zoospores in wet conditions and germ tubes in dry conditions . . ."

The requirements of most pathogenic fungi for free water or high humidity during the incubation stage is commonly referred to as the incubation period, or moist chamber period necessary for infection. In such trials, plants are usually inoculated with a spore suspension, placed in a moist chamber, and returned to a dry environment at intervals. The period in a moist chamber, or the period of time after inoculation that the plants must remain wet for infection to occur, is the minimum incubation period. Some representative values of minimum incubation periods are: 8 hr for *Cronartium ribicola* basidiospores (Hirt, 1942); 5 hr for *Monilinia fructicola* conidia (Klos, 1976); 2 hr for *Phytophthora infestans* sporangia (Cox and Large, 1960); 3 hr for *Peronospora destructor* sporangia (Yarwood, 1943); 6.5 hr for *Phakospora pachyrhiza* uredospores (Melching and Bromfield, 1975); 10 hr for *Piricularia oryzae* conidia (Kahn and Libby, 1958); 8 hr for *Puccinia graminis* uredospores (Kochman and Brown, 1976); 3 hr for *Uromyces phaseoli* uredospores (Yarwood, unpublished); and 9 hr for *Venturia inaequalis* conidia (Kenaga, 1970). With some, if not most, intercellular species, including *Peronospora destructor* (Yarwood 1943), the moist chamber period corresponds roughly to the period of vulnerability to fungicides—that is, as soon as the fungus reaches a stage where it is not inactivated by a dry atmosphere, it is also not inactivated by a fungicide.

This is not the case with subcuticular pathogens. With *Venturia inaequalis* the fungus becomes resistant to the drying action of air in about 9 hr after inoculation, but is not resistant to lime sulfur until about 50 hr after inoculation (Mills, 1947).

These minimum values of incubation periods in a moist chamber do not adequately represent the dosage effect of time and its interaction with temperature and host susceptibility. In 6 trials with bean rust *(Uromyces phaseoli)* at 16°–26°C, and based on 16–1555 lesions counted on the 24-hr controls, the percentage of maximum infection was 0 when the plants were removed from the moist chamber at 1 and 2 hr after inoculation; 0.02% of maximum at 3 hr, 4% at 4 hr, 16% at 6 hr, 59% at 8 hr, 72% at 10 hr, 95% at 13 hr, and 100% at 16, 20, and 24 hr. Temperature is, of course, important. With *Venturia inaequalis* the minimum incubation period was about 2 days at 3°, 28 hr at 6°, 14 hr at 10°, 11 hr at 13°, 9 hr at 21°, and 11 hr at 25°C (Mills, 1947). With bean rust, in one test, the number of lesions resulting on 4 cm² of leaf after 6 hr in a moist chamber was 7 at 16°, and 391 at 27°, while at 14 hr the number of lesions at 16° and 27°C was approximately equal. With *Elsinoe ampelina* the wetness period for infection was 7–10 hr at 12° and 3–4 hr at 21°C (Brook, 1973). With *Pyrenophora trichostoma* on wheat, Horsford (1975) found that a wet period of only 6 hr was necessary for infection of the susceptible variety DN 495, but with increasing resistance the necessary wet period increased to 12 hr with the variety Waldron and 54 hr with Duri. I failed to detect any such trend with bean rust *(Uromyces phaseoli)* on susceptible and resistant varieties of bean.

Water is necessary for most infections to occur, but water can also be inhibitory. To test the effect of excess water, or water in the absence of or shortage of air, bean leaves were inoculated with rust and immersed vertically in water for 8 or more hours before returning to standard glasshouse conditions. When the leaves were dusted with dry uredospores in the late afternoon, half-immersed in water, and returned to the glasshouse bench next morning, heavy infection occurred over the entire immersed portion, but with slightly decreasing infection with increasing depth of water up to a depth of at least 10 cm of water (Fig. 1). If, on the other hand, the leaves were inoculated with a spore suspension and similarly immersed, no or very little infection occurred beyond an immersion depth of about 2 cm. When similar inoculations were made in the morning and the immersed leaves exposed to natural light, heavy infection occurred whether inoculation was by spraying or dusting. My interpretation is that when the leaf with dry spores was immersed, the difficult-to-wet spores (Yarwood, 1969a) carried enough air (oxygen) for germination and infection, whereas

Fig. 1. Infection of bean rust on leaves immersed in water. Attached leaves were dusted with dry spores over the entire lower surfaces of leaves and the leaves immersed vertically in water with the distal end down for about two-thirds of the length of the leaves at 2130 hr, March 17, removed from water at 0800 hr, March 18, and photographed April 2.

the wetted spores applied as a spore suspension did not. But when inoculation and incubation were in the light period, the photosynthesis of the leaf generated sufficient oxygen for germination and infection. When cucumber leaves were incubated with *Pseudoperonospora cubensis* by the above immersion methods, infection occurred to a depth of only about 4 cm. That is, air was more essential for *Pseudoperonospora* than for *Uromyces*.

Powdery mildews are dry weather fungi (Yarwood, 1950b; Reuveni and Rotem, 1973). They may be the only fungi of which the spores are injured by contact with water (Corner, 1935; Arya and Ghemawat, 1954; Hashioka, 1937; Rogers, 1959; Yarwood, 1936b; Ward and Manners, 1974; and Perera and Wheeler, 1975). They usually germinate better on dry glass than on the surface of water; in water the spores may disintegrate in a matter of minutes (Yarwood, 1952a). A water suspension of *Erysiphe polygoni* or *E. cichoracearum* may serve as a satisfactory inoculum, if used soon after preparation, but if kept for more than an hour, its infectivity is lost (Yarwood, 1951a). Water sprays may control powdery mildews (Anonymous, 1861; Yarwood, 1939a; McClellan, 1942; Rogers, 1959), but the heavy syringing necessary was not considered practical by McClellan.

The control of powdery mildews by spraying with water is probably largely a mechanical effect of the impact of water, since immersing mildew-infected *(Sphaerotheca fuliginea)* cucumber leaves in water for up to 2 days did not clearly reduce the mildew, and in many cases clearly increased it.

Powdery mildews are unique among the air-borne fungi in that they do not require free moisture or high humidity for infection (Yarwood, 1936b; Delp, 1954; Rogers, 1959). Even with the few species which require high humidity for germination on glass (Longree, 1939) no increase in infection is detected from incubation in moist chambers. Longree has interpreted this as due to the humidity approaching 100% RH at the leaf surface, but Yarwood and Hazen (1944) believe leaf surfaces in a normal dry environment are normally far below 100% RH and that the infection under these conditions is more likely due to host stimulation of spore germination (Jhooty and McKeen, 1960).

The ascospore stage of Erysiphaceae, like many other fungi, requires free moisture for spore discharge and presumably for spore germination (Cherewick, 1944; Yarwood and Gardner, 1972).

Schmiedeknecht (1958) refers to *Pseudopezia medicaginis* as a xerophytic fungus, but this may not be adequately confirmed. The contrasting effect of moisture on downy mildews, powdery mildews, and rusts was indicated by the observations of Seymour in 1887.

## C. VIRUSES

Viruses apparently do not need free water or high humidity at any stage of the infection process. Ecologically, viruses are more abundant in dry (low rainfall) than in wet areas and seasons (Attafuah and Tinsley, 1958; Wellman, 1972; additional references in Yarwood, 1957). All or most viruses are wound pathogens, and wounds are probably always initially "wet." Growing plants at high soil moisture (Tinsley, 1953) or soaking leaves in water (Yarwood, 1959a) predisposes plants to virus infection, but beyond this, water appears to inhibit the infection process.

Schultz and Folsom (1923) and Saint (1948) have reported that incubating inoculated plants in a moist chamber after inoculation increased transmission, but this has apparently not been adequately confirmed. Artificial inoculations are usually made by rubbing with a water suspension of virus plus abrasive without any special attempt to remove the water, and heavy infection commonly results (Rawlins and Tompkins, 1934; Matthews, 1970). But under many conditions the quicker the free liquid is removed, the heavier the infection. Increases of up to 3000-fold have resulted from drying the leaves within about 3 sec of inoculation, by means of blotters or by directing a blast of dry air against the inoculated leaves (Yarwood, 1955, 1973). The increase due to quick drying was greater with cucumber mosaic virus than with any other virus tested; greater with cowpea than any other indicator host tested; greater with young than with old indicator plants; greater with concentrated than with dilute inoculum; greater with inoculum from terminal than basal leaves of the donor plant; greater for plants placed in a dark moist chamber than left in the normal light; and greater with caffeine, $K_2SO_3$, and $Mg_2Si_3O_8$ in the inoculum than with any other supplement. To avoid the injurious effect of water on infection, Hollings (1955) and Ragetti et al. (1973) have devised methods of dry inoculation. Surprisingly, washing leaves with water immediately after inoculation, which is superficially opposite to quick drying, also increased infection (Holmes, 1929; Yarwood, 1963, 1973). I believe this is due to the removal of inhibitors by the washing process, and if washing was continued beyond 10 sec, infection was reduced.

## D. GRAPHIC PRESENTATION

Each pathogen has its own life history, its requirements for water, and responses to water, but these are correlated somewhat with the taxonomic positions of the pathogens. In Fig. 2, modified from Yarwood (1956) where much of the evidence is presented, I propose a grouping of foliage pathogens according to their moisture requirements at different

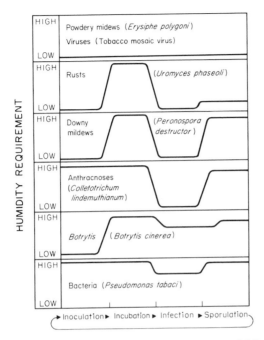

Fig. 2. Diagrammatic representation of humidity requirements of foliage pathogens.

stages of the infection process. They are arranged from those considered the most primitive (high water requirement), to the most advanced (requiring least water). In the most primitive or bacterial type, typified by *Pseudomonas tabaci*, high humidity especially favors inoculation, incubation, and sporulation (release of bacteria from the host) but is needed less for infection (development after entry). With *Botrytis* high humidity is not essential for inoculation (dissemination) but is essential for incubation, and desirable for infection and sporulation. Many other Moniliales may fall in this category. Anthracnoses and other fungi with acervuli and pycnidia, mostly need water for splash dissemination and incubation, and high humidity for sporulation, but not for infection. Downy mildews and rusts need high humidity for incubation, but do not need high humidity for inoculation or infection. Their moisture requirements differ primarily at sporulation, where high humidity is necessary for downy mildews, but not for rusts. Stated in another way, downy mildews need two periods of high humidity to complete their asexual life cycles (Yarwood, 1943; Viranyi, 1975), whereas the uredinial stages of rusts need only one. Rotem and Cohen (1974) believed that epidemics of late blight were favored by low daytime relative humidity. Rusts have been characterized as dew-favored

diseases (Chester, 1946; Yarwood, 1951b) while downy mildews and anthracnoses are rain-favored diseases. Powdery mildews and viruses are
commonly considered dry weather diseases (Yarwood, 1950b, 1955,
1963), though both will tolerate considerable wet weather, and free water
appears essential for the dissemination of ascospores.

## VI. SOURCES OF MOISTURE

A. ATMOSPHERIC HUMIDITY

Most organic and inorganic materials, including living leaves and
spores, can absorb moisture from humid air (Anonymous, 1872; Thut,
1938; Govaerts and Leclercq, 1946; Haines, 1952). Moisture may be incorporated into leaves or spores and/or accumulate preferentially on the
surface as with dew (Anonymous, 1862; Aitkin, 1887). Some fungus-
diseased tissues have a greater power to accumulate water from humid
air or from soil than do healthy tissues (Yarwood, 1965, 1966) and this
may be manifest as water congestion of tissues, as with infections with
*Phytophthora infestans* and *Pseudoperonospora humuli*, or be invisible to
the eye. Humid air, in the absence of free moisture, is especially favorable
for sporulation of such fungi as *Botrytis* and the Peronosporaceae.

The temperature of the leaf is commonly similar to that of the air, but
is usually greater than that of the air in bright sunlight and less than that of
the air in clear dark weather (Yarwood, 1959b). At high humidities a slight
decrease in the temperature of the air or of the leaf may cause or accelerate deposition of free moisture on leaves or other substrata. For these
reasons some reports of infection at high humidity but in the absence of
free water are commonly suspect (Schein, 1964). Precautions may reduce
the likelihood of condensation, but can probably never eliminate it. With
the exceptions of the powdery mildews and viruses, and in spite of much
evidence to the contrary, I believe we are justified in believing that no
infection of foliage pathogens has been adequately demonstrated to occur
in the absence of free moisture.

Critical to any study of the relation of moisture to infection is the
relative humidity at leaf surfaces. How nearly does the RH measured in a
Stevenson screen for weather records (Yarwood, 1959b) represent the RH
at the leaf surface where the fungus must germinate? I believe, in agreement with Wallin (1967), that it is inadequate. Ramsey *et al.* (1938) used an
evaporimeter and a hygrometer in an environment ranging from 10 to 50%
RH, and found the RH at 0.5–1 mm from the leaf surface to be 20–93%.
Longree (1939), on the basis of the finding that *Sphaerotheca pannosa*
conidia germinated well on glass slides only at 95% RH and above, but

gave good infection on leaves at 21% RH, concluded that the RH at the surface of leaves in a dry atmosphere was about 95%. Yarwood and Hazen (1944) used temperature depression of leaves as a measure of RH and found that in an atmosphere of 51% RH, the RH of various untreated surfaces ranged from 52 to 72%, and for Bordeaux-treated leaves ranged from 63 to 83%. They believed that the high germination of *S. pannosa* on leaves was more likely caused by host stimulation of germination (Yarwood, 1936a; Mence and Hildebrandt, 1966) than by the high humidity at leaf surfaces. Subsequently Rogers (1959) observed relative humidities of 58–65% at 0.75 mm from the leaf surface and calculated the RH at the leaf surface as 62–75%, when the RH of the air was 58–60%. Schnathorst (1957) believed the RH at the surface of a lettuce leaf was 33% less than at a free water surface. I believe these relative humidities are inadequate for germination of spores of any plant pathogens other than Erysiphaceae.

Humidity can affect infection by its effect on the host, without a direct effect on the pathogen. High humidity may favor water congestion which may favor infection by bacteria (Braun and Johnson, 1939), and of course low humidity can prevent infection by preventing water congestion. High humidity can favor formation of suberin and wound periderm, which may prevent infections or wall off infections (Artschwager and Starret, 1931; Butin, 1955).

## B. RAIN

For pathogens which require free moisture for infection, in most cases, it seems to make little difference to the pathogen where the moisture comes from, but different sources of moisture may favor certain diseases over others. The major sources of free water are rain, irrigation water, dew, fog, guttation, intercellular water, and intracellular water.

Rain is the most firmly established source of water for fungus infections, as it is for plant growth. The best negative evidence for diseases is the absence of certain infections and/or the lack of spread of certain infections in rain-free growing areas or seasons (Muncie, 1917; Mackie *et al.*, 1945; Menzies, 1954; Rotem and Palti, 1969). Areas such as western United States, Egypt, and Palestine are almost free of rain during the summer growing season and are commonly devoid of such diseases as anthracnoses and foliage bacteria, which require splashing water for dissemination and infection. On the other hand, these same environments are especially prone to powdery mildews, which do not need free moisture for infection, and to rusts, for which dew or guttation is adequate. In a rain-free environment, sprinkling irrigation (Ruppel *et al.*, 1975) and especially heavy syringing (Yarwood, 1939a) may reduce powdery mildews, whereas protection from rain in an area with frequent summer rains (Yarwood, 1936b) may actually increase powdery mildews.

## C. IRRIGATION

Sprinkling as a form of irrigation is basically similar to rain in its effect on disease (Curl and Weaver, 1958; Snyder *et al.*, 1965), except that its timing and rate are controlled by man. In a dry environment, plants may dry off in about 30 min after rain or sprinkling is discontinued so that, for example, 2 hr of heavy sprinkling or rain plus 30 min to dry off, do not give a long enough incubation period for most rain-favored infections to develop. On the other hand, if a short period of rain or sprinkling is discontinued in the evening, the plants may remain wet till the following morning, and this ±10-hr incubation period is adequate for many infections to become established.

Since night temperatures are usually lower than day temperatures, the duration of the wet incubation period for the pathogen during the night may need to be longer or shorter than the incubation period during the day, depending on the actual temperatures, the optimum temperature for the pathogen, and the effect of light on the pathogen.

## D. DEW

Dew is the condensed moisture which is deposited when the temperature of surfaces falls below the temperature of the dew point of the air (Aitken, 1887). If cooling by radiation is intense, as during clear, still nights, this dew point can occur at relatively low humidities. For example, at an air temperature of 20°C, 760 mm barometric pressure, and 75% RH, dew would be deposited on leaf and other surfaces when their surfaces temperatures were cooled to 17°C (Marvin, 1941). In Israel, Duvdevani *et al.* (1946) reported that dew occurred on 101 of 122 days from June to September, and I believe this may be typical of other agricultural areas. But dews usually occur at lower temperatures than do rains, and if these temperatures are far below the optimum temperature for the pathogen, as they commonly are, dews may not be as favorable per unit of time and per unit of moisture as are rains. In Berkeley, dew was deposited on onion leaves by 7–10 P.M. during January and February when the air temperature was about 8°C and the relative humidity of the air was as low as 69% (Yarwood, 1939b). Gaumann and Brierley (1950) considered dew the most important source of water for fungus infections, but I believe most plant pathologists would disagree. Nor do I agree with Gaumann's logic

(Gaumann and Brierley, 1950, p. 28), "At a relative humidity of 98% the undercooling of leaves effected by transpiration will lead to condensation round the spores and germ tubes, so that in fact, they develop in actual drops of water . . ." Rather I would expect that as RH approached 100%, transpiration would stop, and there would be no further undercooling due to transpiration. On the other hand, if the undercooling was due to radiation to the sky, condensation could occur at 100% or much lower RH. But if plants can transpire into a saturated atmosphere (Anonymous, 1871) Gaumann may be correct. Dew and rain are considered of sufficient importance in plant pathology that many devices have been developed to record them (Taylor, 1954; Thies and Calpouzos, 1957). Dew as a source of moisture for infections has been emphasized by Bourke (1955), Duvdevani et al. (1946), Ullrich (1958), Smith (1905), Yarwood (1951b), and many others.

## E. Fog

Fog is the small waterdrops which condense and remain suspended in the air, when it becomes supersaturated with water vapor. I have detected 100% relative humidity in natural outdoor environments only in fog, but never in rain. Where it occurs, it might logically be an important source of water for the initiation of infections, but I have found no clear data in support of this. However, Hirt (1942) estimates the relative importance of rain, fog, and dew in the initiation of infection of white pine blister rust as 10:2:1. Bourke (1955) considered fog important for the development of potato late blight.

## F. Guttation

Of the six sources of external water specified here, guttation is probably regarded at least important, and is commonly not mentioned in treatments of moisture and infection. Burgerstein (1920) reported guttation in 330 genera in 115 families. Guttation may occur through hydathodes or through the cuticle (Lausberg, 1935) or through stomata (Bald, 1952). In my trials guttation occurred in French bean, broad bean, cowpea, tobacco, calla, barley, potato, cucumber, mustard, Gomphrena globosa, and Nicotiana glutinosa. With Gomphrena, barley, mustard, and cucumber, it occurred principally at the edges or tips of leaves, while with others it occurred over the entire surface. Infection of cucumber by Pseudoperonospora, and potato by Phytophthora infestans increased guttation. Young infections (1–3 days from inoculation) of Uromyces in bean increased guttation, but old infections (6 or more days from inoculation) decreased guttation. Infections of *Erysiphe* in bean decreased guttation. No guttation was detected with leaves of peach, plum, wild grape, pear, pea, clover, sugar beet, plantain, *Geum*, or poinsettia, but these negative results are of little significance. I believe, in agreement with Kramer (1949), Long (1957), and Yarwood (1952b) that guttation is commonly underrated because it is confused with dew. I believe, in agreement with Smith (1911), Bald (1952), Endo and Amacher (1964), Endo and Oertli (1963), and Rowell (1951) that guttation is an important source of water for fungal infection. Some specific observations are given below.

When young, vigorous, potted beans (*Phaseolus vulgaris* var. Pinto) with unifoliate leaves were inoculated with dry uredospores of bean rust (*Uromyces phaseoli*) on their lower or upper leaf surfaces and left in the dry greenhouse at about 60% RH, no free water appeared on the leaves and no infection resulted. If such plants were placed overnight in a small closed chamber, with or without free water, at a constant temperature of 20°C, free water appeared on the lower leaf surface. If the period in the moist chamber was over about 8 hr, as much rust infection developed as if the leaves were sprayed or dipped in water after inoculation. If the period was less than 8 hr, more infection developed on the leaves to which water was deliberately added. The free water which appeared on the nonwetted leaves is regarded as guttation water.

If the soil in such trials was dried to near the wilting point and the inoculated plants placed in the same types of chamber, no moisture appeared on the leaves and no infection resulted. If the near-wilted plants were sprayed with water, heavy infection resulted even though the soil remained dry. If only one leaf was sprayed with water, guttation water frequently appeared on the opposite leaf and heavy infection appeared on both leaves. That is, there was some translocated effect of added water. Guttation water did not form so abundantly on the upper as on the lower leaf surfaces, or on old as on young leaves.

If young detached bean leaves were placed with one surface on water in closed petri dishes, water formed on the leaf surface exposed to air. This water is regarded as guttation due to leaf pressure. With medium and old leaves treated as above, no water appeared on the surface exposed to air. If the upper surface of medium-aged leaves were abraded with Carborundum, even if on only one side of the midrib, and placed on water, the intercellular spaces did not fill with water, but water appeared over the entire lower surface exposed to air. If the petiole of an abraded leaf was placed in water, and both leaf surfaces exposed to air in a moist chamber, moisture appeared on the lower leaf surface. The guttation water in such trials was up to 6 mg/cm$^2$ in 12 hr (Fig. 3). A comparable maximum value for dew is 19 mg/cm$^2$ (Ullrich, 1958). Guttation also favored infection of

Fig. 3. Guttation on lower surface of primary leaves of pinto bean. The upper left side of the right leaf was rubbed with a suspension of Celite in water at 1400 hr, May 27, at 7 days from seeding. Both leaves were then incubated with their upper surfaces on water in a closed petri dish. Photographed at 0800 hr, May 28. The abrasion of one-half of the right leaf has induced about equal guttation over entire lower surface whereas there is no apparent guttation on the nonabraded leaf. This guttation on the right leaf is greater than usually found on normal plants.

*Colletotrichum* on French bean (Fig. 4) and *Botrytis* on broad bean (Yarwood, 1952b).

Fog, guttation, and dew are readily confused with each other, and it is usually difficult to decide which is the source of water on leaves. The deposit of water on plants in early morning after a cool foggy night could be due to any one of these, or some combination. The very heavy deposit of water drops on the pubescent leaves of *Plantago lanceolata*, but not on the glabrous leaves, is a good example and the only example I know of trichomes favoring such a deposit. However, whether these water drops are the result of dew, fog, or guttation has not been determined.

## G. Intercellular and Intracellular Water

The intercellular spaces of leaves are normally free of liquid water, and generally well below 100% RH (Thut, 1938). These intercellular spaces may become filled with liquid water (water-soaked or water-congested) as a result of natural or artificial root pressure (Kramer, 1949), high atmospheric humidity (Braun and Johnson, 1939), splashing rain (Diachun *et al.*, 1942, 1944), a strong jet of water, injection with a hypodermic needle (Williams, 1967), vacuum infiltration (Cohen, 1951), mechanical pressure (Yarwood, 1953), heat, immersion in water in dark-

Fig. 4. Relation of guttation to anthracnose on Bountiful bean. The left half of the right leaf of a plant 9 days from seeding was rubbed on the upper surface with a water suspension of Celite, at 0800 hr, April 11. The lower surfaces of both leaves were inoculated at 0805 hr, April 11 with a spore suspension of *Colletotrichum lindemuthianum* and the leaves allowed to dry. Then the leaves were detached at the pulvini and placed with their upper surfaces on water in a closed petri dish at 13°C. Photographed to show lower leaf surface April 17. The abrasion of one-half of one leaf has favored infection over the entire leaf.

ness, or the result of certain infections (Yarwood, 1966). I have found that the intercellular water resulting from induced water congestion of leaves amounts to up to 92% of the original green weight of broad bean leaves, and varied from 31 to 64% for potato, French bean, tobacco, cucumber, cowpea, and corn.

Water congestion from root pressure is favored by young leaves, by darkness, by humid air, moist soil, nitrogen fertilization, and intermediate temperatures. Heating of leaves before incubating plants in a moist chamber may increase or decrease water congestion, depending on dosage of heat and species. Heating of corn leaves 10–90 sec at 50°C favored water congestion. Water congestion and guttation are frequently favored by the same environment, and the same treatment may bring about both congestion and guttation, but water congestion may occur without guttation, and guttation may occur without congestion.

Downy mildew of cucumber may increase water congestion several-fold (Yarwood, 1966). Bacteria and some virus infections may increase water congestion, but infections by rusts and powdery mildews usually decrease water congestion. If, however, bacteria are present in the rust infections, water soak may result (Yarwood, 1969b). Water congestion before inoculation favors infection with several bacterial infections (Johnson, 1947); Johnson also reports that water congestion favors infection by *Colletotrichum lindemuthianum*, *Puccinia graminis*, *Puccinia sorghi*,

*Puccinia helianthi*, and *Erysiphe graminis*. Since several, if not all, powdery mildews are favored by low soil moisture (Yarwood, 1949), I am skeptical of any powdery mildew being favored by water congestion; Cohen (1951) found that water congestion actually prevented infection with bean rust *Uromyces phaseoli*, as I have confirmed.

Intracellular water is believed always present in the vacuoles of living cells and is presumably necessary for life or for any infection, but variations in intracellular water of the host affecting disease seem not to have been studied except as reflected in effects of relative humidity of the air and soil moisture.

## VII. MANIPULATION OF WATER TO CONTROL DISEASE

Regulation of water to control disease has important but limited application. If rain, dew, fog, guttation, or irrigation could be prevented, especially at night, or if the period of wetness reduced to about 2 hr per day, most aerial infections of bacteria and fungi could be avoided. This reduction in disease is accomplished by glasshouse culture (Chabot, 1946), by growing plants in rain-free summers (Mackie *et al.*, 1945), and by ridge culture and controlled irrigation. Watering and dragging in the early morning will speed the drying of golf greens and reduce *Rhizoctonia* (Monteith and Dahl, 1932). Fanning is also successful (Monteith and Reid, 1936).

Prevention of infection by addition of water has less potential application than disease control by reduction of water. Growing roses in intermittent mist (Rogers, 1959) will reduce rose mildew, and syringing of plants with water in the evening will reduce several powdery mildews (Yarwood, 1939a). Soaking barley seeds in water reduces loose smut (Tyner, 1953). Sprinkling reduces *Botrytis* of tomatoes (Kloke, 1963) and washing of cherries reduces *Monilinia* and *Botrytis* (Yarwood and Harvey, 1952). High soil moisture reduces *Streptomyces* on potatoes (Teichert and Janke, 1973) and several powdery mildews (Yarwood, 1949). The last must be some kind of translocated effect, since the mildew infection is on the leaves. None of the above methods seems to have become a major method of disease control.

### REFERENCES

Aitken, J. (1887). On dew. *Trans. R. Soc. Edinburgh* **33**, 9–64.
Altman, P. L., and Dittmer, D. S., eds. (1962). "Growth." Fed. Am. Soc. Exp. Biol., Washington, D. C.
Anderson, P. J. (1918). Rose canker and its control. *Mass., Agric. Exp. Stn., Bull.* **183**, 1246.
Anonymous (1861). Mildew upon the grape. *Calif. Farmer* **15**, 148.

Anonymous. (1862). (Amount and function of dew on plants.) *Gard. Chron.* **22**, 427.

Anonymous. (1871). (Humidity and transpiration.) *Gard. Chron.* **31**, 70–138.

Anonymous. (1872). (Absorption of water by leaves.) *Gard. Chron.* **32**, 1128.

Armolik, N., and Dickson, J. G. (1956). Minimum humidity requirement for germination of conidia associated with storage of grain. *Phytopathology* **46**, 462–465.

Arthur, J. C. (1929). "The Plant Rusts." Wiley, New York.

Artschwager, E., and Starrett, R. (1931). Suberization and wound periderm formation in sweet potato and gladiolus as affected by temperature and humidity. *J. Agric. Res.* **43**, 353–364.

Arya, H. C., and Ghemawat, M. S. (1954). Occurrence of powdery mildew of wheat in the neighborhood of Jodpur. *Indian Phytopathol.* **6**, 123–160.

Aso, K. (1900). The chemical composition of the spores of *Aspergillus oryzae. Bull. Coll. Agric., Tokyo Imp. Univ.* **4**, 81–96.

Attafuah, A., and Tinsley, T. W. (1958). Viruses diseases of *Adansonia digitata* L. Bomboreae and their relation to cacao in Ghana. *Ann. Appl. Biol.* **46**, 20–22.

Bald, J. G. (1952). Stomatal droplets and the penetration of leaves by plant pathogens. *Am. J. Bot.* **39**, 97–99.

Barnes, M., and Parker, M. S. (1966). The increase in size of mould spores during germination. *Trans. Br. Mycol. Soc.* **49**, 487–494.

Berwith, C. E. (1936). Apple powdery mildew. *Phytopathology* **26**, 1071–1073.

Bohn, G. H., and Maliot, J. C. (1945). Inoculation experiments with *Pseudomonas ribicola. Phytopathology* **35**, 1008–1016.

Bonner, J. T. (1948). A study of temperature and humidity requirements of *Aspergillus niger. Mycologia* **40**, 728–738.

Boughey, A. S. (1949). The ecology of fungi which cause economic plant diseases. *Trans. Br. Mycol. Soc.* **32**, 179–189.

Bourke, P. M. A. (1955). (Potato blight in the climate of La Serena.) *Rev. Appl. Mycol.* **37**, 303 (abstr.). 1958.

Braun, A. C., and Johnson, J. (1939). Natural water soaking and bacterial infection. *Phytopathology* **39**, 23.

Brodie, H. J. (1945). Further observations on the mechanism of germination of the conidia of various species of powdery mildew at low humidity. *Can J. Sect. Res., C* **23**, 198–211.

Brodie, H. J., and Neufeld, C. C. (1942). The development and structure of conidia of *Erysiphe polygoni* DC and their germination at low humidity. *Can. J. Res., Sect. C* **20**, 41–62.

Bromfield, K. R. (1967). Some uredospore characteristics of importance in experimental epidemiology. *Plant Dis. Rep.* **51**, 248–252.

Brook, P. L. (1973). Epidemiology of grapevine anthracnose caused by *Elsinoe ampelina. N. Z. J. Agric. Res.* **16**, 333–342.

Burchill, R. T. (1966). Discussion. *In* "The Fungus Spore" (M. F. Madelin, ed.), p. 286. Butterworth, London.

Burgerstein, A. (1920). "Die Transpiration der Pflanzen II." Fischer, Jena.

Butin, H. (1955). Uber den Einfluss der Wassergehaltes der Pappel auf ihre Resistenz gegenuber *Cytospora chrysosperma* (Pers) Fr. *Phytopathol. Z.* **24**, 245–264.

Chabot, E. (1946). "Greenhouse Gardening for Everyone." Barrows, New York.

Cherewick, W. J. (1944). Studies on the biology of *Erysiphe graminis* DC. *Can. J. Res., Sect. C* **22**, 52–68.

Chester, K. S. (1946). "The Cereal Rusts." Chronica Botanica, Waltham, Massachusetts.

Chowdhury, S. (1937). Germination of fungal spores in relation to atmospheric humidity. *Indian J. Agric. Sci.* **7**, 653–657.

Clayton, C. N. (1942). The germination of fungal spores in relation to controlled humidity. *Phytopathology* **32**, 921–943.

Clayton, E. E. (1936). Water soaking of leaves in relation to development of wildfire disease of tobacco. *J. Agric. Res.* **52**, 239–269.

Cochrane, V. W. (1958). "Physiology of Fungi." Wiley, New York.

Cohen, M. (1951). Increased resistance to bean rust associated with water infiltration. *Phytopathology* **41**, 937.

Colhoun, J. (1973). Effect of environmental factors on plant disease. *Annu. Rev. Phytopathol.* **11**, 343–364.

Corner, E. J. H. (1935). Observations on resistance to powdery mildews. *New Phytol.* **34**, 180–200.

Cox, A. E., and Large, E. C. (1960). "Potato Blight Epidemics." U. S. Dep. Agric., Washington, D. C.

Cramer, E. (1894). Die Zuammensitzung der sporen von *Penicillium glaucum* und ihre Beziehung zu der Widerstandfahigkeit derselben gegen aussere Einflusse. *Arch. Hyg.* **20**, 197–210.

Curl, E. A., and Weaver, H. A. (1958). Diseases of forage crops under sprinkler irrigation in the Southwest. *Plant Dis. Rep.* **42**, 637–644.

DeBary, A. (1863). Recherches sur le développement de quelques champignons parasites *Ann. Sci. Nat., Bot. Biol. Veg.* [4] **20**, 5–148.

Delp, C. J. (1954). Effect of temperature and humidity on the grape powdery mildew fungus. *Phytopathology* **44**, 615–626.

Diachun, S., Valleau, W. D., and Johnson, E. M. (1942). Relation of moisture to invasion of tobacco leaves by *Bacterium tabacum* and *Bacterium ungulatum*. *Phytopathology* **32**, 379–387.

Diachun, S., Valleua, W. D., and Johnson, E. M. (1944). Invasion of water-soaked tobacco leaves by bacteria, solutions, and tobacco mosaic virus. *Phytopathology* **34**, 250–253.

Doran, W. L. (1922). Effect of external and internal factors on the germination of fungus spores. *Bull. Torrey Bot. Club* **49**, 313–340.

Drandarevski, C. A. (1969). Untersuchungen uber den echten Reben mehltau *Erysiphe betae* (Vanha). *Phytopathol. Z.* **65**, 124–154.

Duggar, B. M. (1909). "Fungus Diseases of Plants." Ginn, Boston, Massachusetts.

Durrell, L. W. (1918). Factors influencing the uredospore germination of *Puccinia coronata*. *Phytopathology* **8**, 81–82.

Duvdevani, S., Reichert, I., and Palti, J. (1946). The development of downy and powdery mildew of cucumbers as related to dew and other environmental factors. *Palest. J. Bot., Rehovot Ser.* **5**, 127–151.

Ekundayo, J. A., and Carlisle, M. J. (1964). The germination of sporangiospores of *Rhizopusarrhizus;* spore swelling and germ tube emergence. *J. Gen. Microbiol.* **35**, 261–269.

Endo, R. M., and Amacher, R. H. (1964). Influence of guttation fluid on infection structures of *Helminthosporium sorokinianium*. *Phytopathology* **54**, 1327–1334.

Endo, R. M., and Oertli, J. J. (1963). Stimulation of fungal infection of bentgrass. *Nature (London)* **201**, 313.

Fernando, M., and Stevenson, G. (1952). Effect of the condition of potato tissue as modified by temperature and water content, upon attack by certain organisms and their pectinase enzymes. *Ann. Bot. (London)* [N. S.] **16**, 103–114.

Foister, C. E. (1935). The relation of weather to fungus and bacterial diseases of plants. *Bot. Rev.* **1**, 497–516.

Foister, C. E. (1946). The relation of weather to fungus diseases of plants II. *Bot. Rev.* **12**, 548–591.

Foster, J. W. (1949). "Chemical Activities of Fungi." Academic Press, New York.

Frampton, V. L., and Longree, K. (1941). The vapor pressure gradient above a transpiring leaf. *Phytopathology* **31**, 1040–1042.

Gaumann, E., and Brierley, W. B. (1950). "Principles of Plant Infection." Crosby Lockwood, London.

Glaz, E. T. (1955). Researches about the viability and preservation of ergot conidia of *Claviceps purpurea* (Fr) Tul. grown in submerged culture. *Acta Microbiol. Acad. Sci. Hung.* **3**, 315–325.

Govaerts, J., and Leclercq, J. (1946). Water exchange between insects and air moisture. *Nature (London)* **157**, 483.

Graf-Marin, A. (1934). Studies on powdery mildew of cereals. *N. Y., Agric. Exp. Stn., Ithaca, Mem.* **157**, 1–48.

Grainger, J. (1947). The ecology of *Erysiphe graminis* DC. *Trans. Br. Mycol. Soc.* **31**, 54–65.

Gregg, N. (1952). Enzyme secretion by strains of *Bacterium carotovorum* and other pathogens in relation to parasitic vigor. *Ann. Bot. (London)* [N. S.] **19**, 235–250.

Griffin, D. M. (1972). "Ecology of Soil Fungi." Syracuse Univ. Press, Syracuse, New York.

Haines, F. M. (1952). The absorption of water by leaves in an atmosphere of high humidity. *J. Exp. Bot.* **3**, 95–98.

Hart, M. P., and MacLeod, D. M. (1955). An apparatus for determining the effects of temperature and humidity on germination of fungus spores. *Can. J. Bot.* **33**, 289–292.

Hashioka, Y. (1937). Relation of temperature and humidity to *Sphaerotheca fuliginea* (Schlecht) Poll. with special reference to germination, viability and infection. *Trans. Nat. Hist. Soc. Formosa* **27**, 129–145.

Hawker, L. E., and Madelin, M. F. (1976). The dormant spore. *In* "The Fungus Spore" (D. J. Weber and W. M. Hess, eds.), pp. 1–70. Wiley, New York.

Hermansen, J. E. (1972). Successful low temperature storage of conidia of *Erysiphe graminis* produced under dry conditions. *Friesia* **10**, 86–88.

Hirt, R. R. (1942). The relation of certain meteorological conditions to the infection of eastern white pine by the blister rust fungus. *N. Y. State Coll. For. Syracuse Univ., Tech. Publ.* **59**, 1–65.

Hollings, M. (1955). *Chenopodium amaranticolor* as a test plant for plant viruses. *Plant Pathol.* **5**, 56–60.

Holmes, F. O. (1929). Inoculation methods in tobacco mosaic studies. *Bot. Gaz. (Chicago)* **87**, 56–63.

Horsford, R. M. (1975). Varietal resistance and fungal pathogenicity related to wet period. *Proc. Am. Phytopathol. Soc.* **2**, 59.

Ingold, C. T. (1954). Fungi and water. *Trans. Br. Mycol. Soc.* **37**, 97–107.

Jhooty, J. S. (1971). Germination of powdery mildew conidia *in vitro* on host and nonhost leaves. *Indian Phytopathol.* **24**, 67–73.

Jhooty, J. S., and McKeen, W. E. (1960). The influence of host leaves on the germination of asexual spores of *Sphaerotheca macularis* (Wallr. ex. Fr.) Cooke. *Can. J. Microbiol.* **11**, 539–545.

Jhooty, J. S., and McKeen, W. E. (1965). Water relations of asexual spores of *Sphaerotheca macularis* (Wallr. ex Fr.) Cooke and Erysiphe polygoni DC. *Can. J. Microbiol.* **11**, 531–545.

Johnson, D., Weber, D. J., and Hess, W. M. (1976). Lipids from conidia of *Erysiphe graminis tritici* (powdery mildew). *Trans. Br. Mycol. Soc.* **66**, 35–43.

Johnson, E. M., and Valleau, W. D. (1935). Mosaic from tobacco one to fifty-two years old *Ky., Agric. Exp. Stn., Bull.* **361**, 264–271.

Johnson, J. (1947). Water congestion and fungus parasitism. *Phytopathology* **37**, 403–417.

Kahn, R. P., and Libby, J. L. (1958). The effect of environmental factors and plant age on the infection of rice by the blast fungus *Piricularia oryzae*. *Phytopathology* **48**, 25–30.

Keitt, G. W., and Jones, L. K. (1926). Studies on the epidemiology and control of apple scab. *Wis., Agric. Exp. Stn., Res. Bull.* **73**, 1–104.

Kenaga, C. B. (1970). "Principles of Phytopathology." Balt, Lafayette, Indiana.

Kimbrough, J. W. (1963). The development of *Pleochaeta polychaeta* (Erysiphaceae). *Mycologia* **55**, 608–618.

Kloke, A. (1963). The influence of sprinkling and fertilizing on parasitic and nonparasitic diseases of tomato. *Mitt. Biol. Bundesanst. Land- Forstwirtsch., Berlin-Dahlem* **108**, 58–65.

Klos, E. J. (1976). A, B, C's of brown rot control. *Rep. Mich. State Hortic. Soc.* **105**, 94–98.

Kochman, J. K., and Brown, J. F. (1976). Host and environmental effect on the penetration of oats by *Puccinia graminis avenae* and *Puccinia coronata avenae*. *Ann Appl. Biol.* **82**, 251–258.

Kramer, P. J. (1949). "Plant and Soil Water Relationships." McGraw-Hill, New York.

Lapwood, D. H. (1957). On the parasitic action of certain bacteria in relation to the capacity to secrete pectolytic enzymes. *Ann. Bot. (London)* [N. S.] **21**, 167–184.

Large, E. C. (1940). "The Advance of the Fungi." Cape, London.

Lausberg, T. (1935). Quantitative Untersuchungen uber die kuticulare Exkretion des Laublattes. *Jahrb. Wiss. Bot.* **81**, 769–806.

Lesage, P. (1895). Recherches expérimentales sur la germination des spores de *Penicillium glaucum. Ann. Sci. Nat., Bot. Biol. Veg.* [8] **1**, 309–322.

Leveille, J. H. (1851). Des espèces qui composent le genre *Erysiphe. Ann. Sci. Nat., Bot. Biol. Veg.* [3] **15**, 104–179.

Long, I. F. (1955). Dew and guttation. *Weather* **10**, 128.

Longree, K. (1939). The effect of temperature and relative humidity on the powdery mildew of roses. *N. Y., Agric. Exp. Stn., Ithaca, Mem.* **273**, 1–43.

McClellan, W. D. (1942). Control of powdery mildew of roses in the greenhouse. *N. Y., Agric. Exp. Stn., Ithaca, Bull.* **785**, 1–39.

McKeen, W. E. (1970). Lipid in *Erysiphe graminis hordei* and its possible role during germination. *Can. J. Bot.* **45**, 1489–1496.

McKeen, W. E., Mitchell, N., and Smith, R. (1967). The *Erysiphe cichoracearum* conidium. *Can. J. Bot.* **45**, 1489–1496.

Mackie, W. W., Snyder, W. C., and Smith, F. L. (1945). Production in California of snap bean seed free from blight and anthracnose. *Calif., Agric. Exp. Stn., Bull.* **689**, 1–23.

Mandels, G. R., and Darby, R. T. (1953). A rapid cell colume assay for fungitoxicity using fungus spores. *J. Bacteriol.* **65**, 16–26.

Marchant, R., and White, M. F. (1966). Spore swelling and germination in *Fusarium culmorum. J. Gen. Microbiol.* **42**, 237–244.

Marvin, C. F. (1941). Psychrometric tables. *U. S. Dep. Commer., Weather Bur., Bull.* **235**, 1–87.

Matthews, R. E. F. (1970). "Plant Virology." Academic Press, New York.

Mazur, P. (1968). Survival of fungi after freezing and dessication. *In* "The Fungi" (G. C. Ainsworth and A. S. Sussman, eds.), Vol. 3, pp. 325–394. Academic Press, New York.

Melching, J. S., and Bromfield, K. R. (1975). Factors influencing spore germination and infection by *Phakospora pachyrhizi* and intensification and spread of soybean rust under controlled conditions. *Proc. Am. Phytopathol. Soc.* **2**, 125.

Mence, M. J., and Hildebrandt, A. C. (1966). Resistance to powdery mildew in rose. *Ann. Appl. Biol.* **58**, 309–320.

Menzies, J. D. (1954). Effect of sprinkler irrigation in an arid climate on the spread of bacterial diseases of bean. *Phytopathology* **44**, 553–556.

Mills, W. D. (1947). How to use the time-temperature chart of apple scab control. *Proc., N. Y. State Hortic. Soc.* **92**, 199–202.

Monteith, J., and Dahl, A. S. (1932). Turf diseases and their control. *USGA Green Sect., Bull.* **12**, 87–186.

Monteith, J., and Reed, M. E. (1936). Control of brown patch in turf by fanning. *Phytopathology* **26**, 102.

Muncie, J. A. (1917). Experiments on the control of anthracnose and bean blight. *Mich., Agric. Exp. Stn., Bull.* **38**, 1–50.

Nour, M. A. (1958). Studies on *Leveillula taurica* (Lev.) Arn. and other powdery mildews. *Trans. Br. Mycol. Soc.* **41**, 17–38.

Nutman, F. J., Roberts, F. M., and Clarke, R. T. (1963). Studies on the biology of *Hemileia vastatrix* Berk. and Br. *Trans. Br. Mycol. Soc.* **46**, 27–48.

Panopoulos, N. J., and Schroth, M. N. (1974). Role of flagellar motility in the invasion of bean leaves by *Pseudomonas phaseolicola. Phytopathology* **64**, 1389–1397.

Perera, R. G., and Wheeler, B. E. J. (1975). Effect of water drops on the development of *Sphaerotheca pannosa* on rose leaves. *Trans. Br. Mycol. Soc.* **64**, 313–319.

Peries, O. S. (1962). Studies on strawberry mildew caused by *Sphaerotheca macularis* (Wallr. ex Fries) Jaczewski. *Ann. Appl. Biol.* **50**, 211–233.

Powell, J. M. (1974). Environmental factors affecting germination of *Cronartium camandrae* aeciospores. *Can. J. Bot.* **52**, 659–667.

Ragetli, H. W. S., Weintraub, M., and Elder, M. (1973). Effective mechanical inoculation of plant viruses in the absence of water. *Can. J. Bot.* **51**, 1977–1981.

Ramsey, J. A., Butler, E. G., and Sang, J. A. (1938). The humidity gradient at the surface of a transpiring leaf. *J. Exp. Bot.* **15**, 255–265.

Rawlins, T. E., and Tompkins, C. M. (1934). The use of carborundum as an abrasive in plant virus inoculations. *Phytopathology* **24**, 1147.

Reuveni, R., and Rotem, J. (1973). Epidemics of *Leveillula taurica* on tomatoes and peppers as affected by the conditions of humidity. *Phytopathol. Z.* **76**, 153–157.

Richmond, D. V., and Somers, E. S. (1963). Studies on the fungitoxicity of captan. III. *Ann. Appl. Biol.* **52**, 327–336.

Riker, A. J., and Riker, R. S. (1936). "An Introduction to Research on Plant Diseases." J. S. Swift, St. Louis, Missouri.

Rogers, M. N. (1959). Some effects of moisture and host plant susceptibility on the development of powdery mildew roses caused by *Sphaerotheca pannosa* var. *rosae. N. Y., Agric. Exp. Stn., Ithaca, Mem.* **363**, 1–37.

Rotem, J. (1968). Thermoxerophytic properties of *Alternaria porri* f sp. *solani. Phytopathology* **58**, 1284–1287.

Rotem, J., and Cohen, Y. (1974). Epidemiological patterns of *Phytophthora infestans* under semi-arid conditions. *Phytopathology* **64**, 711–714.

Rotem, J., and Palti, J. (1969). Irrigation and plant diseases. *Annu. Rev. Phytopathol.* **7**, 267–288.

Rowell, J. B. (1951). Observations on the pathogenicity of *Rhizoctonia solani* on bentgrass. *Plant Dis. Rep.* **35**, 240–242.

Ruppel, E. G., Hills, F. J., and Mumford, D. L. (1975). Epidemiological observations on the sugarbeet powdery mildew epiphytotic in Western U. S. A. in 1974. *Plant Dis. Rep.* **59**, 283–286.

Saint, S. J. (1948). Annual report of the British West Indies Sugar Cane Breeding Station Barbados for the year ending Sept. 30, 1945. *Rev. Appl. Mycol.* **27**, 157 (abstr.).

Schein, R. D. (1964). Comments on the moisture requirements of fungus germination. *Phytopathology* **54**, 1427.

Schmiedeknecht, M. (1958). *Pseudopeziza medicaginis* (Lit) Sac., ein zerophilir pflanzenpathogener ascomycet. *Naturwissenschaften* **21**, 525.

Schnathorst, W. C. (1957). Microclimates and their significance in the development of powdery mildew of lettuce. *Phytopathology* **47**, 533.

Schnathorst, W. C. (1960). Effect of temperature and moisture stress on the lettuce powdery mildew fungus. *Phytopathology* **50**, 304–308.

Schneider, R. (1953). Untersuchungen uber fruchtigkeitsanspruchen parasitischer Pilze. *Phytopathol. Z.* **21**, 63–78.

Schultz, E. S., and Folsom, D. (1923). Transmission, variation and control of certain degeneration diseases of Irish potato. *J. Agric. Res.* **25**, 45–118.

Seymour, A. B. (1887). Relation of moisture of plant disease. *Bot. Gaz. (Chicago)* **12**, 229.

Sharp, E. L., and Smith, F. G. (1957). Further study of the preservation of Puccinia uredospores. *Phytopathology* **47**, 423–429.

Shaw, L. (1935). Intercellular humidity in relation to fire blight susceptibility in apple and pear. *N. Y., Agric. Exp. Stn., Ithaca, Mem.* **181**, 1–40.

Shaw, M. (1964). The physiology of rust uredospores. *Phytopathol. Z.* **50**, 159–180.

Smiley, E. M. (1920). Phyllosticta blight of snapdragon. *Phytopathology* **10**, 232–248.

Smith, E. F. (1911). "Bacteria in Relation to Plant Diseases," Vol. II. Carnegie Inst., Washington, D. C.

Smith, R. E. (1905). Asparagus and asparagus rust in California. *Calif., Agric. Exp. Stn., Bull.* **165**, 1–99.

Snow, D. (1949). The germination of mold spores at controlled humidities. *Ann. Appl. Biol.* **36**, 1–13.

Snyder, W. C., Grogan, R. G., Bardin, R., and Schroth, M. N. (1965). Overhead irrigation encourages wet weather diseases. *Calif. Agric.* **19**, 11.

Somers, E., and Horsfall, J. G. (1966). The water content of powdery mildew conidia. *Phytopathology* **56**, 1031–1035.

Stakman, E. C., and Harrar, J. G. (1957). "Principles of Plant Pathology." Ronald, New York.

Stapleton, G. E., and Hollaender, A. (1952). Mechanism of lethal and mutagenic action of ionizing radiation on *Aspergillus terreus. J. Cell. Comp. Physiol.* **39**, Suppl. 1, 101–113.

Stock, F. (1931). Untersuchungen uber Keimung und Keimschlauchwachstum der Uredosporen einiger Getreideroste. *Phytopathol. Z.* **3**, 231–302.

Stoll, K. (1941). Untersuchungen uber den Apfelmehltau *Podosphaera leucotricha* (E. & Ev.) Salm. *Forschungdienst* **11**, 59–70.

Sumi, M. (1928). Uber die chemischen Bestandteile der sporen von *Aspergillus oryzae. Biochem. Z.* **195**, 161–174.

Sussman, A. S. (1968). Longevity and survivability of fungi. *In* "The Fungi" (G. C. Ainsworth and A. S. Sussman, eds.), pp. 447–486. Academic Press, New York.

Suzuki, H. (1975). Meterological factors in the epidemiology of rice blast. *Annu. Rev. Phytopathol.* **13**, 239–256.

Tarr, S. A. J. (1972). "The Principles of Plant Pathology." Winchester, New York.

Taylor, C. F. (1954). A device for recording the duration of dew deposits. *Phytopathology* **44**, 390.

Teichert, D., and Janke, C. (1973). Erganzende Beobachtungen zur Bekampfungsmoglichkeit des Kortoffelschortes durch Bereghung. *Nachr. Dtsch. Pflanzenschutdienstes DDR* **27**, 157–161.

Thies, T., and Calpouzos, L. (1957). A seven day instrument for recording periods of rainfall and dew. *Phytopathology* **47**, 746–747.

Thomas, H. E., and Ark, P. A. (1934). Nectar and rain in relation to fire-blight. *Phytopathology* **24**, 682–685.

Thut, H. F. (1938). Relative humidity variations affecting transpiration. *Am. J. Bot.* **25**, 589–595.

Tinsley, T. W. (1953). The effect of varying the water supply of plants on their susceptibility to infection with viruses. *Ann. Appl. Biol.* **40**, 750–760.

Turel, F. L. M., and Ledingham, G. A. (1959). Utilization of labelled substrates by the mycelium and uredospores of flax rust. *Can. J. Microbiol.* **5**, 537–545.

Tyner, L. E. (1953). The control of loose smut of barley and wheat by Spergon and by soaking in water at room temperature. *Phytopathology* **43**, 313–316.

Ullrich, J. (1958). Die Tau - und Regenbenetzung von Kartoffelbestanden. Ein Beitrag zur Epidemiologie der Krautfaule *(Phytophthora infestans)*. *Angew. Bot.* **32**, 125–146.

Viranyi, F. (1975). Studies on the biology and ecology of onion downy mildew *(Peronospora destructor)* Berk. (Fries) in Hungary. *Acta Phytopathol. Acad. Sci. Hung.* **10**, 321–328.

Waggoner, P. E. (1965). Microclimate and plant disease. *Annu. Rev. Phytopathol.* **3**, 103–126.

Walkinshaw, C. A., Hyde, J. M., and van Zandt, J. (1967). Fine structure of quiescent and germinating aeciospores of *Cronartium fusiforme*. *J. Bacteriol.* **94**, 245–254.

Wallin, R. H. (1967). Ground level climate and plant disease epiphytotics. *In* "Ground Level Climatology" (R. H. Shaw, ed.), Publ. No. 86, pp. 149–163. Am. Assoc. Adv. Sci., Washington, D. C.

Ward, S. V., and Manners, J. G. (1974). Environmental effects on the quantity and viability of conidia produced by *Erysiphe graminis*. *Trans. Br. Mycol. Soc.* **62**, 119–128.

Weimer, J. L. (1917). Three cedar rust fungi. Their life histories and the diseases they produce. *Cornell, Agric. Exp. Stn., Bull.* **390**, 509–549.

Wellman, F. L. (1972). "Tropical American Plant Disease." Scarecrow, Metuchen, New Jersey.

Whetzel, H. H., Hesler, L. R., Gregory, C. T., and Rankin, W. H. (1925). "Laboratory Outlines in Plant Pathology." Saunders, Philadelphia, Pennsylvania.

Williams, A. S. (1967). Bacteria and fungi as causal agents in disease. *In* "Sourcebook of Laboratory Exercises in Plant Pathology," pp. 8–10. Freeman, San Francisco, California.

Williams, P. H., and Keen, N. T. (1967). Relation of cell permeability alterations to water congestion in cucumber angular leaf spot. *Phytopathology* **57**, 1378–1385.

Yarwood, C. E. (1936a). The diurnal cycle of the powdery mildew, *Erysiphe polygoni*. *J. Agric. Res.* **52**, 645–657.

Yarwood, C. E. (1936b). The tolerance of *Erysiphe polygoni* and certain other powdery mildews to low humidity. *Phytopathology* **26**, 845–859.

Yarwood, C. E. (1939a). Control of powdery mildews with a water spray. *Phytopathology* **29**, 288–290.

Yarwood, C. E. (1939b). Relation of moisture to infection with some downy mildews and rusts. *Phytopathology* **29**, 933–945.

Yarwood, C. E. (1943). Onion downy mildew. *Hilgardia* **14**, 595–691.

Yarwood, C. E. (1949). Effect of soil moisture and nutrient concentration on the development of bean powdery mildew. *Phytopathology* **39**, 780–788.

Yarwood, C. E. (1950a). Water content of fungus spores. *Am. J. Bot.* **37**, 636–639.

Yarwood, C. E. (1950b). Dry weather fungi. *Calif. Agric.* **4**, 7–12.

Yarwood, C. E. (1951a). Fungicides for powdery mildews. *Proc. Int. Congr. Crop Prot.*, *2nd, 1951* pp. 1–22.

Yarwood, C. E. (1951b). Defoliation by a rain-favored, a dew-favored, and a shade-favored disease. *Phytopathology* 41, 194–195.

Yarwood, C. E. (1952a). Some water relations of *Erysiphe polygoni* conidia. *Mycologia* 44, 506–522.

Yarwood, C. E. (1952b). Guttation due to leaf pressure favors fungus unfections. *Phytopathology* 42, 520.

Yarwood, C. E. (1953). Pressure effects in fungus and virus infections. *Phytopathology* 43, 70–72.

Yarwood, C. E. (1955). Deleterious effects of water in plant virus inoculations. *Virology* 1, 268–285.

Yarwood, C. E. (1956). Humidity requirements of foliage pathogens. *Plant Dis. Rep.* 40, 318–321.

Yarwood, C. E. (1957). Mechanical transmission of plant viruses. *Adv. Virus Res.* 4, 243–278.

Yarwood, C. E. (1959a). Virus susceptibility increased by soaking bean leaves in water. *Plant Dis. Rep.* 43, 841–844.

Yarwood, C. E. (1959b). Microclimate and infection. *In* "Plant Pathology 1908–1958" (C. S. Holton *et al.*, eds.), pp. 548–556. Univ. of Wisconsin Press, Madison.

Yarwood, C. E. (1963). The quick drying effect in plant virus inoculations. *Virology* 20, 621–628.

Yarwood, C. E. (1965). Selective hygroscopicity of diseased leaves. *Phytopathology* 55, 1373–1374.

Yarwood, C. E. (1966). Selective accumulation of water by diseased leaves. *Phytopathology* 56, 152.

Yarwood, C. E. (1969a). Selective retention of uredospores by leaves. *Phytopathology* 58, 359–361.

Yarwood, C. E. (1969b). Association of rust and halo blight on beans. *Phytopathology* 59, 1302–1305.

Yarwood, C. E. (1973). Quick drying vs. washing in virus inoculations. *Phytopathology* 63, 72–76.

Yarwood, C. E., and Gardner, M. W. (1972). Ascospore discharge by *Erysiphe trina*. *Mycologia* 64, 799–805.

Yarwood, C. E., and Harvey, H. T. (1952). Reduction of cherry decay by washing. *Plant Dis. Rep.* 36, 389.

Yarwood, C. E., and Hazen, W. E. (1944). The relative humidity at leaf surfaces. *Am. J. Bot.* 31, 129–135.

Yarwood, C. E., Sidky, S., Cohen, M., and Santilli, V. (1954). Temperature relations of powdery mildews. *Hilgardia* 22, 603–622.

Yarwood, C. E., Resconich, E. C., Ark, P. A., Schlegel, D. E., and Smith, K. M. (1961). So-called beet latent virus is a bacterium. *Plant Dis. Rep.* 45, 85–89.

Yossifovitch, M. (1923). Contribution à l'étude de l'odium de la vigne et de son traitement. *Bull. Soc. Bot. Fr.* 70, 574.

Zaracovitis, C. (1964). Factors in testing fungicides against powdery mildews. *Ann. Inst. Phytopathol. Benaki* 6, 73–106.

Zaracovitis, C. (1966). The germination in vitro of conidia of powdery mildew fungi. *In* "The Fungus Spore" M. F. Madelin ed. pp. 273–286. Butterworth, London.

Zellner, J. (1911). Zur chemie der hoher Pilze. VIII. *Monatsh. Chem.* 32, 1065–1074.

Zobl, K. (1950). Uber die Bezichungen zwischen chemischer Zusammensetzung von Pilzsporen und ihren Verhalten gegen Erhitzen. *Sydowia* 4, 175–184.

# CHAPTER 6

# EFFECT OF SOIL MOISTURE ON SURVIVAL AND SPREAD OF PATHOGENS

## D. M. Griffin

DEPARTMENT OF FORESTRY, AUSTRALIAN NATIONAL UNIVERSITY, CANBERRA, AUSTRALIA

## I. INTRODUCTION

Beginning in 1963, various aspects of the influence of soil water on fungal activity and pathogenicity have been periodically reviewed (Griffin, 1963, 1969, 1970, 1972; Cook and Papendick, 1970b, 1972; Cook, 1973; Baker and Cook, 1974; Papendick and Campbell, 1975; Schoeneweiss, 1975). Research during the last decade has greatly clarified physical aspects of the topic, but our understanding of the physiological and biochemical basis of varying fungal responses to change in the water regime is still very superficial. This chapter will therefore pay particular attention to physiological issues, largely to emphasize topics requiring additional research.

A change in the status of water in soil can affect microbial activity via many routes (Griffin, 1970). No account will be taken here of reports in which soil moisture has been shown to have an effect on microbial activity yet the precise mode of influence has not been established with reasonable certainty, e.g., through alteration of matric potential or solute diffusion. Fungal pathogens will be the main focus of attention, but the effects

of soil moisture on saprophytic fungi and bacteria will be considered because of important antagonistic interactions between the groups. Further, data from pure culture studies of microorganisms will be discussed because of their importance in providing insight into their activity in soil.

In order to survive, a fungus may utilize endogenous reserves and exist in a dormant state. Alernatively, it may utilize exogenous nutrients and either grow and colonize new substrates or enter a state called "saprophytic survival" in which hyphal extension and substrate colonization are very limited yet respiration of exogenous carbohydrates still occurs (Garrett, 1970; Griffin, 1972; Gray, 1976). In most cases the cycle of change between dormancy and growth also involves sporulation and germination. The measurement of dormancy or saprophytic survival is essentially a measure of longevity, whereas that for saprophytic or parasitic growth is primarily a measure of new substrate colonized or utilized, or new cell material synthesized. Spread of microorganisms may occur actively through growth or motility of propagules or passively through movement in soil or water. Bowen and Rovira (1976) reviewed the spread of microorganisms in connection with root colonization.

The various ways in which a change in water content of soil may affect microorganisms have been discussed by Griffin (1970, 1972) and by Marshall (1975). These include changes in (1) water potential including matric potential and osmotic potential (*sensu* Gardner, 1968) or solute potential (*sensu* Taylor, 1968); (2) solute diffusion; (3) movement of microbial propagules; and, (4) concentration and diffusion rates of gases in the soil voids.

The last mode of influence lies outside the scope of this review but aspects of it have been reviewed elsewhere (Griffin, 1968, 1972; Robinson, 1973).

## II. FUNGAL RESPONSE TO WATER POTENTIAL

### A. MATRIC AND SOLUTE POTENTIALS

Within the soil system water potential ($\psi$) may be considered to be the sum of matric potential ($\psi_m$) and solute potential ($\psi_s$). Both can be conveniently measured in bar units, which can easily be converted to S.I. units (1 bar $= 10^5$ Pa). As constraint within a physical matrix and the presence of a solute both reduce the free energy of water in soil, $\psi$ is in all cases negative. Furthermore

$$\psi_m = 10.65T \log p/p_0$$

where $T$ is temperature (°K) and $p/p_0$ is relative vapor pressure of water and

$$\psi_s = 24.7\ vm\ \phi\ \text{(at 25°C)}$$

where $v$ is ions molecule$^{-1}$ (taken as 1 for nonionic solutes), $m$ is molality, and $\phi$ the osmotic coefficient at $m$ and $T$. In most soils, the contribution of $\psi_m$ greatly exceeds that of $\psi_s$ in reducing $\psi$. Within hyphae, or plant cells, the bounding wall usually exerts a turgor pressure ($\psi_p$) so that $\psi$ of the cell is increased (less negative) by the amount of $\psi_p$. Thus $\psi = \psi_s + \psi_m + \psi_p$. For microbial cell walls, $\psi_m$ is thought to be negligible.

Because of the opacity of most matrices, particularly soil, it is easier to determine fungal response to $\psi_s$ than to $\psi_m$. Solutes, however, may directly influence growth in ways individual to the particular solute, as well as through $\psi_s$. Thus carbohydrates may be metabolized and ions may be toxic. In experiments on effects of $\psi_s$ on microbial activity, it must therefore be established that the effects are indeed due to $\psi_s$ and not to specific solute factors. If a variety of solutes produce the same effect at comparable values of $\psi_s$ then it can be assumed that the effect is indeed due to $\psi_s$. A considerable body of evidence now demonstrates that the effects of commonly used solutes are exerted primarily through $\psi_s$ (Fig. 1),

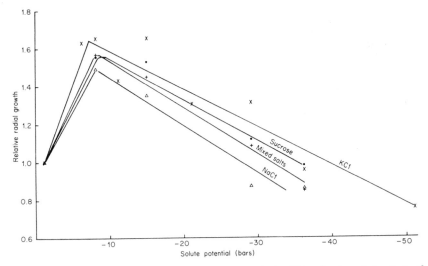

Fig. 1. Relationship between colony radial growth during 72 hr after inoculation to $\psi_s$ of the medium for *Fusarium culmorum* grown on basal medium agar at 20°C, $\psi_s$ modified by addition of KCl ($\times$), NaCl ($\triangle$), mixed salts (NaCl, KCl, and Na$_2$SO$_4$) ($+$) or sucrose ($\bullet$). (After Cook *et al.*, 1972.)

although the fungal response varies slightly from solute to solute, indicating secondary specific solute effects (Sommers *et al.*, 1970; Dubé *et al.*, 1971; Bruehl *et al.*, 1972; Bruehl and Manandhar, 1972; Cook *et al.*, 1972; Manandhar and Bruehl, 1973; Wong and Griffin, 1974; Taylor, 1974; Cook and Christen, 1976; Wearing, 1976). Even where it has been shown that there are statistically significant differences between the effects of the commonly used solutes, the general forms of the curves relating fungal growth to $\psi_s$ for each solute are very similar (Wilson, 1973). An exception is provided by *Sclerotinia borealis*, where the growth response to sucrose and to ions is quite different (Bruehl and Cunfer, 1971). This may be related to the abnormalities in respiration reported by Ward (1966). Polyethylene glycols of various molecular weights have been used as supposedly inert solutes (Hoch and Mitchell, 1973; Mexal and Reid, 1973), but these compounds greatly reduce oxygen diffusion (Mexal *et al.*, 1975) and so introduce a disturbing variable that almost certainly accounts for some of the unexpected responses following their use.

Care may need to be taken in the precise interpretation of data derived from growth of microorganisms on agar (Gardner *et al.*, 1972). This gel provides both $\psi_m$ and $\psi_s$ components of $\psi$ and also retards solute diffusion as it dries.

## B. Water Potential, Hyphal Growth, and Respiration

Saprophytic growth and activity, whether in pure culture or in the competitive environment of the soil, can be measured in three main ways. In each, it is necessary to distinguish between the effect of the environmental factor on the lag phase and on steady growth rate.

### 1. Specific Growth Rate and Colony Dry Weight

The specific growth rate is a sensitive measure of increase in colony dry weight and thus of protoplasmic synthesis and is given by

$$a = dW/Wdt = \ln2/t_d$$

where $a$ = specific growth rate, $W$ = dry mass of culture at times $t$, and $t_d$ = doubling or mean effective generation time. Trinci (1969) has studied the relationships between the rate of radial growth of a colony and specific growth rate for a number of fungi and has shown that they are by no means always equivalent measures of growth. No data are available for specific growth rates when $\psi$ is the controlling variable, but the related measure of colony dry weight after a certain period of growth has been reported (Mozumder and Caroselli, 1966; Taylor, 1974; Wearing, 1976; Griffin, 1977).

The use of colony dry weight as a measure of growth raises a number of theoretical and practical issues that do not appear to have been given adequate consideration. If a fungus is to continue to grow, it must obtain water from its environment to form new protoplasm. In theory, water might be obtained by movement along an appropriate osmotic gradient or by some "pump" mechanism whereby the organism does work against a water potential gradient. In fungi, there is no certain evidence for the latter, although such a mechanism has been postulated (Links *et al.*, 1957; Ekundayo and Carlile, 1964; Davidson, 1974). For water to move along a potential gradient, $\psi$ of the cell must be lower than that of the environment.

In turn, because of the positive potential imposed by turgor of the taut wall ($\psi_p$), $\psi_s$ of the cytoplasm must be lower than $\psi$ of the cell as a whole. Without turgor, growth ceases (Robertson, 1968). However, $\psi_p$ appears to vary, with species and external water potential, between 2.6 and 18.3 bars (Table I). The creation of the low cytoplasmic $\psi_s$ needed to produce the appropriate potential gradient will usually require energy. The potential difference between environment and cell will generally be

TABLE I

RELATIONSHIPS BETWEEN WATER POTENTIAL ($\psi$), SOLUTE POTENTIAL ($\psi_s$), AND TURGOR POTENTIAL ($\psi_p$) FOR FUNGAL HYPHAE

| Fungus | $\psi^a$ | $\psi_s$ | $\psi_p$ | Reference |
|---|---|---|---|---|
| *Aphanomyces euteiches* | −0.2 | −3.3 | 3.1 | Hoch and Mitchell (1973) |
| | −1.4 | −8.0 | 6.6 | |
| | −4.2 | −7.7 | 3.5 | |
| | −9.2 | −10.3 | 1.1 | |
| | −13.8 | −14.0 | 0.2 | |
| *Aspergillus niger* | −0.3 | −3.6 | 3.3 | Park and Robinson (1966) |
| *Aspergillus wentii* | −6.2 | −21.5 | 15.3 | Adebayo *et al.*, (1971) |
| | −16.2 | −31.8 | 15.5 | |
| | −31.2 | −44.3 | 13.1 | |
| *Fusarium oxysporum* | −3.0 | −5.6 | 2.6 | Robertson (1958) |
| *Mucor hiemalis* | −1.2 | −5.3 | 4.1 | Adebayo *et al.* (1971) |
| | −6.2 | −12.9 | 6.5 | |
| | −16.2 | −24.0 | 7.8 | |
| | −31.2 | −42.5 | 11.3 | |
| *Neurospora crassa* | −3.0 | −18.0 | 15.0 | Robertson and Rizvi (1968) |

[a] Water potential ($\psi$) of fungus is assumed to be very nearly equal to $\psi_s$ of external nutrient solution.

small, however, or the rate of water intake will exceed the rate of cell formation and wall rupture will eventually occur.

The lowered $\psi_s$ of the protoplasm might be achieved in one of three ways. The fungus may be growing in a basal nutrient solution (Sommers *et al.*, 1970) of reduced $\psi_s$ because of the presence of high concentration of, say, potassium chloride. A low internal $\psi_s$ could theoretically be achieved either by synthesis of carbohydrates of relatively high molecular weight or by passive or active movement of the major solute (potassium chloride) through the hyphal wall into the protoplasm. [It is often thought that, if a membrane is partly or wholly permeable to a solute, this somehow negates the effect of the solute. Although the osmotic *pressure* on the cell membrane will be reduced by such diffusion, the protoplasm will experience the same solute *potential* (Griffin, 1972).] The latter mechanism is indicated for both bacteria (Christian and Waltho, 1964) and fungi (Davidson, 1974; D. M. Griffin, unpublished). In the alga *Platymonas subcordiformis*, however, mannitol appears to be responsible for $\psi_s$ balance in the presence of external sodium chloride (Kirst, 1975). Polyols, especially arabitol, are important as compatible solutes in cells of sugar-tolerant yeasts grown in the presence of carbohydrates or polyethylene glycol (Brown and Simpson, 1972; Brown, 1974).

Second, the fungus may be growing in the same nutrient solution, to which polyethylene glycol has been added to reduce $\psi_s$ external to the fungus. In this instance, the membranes not being permeable to polyethylene glycol, $\psi_s$ of the protoplasm might be lowered by either cytoplasmic synthesis of carbohydrates or by cytoplasmic accumulation of minor solutes, generally ions against an ionic gradient from the basal medium. The latter occurs in *Fomes lividus* (D. M. Griffin, unpublished data), potassium ions in particular being accumulated, and such a process must consume energy.

Finally, if the fungus is growing in an environment of low ionic concentration, such as soil or wood, and of low $\psi_m$, reduction in internal $\psi_s$ will presumably be effected by cytoplasmic synthesis of carbohydrates of high molecular weight but may depend on accumulation of scarce ions. If the latter is the case, and there is no evidence, growth might be determined by the rate at which ions can be accumulated.

Regardless of mechanism, the density of unit volume of active hyphae must be greater the lower $\psi$. Furthermore, the exact density at a given $\psi$ will depend on the molecular weight and ionic number of the major internal solute, and thus perhaps on the external solute. It thus seems inevitable that the dry weight of colonies grown at different values of $\psi$ will be determined not only by their growth (in the sense of increase in volume) but also by their density (gm cm⁻³), the latter depending not only on $\psi$ but also on the major internal solute. Dry weight thus becomes a complex comparative measure of growth.

Apart from these problems, there exists a major practical one. To obtain colony dry weight, the mycelium must first be separated from the growth medium. If small volumes of solution, containing high concentrations of solutes, remain trapped between the hyphae, colony dry weight will be erroneously increased. This is particularly important where the major external solute (e.g., polyethylene glycol) is of very high molecular weight. Prolonged or vigorous washing, however, will remove solutes from the cytoplasm and so erroneously decrease dry weight (Hoch and Mitchell, 1973; Wearing, 1976).

The few data published show that dry weight response of fungi to $\psi$ is similar to that for radial growth (Mozumder and Caroselli, 1966; Taylor, 1974; Wearing, 1976; Griffin, 1977) with the former showing slightly greater sensitivity to reduced solute potential (Fig. 2).

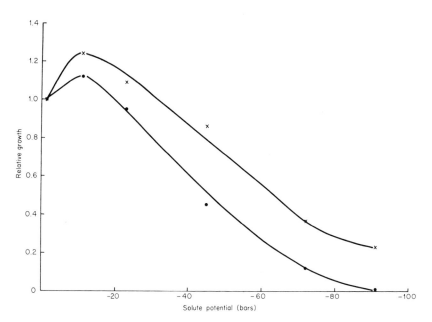

Fig. 2. Relationships between colony radial growth ($\times$), colony dry weight ($\bullet$), and $\psi_s$ of Czapek-Dox medium for *Gibberella zeae* grown at 25°C, $\psi_s$ modified by addition of KCl. (Data from Wearing, 1976.)

## 2. Rate of Colony Radial Growth

In pure culture studies, the rate of colony radial growth (linear hyphal extension rate) is the most frequently used measure of fungal growth. It neglects, however, to take account of differences in hyphal branching and colony structure that are comprehended in measures of dry weight.

There is now a large amount of data on the response of colony radial growth rate to changing $\psi_s$ and a representative sample is depicted in Fig. 3. *Clavulina amethystina* is representative of a number of extremely sensitive homobasidiomycetes (Tresner and Hayes, 1971; Wilson, 1973), including many wood-decay fungi (Griffin, 1977). No soil-borne plant pathogen of equal sensitivity is known. *Phytophthora cinnamomi* is representative of many phycomycete fungi and also of such pathogens as *Gaeumannomyces graminis* and *Rhizoctonia solani*. *Fusarium moniliforme* is representative of a number of *Fusarium* spp., *Verticillium albo-atrum*, *Cochliobolus sativus*, and of some species of *Aspergillus* and *Penicillium*. Other species of the last two genera are, however, far more tolerant of reduced $\psi_s$, growing at half-maximal rate at −200 bars (Ayerst, 1969; Griffin, 1972; Cook and Christen, 1976). Actinomycetes appear to be similar to *Phytophthora* in their responses although some are certainly more tolerant, showing some growth at −60 to −150 bars (Griffin, 1972; Williams *et al.*, 1972; Wong and Griffin, 1974).

Growth from mycelial inoculum imbedded in agar is characterized by

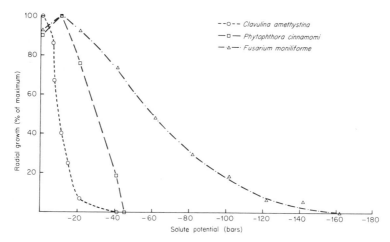

Fig. 3. Relationship between colony radial growth, during 168 hr after inoculation, to $\psi_s$ of the medium for three fungi grown on basal medium agar (Sommers *et al.*, 1970) at 25°C, the potential modified by addition of KCl. (Data from Wilson, 1973.)

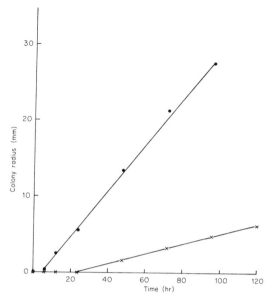

Fig. 4. Relationships between colony radius and time for *Phytophthora cinnamomi* when grown at 25°C on basal medium agar [-1.2 bars $\psi_s$ (●)] or on basal medium agar adjusted to $\psi_s = -41.2(\times)$ by addition of KCl. (Data from Wilson, 1973.)

a lag phase followed by a uniform rate of radial growth (Fig. 4). Reduced $\psi_s$ increases the duration of the lag phase and also reduces the radial growth rate (Scott, 1957; Ayerst, 1969; Wilson, 1973), the latter effect predominating. An anomalously long lag phase has been reported when polyethylene glycol was used as a solute (Mexal and Reid, 1973).

Fungi appear to be more sensitive to $\psi_m$ than to $\psi_s$ (Fig. 5). The effect varies with species, but in some the minimum $\psi_s$ permitting growth is even twice as low as that for $\psi_m$ (Adebayo and Harris, 1971; Cook *et al.*, 1972; Manandhar and Bruehl, 1973; Wearing, 1976). It is likely that reduction in solute diffusion, concurrent with reduction in $\psi_m$ is a major factor in this difference (Adebayo and Harris, 1971; Wearing, 1976) and in similar differences reported for leaf infection (Grogan and Abawi, 1975), but the lower nutrient status of many matric systems may be significant.

The response of many microorganisms to water as an environmental factor is subject to interaction with many other factors. Thus extensive data show that growth of microorganisms is possible at lower values of $\psi$ on substrates of high nutritional status (Ayerst, 1968; Wearing, 1976). In pure culture, strong interactions between $\psi$ and temperature in their effects on growth have also been shown. In general, the optimal temper-

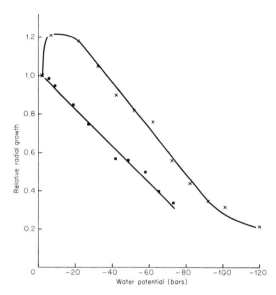

Fig. 5. Relationship between colony radial growth, during 72 hr after inoculation, to $\psi_s$ (×) and $\psi_m$ (■) for *Gibberella zeae*. (Data from Wearing, 1976.)

ature increases by about 5°C as $\psi$ decreases over the range permitting growth (Ayerst, 1968, 1969; Bruehl and Manandhar, 1972; Manandhar and Bruehl, 1973; Cook and Christen, 1976; Wearing, 1976). In soil, a similar interaction has been demonstrated, with *Aspergillus* spp. more active than *Penicillium* spp. at all potentials at temperatures of 30° and 35°C but only at potentials less than −220 bars at 15° and 20°C (Chen and Griffin, 1966b). The only data reported indicate no such interaction between pH and $\psi$ (Manandhar and Bruehl, 1973).

A slight reduction in $\psi$, but not $\psi_m$, frequently increases radial growth rate of fungi and streptomycetes (Fig. 5) (Griffin, 1972; Cook *et al.*, 1972; Bruehl and Manandhar 1972; Cook, 1973; Manandhar and Bruehl, 1973; Wong and Griffin, 1974; Grogan and Abawi, 1975). The reason for this is not known.

With *Fusarium graminearum*, *F. culmorum*, and *Gaeumannomyces graminis* var *tritici*, the results of *in vitro* experiments on response to reduced $\psi_s$ parallel the known effects of soil water on disease (Papendick and Cook, 1974; Cook and Christen, 1976). Similar experiments with *Fusarium oxysporum* f. sp. *vasinfectum* and *Verticillium albo-atrum* failed to reveal any such correlation (Manandhar and Bruehl, 1973) between *in vitro* growth and field pathogenicity.

## 3. Respiration

Yet another measure of microbial activity, and hence of growth in many cases, is provided by the rate of respiration. In the terrestrial filamentous fungi investigated, the respiration accompanying production of unit growth increased as $\psi_s$ decreased (Wilson and Griffin, 1975b). This rate of increase in respiration per unit growth as $\psi_s$ declined was greatest in fungi, such as *Phytophthora cinnamomi*, that have relatively high minima for $\psi_s$ permitting growth, and was least in xerotolerant fungi such as *Penicillium canescens* (Fig. 6). It is tempting to postulate that some fungi are restricted to higher potentials because of their inefficient use of respiratory energy at lower potentials. An analogous situation exists in the marine ascomycete *Lulworthia medua* in which respiration becomes uncoupled from biomass increase in fresh water (Davidson, 1974). In yeasts, yet another response was found. Respiration in the sugar-tolerant *Saccharomyces rouxii* was unaffected by the lowered $\psi_s$ caused by high concentrations of carbohydrates but that in the nontolerant *S. cerevisiae* was reduced (Brown, 1975). The interrelationships between growth, respiration, ionic uptake, and osmoregulation in fungi remain largely unknown and richly deserve precise investigation.

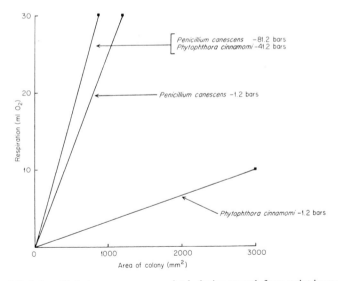

Fig. 6. Relationship between oxygen respired, during growth from point inoculum, to a colony of given area, area of colony, and $\psi_s$ (adjusted with KCl) for *Penicillium canescens* and *Phytophthora cinnamoni*. (After Wilson and Griffin, 1975b.)

## 4. Other Measures of Activity

Metabolism is reflected not only in increased biomass but also in production of chemicals, such as antibiotics, that are released into the surrounding medium. Production of antibiotics, *in vitro* by both *Streptomyces* sp. and *Cephalosporium gramineum*, is sensitive to $\psi_s$, being maximal for the former at potentials between $-20$ and $-35$ bars; for the latter between $-22$ and $-55$ bars. In each case, maximum production was at a potential at which growth was less than half-maximal (Bruehl *et al.*, 1972; Wong and Griffin, 1974). Production of penicillins by *Penicillium chrysogenum* in maize seeds has also been reported at $\psi$ between $-28$ and $-70$ bars (Hill, 1972). The production of antibiotics at these relatively low potentials is probably of considerable ecological significance, for the greatest inhibition *in vitro* of three pathogenic fungi by a streptomycete occurred at $-35$ bars, thus possibly narrowing the potential range of $\psi$ over which these fungi are active in soil (Wong and Griffin, 1974).

Despite all this evidence showing that $\psi$ is a factor of great ecological importance for fungi, it is interesting that mycofloristic lists do not demonstrate this. Thus the flora of desert soils is not conspicuously richer in the number of xerophytic species than an arable or rain forest soil in comparable latitudes (Chen and Griffin, 1966a), nor are the fungi of a Kuwait salt marsh exceptional (Moustafa, 1975). Ecological differentiation is therefore likely to be expressed within a soil, rather than between soils. As a given soil undergoes the normal cycle of wetting and drying, a succession of fungal species will move from dormancy to activity in response to the varying $\psi$ and associated factors.

## C. WATER POTENTIAL AND GROWTH INITIATION

When spores or sclerotia form the inoculum, activity commences by germination. Representative data have been collected on the influence of $\psi$ on the latent period for germination (Griffin, 1963, 1972; Ayerst, 1968, 1969). The latent period increases rapidly as $\psi$ declines, and germination occurs at lower $\psi$ if nutrients are readily available and if other environmental factors, especially temperature, are optimal (Fig. 7) (Ayerst, 1968, 1969; Griffin, 1972; Byrne and Jones, 1975). The practical significance of germination under extreme conditions will vary. Ideally, the relation between $\psi$ and the latent period should be considered in the light of the general ecology of the organism in its natural milieu. For instance, *Aspergillus flavus* within a grain silo will still find its substrate available after the 10 days required for germination at $-108$ bars (Pitt and Christian, 1968) or even after the 4 months at $-306$ bars (Armolik and Dickson, 1956). It is, however, most unlikely that pioneer colonizer such as *Fusarium oxy-*

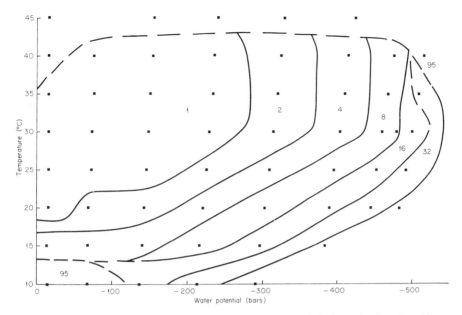

Fig. 7. Relationship between $\psi$, temperature, and latent period of germination of conidia of *Aspergillus chevalieri*. The lines delimit zones in which the minimum periods for germination are 95, 32, 16, 8, 4, 2, or 1 (or less than 1) day. Points indicate observations. (Data from Ayerst, 1969.)

*sporum* will find its substrate within soil still available after the 8 week latent period at – 157 bars (Schneider, 1954).

It is therefore likely that the range of potential permitting germination and continued growth in the far-from-optimal conditions usually present in field soils will be greatly restricted compared to that *in vitro*. Despite the doubtful significance of the extreme data, response of germination to reduced $\psi$ is clearly related to the ecology of the fungus, tolerance being associated with growth in natural substrates of reduced $\psi$ (Griffin, 1963, 1972; Byrne and Jones, 1975).

Little is known about the precise physiological effects of $\psi$ on spore germination but the optimal temperature increases with declining $\psi$ (Fig. 7) (Ayerst, 1969; Mozumder and Caroselli, 1970; Mozumder *et al.*, 1970; Griffin, 1972). In *Pythium* spp. thinning of the oospore wall is a precondition for germination and this is sensitive to $\psi_m$ (Lumsden and Ayres, 1975). Even in such a "water mold," germination occurs at – 15 bars in soil and the data indicate that wet soil conditions favor the activity of the genus, principally by increasing nutrient diffusion and hence availability of nutrients to the germinating spore (Stanghellini and Burr, 1973). The chlamydospores of *Phytophthora drechsleri* appear anomalous in that they germinate at $-98.5$ bars $\psi_s$ yet the minimum $\psi_s$ for continued mycelial growth is $-56$ bars (Cother and Griffin, 1974).

Zoosporangia normally germinate by formation of zoospores. In *Phytophthora drechsleri* and *P. cinnamomi* these were released in soils wetter than $-0.3$ to $-0.4$ bars $\psi_m$ (Reeves, 1975; Duniway, 1975b). Complete zoosporogenesis in *Aphanomyces euteiches* was most abundant at $-0.5$ bar $\psi_s$ or greater and did not occur at less than $-3$ to $-7$ bars $\psi_s$, depending on the solute (Hoch and Mitchell, 1973).

Sclerotia are reported to germinate only at high values of $\psi$ (Abeygunawardena and Wood, 1957) but a previous period of drying may be a necessary precursor for germination (Smith, 1972a,b,c). The effect of $\psi$ on spore germination is discussed further in Chapter 5 of this volume.

## D. WATER POTENTIAL AND REPRODUCTION

Asexual sporulation in ascomycetes and imperfect fungi can probably occur in most species throughout the range of $\psi$ permitting mycelial growth, although sporulation may be very tardy at low values of $\psi$. Ascospore production in cleistothecia is, however, more sensitive and in species investigated minimal values of $\psi$ may be up to 100 bars higher than for vegetative growth (Pitt and Christian, 1968). The higher $\psi$ minima reported by Curran (1971) are almost certainly due to the relatively short duration of her experiments (3 weeks instead of 120 days).

Production of apothecia from sclerotia of *Whetzelinia sclerotiorum* was inhibited by $-6$ bars $\psi_s$. Sclerotia, however, were produced between $-1$ to $-64$ bars $\psi_s$ (Grogan and Abawi, 1975). Sporulation in the Mastigomycotina is far more sensitive than in these other groups. Thus in *Pythium ultimum* the ratio of oospores to zoosporangia produced declined rapidly as $\psi_m$ fell from $-0.2$ to $-0.5$ bars (Bainbridge, 1970). In contrast, oogonia of *Phytophthora cactorum* were produced more abundantly at $-3$ bars than at higher $\psi_m$ (Sneh and McIntosh, 1974).

In *Phytophthora cactorum*, *P. drechsleri*, and *P. cinnamomi*, about $-0.3$ bar $\psi_m$ was optimal for formation of zoosporangia, with production declining in saturated soil because of impeded gas diffusion and also becoming negligible at about $-4$ bars (Sneh and McIntosh, 1974; Duniway, 1975a,b; Reeves, 1975). Chlamydospores of *P. cinnamomi*, however, were produced under all the moisture regimes tested, some being well below $-15$ bars (Reeves, 1975).

## E. WATER POTENTIAL AND SURVIVAL OF INOCULUM

In the cases of dormancy and saprophytic survival, neither the longevity of an individual propagule nor the extreme longevity of the exceptional case are of greatest significance. It is the population that is usually important and Yarwood and Sylvester (1959) have cogently argued that half-life period is the best overall measure of population survival. Unfortunately no such half-life data are available when $\psi$ is the variable.

Earlier work on the influence of water on survival of microorganisms has been reviewed by Cochrane (1968) and Mazur (1968). These reviews show that little is known of the physiological and biochemical factors affecting longevity of propagules. The relevance of the published data to field conditions is also reduced by the fact that longevity of propagules is affected by nutrient composition of the medium on which the propagule originated.

Survival of propagules of some fungi is greatest at high values of $\psi$, asexual spores of members of the Mastigomycotina being noteworthy (Cochrane, 1958.) Even with these, however, the evidence is not uniform. Thus Pratt and Mitchell (1975) did not detect any survival of *Phytophthora megasperma* in soil after 7 months at $-40$ bars, yet *Pythium ultimum* sporangia survived for 11 months in wet or air-dried soil (Stanghellini and Hancock, 1971a). Zoosporangia and oospores of *Phytophthora cactorum* survived better at $-5$ bars than at $-0.2$ bar $\psi_m$ (McIntosh, 1972; Sneh and McIntosh, 1974), probably because of reduced bacterial antagonism. At $-62$ bars, survival in soil of many species of the Zygomycotina was reduced more than that of a number of ascomycetes and imperfect fungi (Shameemullah *et al.,* 1971).

Cochrane (1958) identified a second category of fungi, whose spores survived best at intermediate values of $\psi$. None, however, would normally be considered soil fungi, with the possible exception of *Urocystis tritici*, whose teleutospores respond in this way.

Many fungal spores survive longest at very low values of $\psi$, often far below $-1000$ bars (Clerk and Madelin, 1965; Ledingham, 1970; Grogan and Abawi, 1975).

A final group consists of fungi such as *Aspergillus flavus* (Teitell, 1958) and *Metarrhizium anisopliae* (Clerk and Madelin, 1965) whose spores show a great decrease in survival at $-400$ to $-1000$ bars. A similar decrease in survival at $-145$ to $-390$ bars has also been revealed for hyphae of *Gibberella zeae* (Burgess and Griffin, 1968) and *Cephalosporium gramineum* (Bruehl and Lai, 1968) in wheat straws.

Low survival at low values of $\psi$ is probably simply explained by lethal loss of water. Maximal survival at a similar $\psi$ might be attributed to virtual cessation of respiration and hence to conservation of endogenous reserves. However, explanation is far from clear when survival is best or worst at intermediary $\psi$. The low survival of *Metarrhizium anisopliae* is clearly associated with unexplained peculiarities in respiration (Clerk and

Madelin, 1965) and the same may be true of survival of hyphae in *Gibberella zeae* and *Cephalosporium gramineum*. In the latter cases, however, low survival coincided with intense activity of *Penicillium* spp. and might therefore be attributed to antagonism from these xerophytic fungi.

The survival of sclerotia presents new problems. The rind does not effectively restrict water loss or subsequent gain, but sclerotia can survive exposure to 0% relative humidity for 4 wk (Trevethick and Cooke, 1973). In *Sclerotium rolfsii*, *S. delphinii*, and *S. oryzae*, drying causes substantial leakage, probably from the cells themselves rather than from free space within the sclerotium, on rewetting and this is associated with rapid rotting. No such leakage or rotting occurs in sclerotia of *Botrytis cinerea*, *B. tulipae*, or *Sclerotium cepivorum* (Smith, 1972a,b,c; Coley-Smith *et al.*, 1974; Keim and Webster, 1974).

## III. SOIL WATER CONTENT AND MICROBIAL ACTIVITY

Water content of a soil is most frequently expressed gravimetrically (gm water per gm soil dried for 24 hr at 105°C). A volumetric measure (cm$^3$ water per cm$^3$ soil) can be converted to a gravimetric one by dividing the former by the soil bulk density (mass of solids divided by total volume of solids, liquids, and gases), which in its own right may be a factor of biological importance (Griffin, 1972; Miller and Burke, 1974).

Water content influences microbial activity primarily through its effect on movement through soil, whether this be movement of solutes by diffusion or of microbial cells by Brownian activity or by flagella. In all cases, there are at least two classes of effect. The first is simply geometric and assumes an entirely inert matrix. All matrices, and especially clays, are surface active, however. As water content declines, solute molecules and microbial cells become necessarily closer to these active surfaces and a variety of attractions and repulsions then operate more or less strongly. This latter aspect, with its close association with clay mineralogy, has been discussed by Stotzky (1972), Marshall (1975), and Wong and Griffin (1976a).

The effect of water content on solute diffusion as an important factor in pathology has been demonstrated (Flentje, 1964; Kerr, 1964; Cook and Flentje, 1967; Stanghellini and Hancock, 1971b; Griffin, 1972; Short and Lacy, 1974, 1976), primarily in diseases where diffusion of nutrients from seed or seedling is necessary to stimulate germination of fungal spores.

Movement of microbial cells in soil will occur only if continuous water-filled pathways of the appropriate dimensions are present. Thus bacterial movement through soil is sensitive to water content (Griffin and

Quail, 1968; Hamdi, 1971; Bowen and Rovira, 1976; Wong and Griffin, 1976a,b,), the rate becoming negligible as soils drain to between about $-0.2$ and $-1$ bars $\psi_m$. Such restrictions affect chemical transformations in soil such as nitrification and sulfur oxidation (Griffin, 1972). They are also important in limiting bacterial-induced lysis of germ tubes and hyphae to soils of high $\psi_m$ (Cook and Papendick, 1970a; Sneh and McIntosh, 1974). The limitation on infection of potatoes by *Streptomyces scabies* in soils of high $\psi_m$, associated with prior colonization of the lenticels by antagonistic bacteria, is also relevant (Lapwood, 1966; Lewis, 1970; Lapwood and Adams, 1973; Davis *et al.*, 1974).

In soil, respiration of microorganisms declines with decreasing water content far more rapidly than can be explained in terms of $\psi_m$ alone and once again limited bacterial movement is likely to be an important factor. In soils of lower than about $-7$ bars $\psi_m$, fungal respiration probably predominates (Wilson and Griffin, 1975a), in accord with data of Anderson and Domsch (1973). The implications for the understanding of antagonism between microorganisms at different potentials is clear.

Zoospores are likely to be even more restricted than bacteria by pore geometry, because of the larger size of the former. Thus movement through various soils by zoospores of *Phytophthora cyptogea* was reduced at $-0.001$ to $-0.01$ bar $\psi_m$ and ceased at $-0.01$ to $-0.1$ bar, depending on soil texture (Duniway, 1976). In addition, contact with solid surfaces increases as movement is restricted by drainage to smaller and smaller pores, and this contact in itself reduces zoospore mobility (Ho and Hickman, 1967; Bimpong and Clerk, 1970).

As discussed in Chapter 4 of this volume, spores of all types can be carried or splashed in moving water and such passive spread may be significant (Hirst, 1965; Hickman and Ho, 1966; Dick, 1968), particularly in water flooding over the soil surface. The significance of downward movement of spores through the soil profile is, however, little understood. Such movement must be very common when rain falls onto dry, cracked, clay soils in which fissures to a depth of more than 1 m may exist. Not only spores but soil particles themselves will be washed down. In this way, microbial propagules may be carried into layers more or less favorable to their survival or activity but no reports were found of experiments directed at the specific question of longevity of propagules after water transport.

Downward movement of fungal and actinomycete spores by water through the smaller stable pore system of soils has been studied by several investigators (Hirst, 1965; Griffin, 1972; Ruddick and Williams, 1972; Bitton *et al.*, 1974) but is of doubtful importance. The obvious limitations imposed by pore geometry and spore size apply. Dry-walled spores are not easily carried by water, and none are moved far unless the system is near saturated and the rate of water movement considerable. Even then vertical spread is unlikely to exceed a few centimeters.

## REFERENCES

Abeygunawardena, D. V. W., and Wood, R. K. S. (1957). Factors affecting germination of sclerotia of *Sclerotium rolfsii*. *Trans. Br. Mycol. Soc.* **40**, 221–231.

Adebayo, A. A., and Harris, R. F. (1971). Fungal growth response to osmotic as compared to matric water potential. *Soil Sci. Soc. Am., Proc.* **35**, 465–469.

Adebayo, A. A., Harris, R. F., and Gardner, W. R. (1971). Turgor pressure of fungal mycelia. *Trans. Br. Mycol. Soc.* **57**, 145–151.

Anderson, J. P. E., and Domsch, K. H. (1973). Quantification of bacterial and fungal contributions to soil respiration. *Arch. Microbiol.* **93**, 113–127.

Armolik, N., and Dickson, J. G. (1956). Minimum humidity requirement for germination of conidia of fungi associated with storage of grain. *Phytopathology* **46**, 462–465.

Ayerst, G. (1968). Prevention of biodeterioration by control of environmental conditions. *In* "Biodeterioration of Materials" (A. H. Walters and J. J. Elphick, eds.), pp. 223–241. Elsevier, Amsterdam.

Ayerst, G. (1969). The effects of water activity and temperature on spore germination and growth in some mould fungi. *J. Stored Prod. Res.* **5**, 127–141.

Bainbridge, A. (1970). Sporulation by *Pythium ultimum* at various soil moisture tensions. *Trans. Br. Mycol. Soc.* **55**, 485–488.

Baker, K. F., and Cook, R. J. (1974). "Biological Control of Plant Pathogens." Freeman, San Francisco, California.

Bimpong, C. E., and Clerk, G. C. (1970). Motility and chemotaxis in zoospores of *Phytophothora palmivora* (Butl.). *Ann. Bot. (London)* [N. S.] **34**, 617–624.

Bitton, G., Lahav, N., and Henis, Y. (1974). Movement and retention of *Klebsiella aerogenes* in soil columns. *Plant Soil* **40**, 373–380.

Bowen, G. D., and Rovira, A. D. (1976). Microbial colonization of plant roots. *Annu. Rev. Phytopathol.* **14**, 121–144.

Brown, A. D. (1974). Microbial water relations. Features of the intracellular composition of sugar-tolerant yeasts. *J. Bacteriol.* **118**, 769–777.

Brown, A. D. (1975). Microbial water relations. Effects of solute concentration on the respiratory activity of sugar-tolerant and non-tolerant yeast. *J. Gen. Microbiol.* **86**, 241–249.

Brown, A. D., and Simpson, J. R. (1972). Water relations of sugar-tolerant yeasts: the role of intracellular polyols. *J. Gen. Microbiol.* **72**, 589–591.

Bruehl, G. W., and Cunfer, B. (1971). Physiologic and environmental factors that effect the severity of snow mold of wheat. *Phytopathology* **61**, 792–799.

Bruehl, G. W., and Lai, P. (1968). Influence of soil pH and humidity on survival of *Cephalosporium gramineum* in infested wheat straw. *Can. J. Plant Sci.* **48**, 245–252.

Bruehl, G. W., and Manandhar, J. B. (1972). Some water relations of *Cercosporella herpotrichoides*. *Plant Dis. Rep.* **56**, 594–596.

Bruehl, G. W., Cunfer, B., and Toivianinen, M. (1972). Influence of water potential on growth, antibiotic production and survival of *Cephalosporium gramineum*. *Can. J. Plant Sci.* **52**, 417–423.

Burgess, L. W., and Griffin, D. M. (1968). The recovery of *Gibberella zeae* from wheat straws. *Aust. J. Exp. Agric. Anim. Husb.* **8**, 364–370.

Byrne, P., and Jones, E. B. G. (1975). Effect of salinity on spore germination of terrestrial and marine fungi. *Trans. Br. Mycol. Soc.* **64**, 497–503.

Chen, A. W., and Griffin, D. M. (1966a). Soil physical factors and the ecology of fungi. V. Further studies in relatively dry soils. *Trans. Br. Mycol. Soc.* **49**, 419–426.

Chen, A. W., and Griffin, D. M. (1966b). Soil physical factors and the ecology of fungi. VI. Interaction between temperature and soil moisture. *Trans. Br. Mycol. Soc.* **49**, 551–561.

Christian, J. H. B., and Waltho, J. A. (1964). The composition of *Staphylococcus aureus* in relation to the water activity of the growth medium. *J. Gen. Microbiol.* **35**, 205–213.

Clerk, G. C., and Madelin, M. F. (1965). The longevity of conidia of three insect-parasitizing hyphomycetes. *Trans. Br. Mycol. Soc.* **48**, 193–210.

Cochrane, V. W. (1958). "Physiology of Fungi." Wiley, New York.

Coley-Smith, J. R., Ghaffar, A., and Javed, Z. U. R. (1974). The effect of dry conditions on subsequent leakage and rotting of fungal sclerotia. *Soil Biol. & Biochem.* **61**, 307–312.

Cook, R. J. (1973). Influence of low plant and soil water potentials on diseases caused by soilborne fungi. *Phytopathology* **63**, 451–458.

Cook, R. J., and Christen, A. A. (1976). Growth of cereal root-rot fungi as affected by temperature-water potential interactions. *Phytopathology* **66**, 193–197.

Cook, R. J., and Flentje, N. T. (1967). Chlamydospore germination and germling survival of *Fusarium solani* f. *pisi* in soil as affected by soil water and pea seed exudation. *Phytopathology* **57**, 178–182.

Cook, R. J., and Papendick, R. I. (1970a). Soil water potential as a factor in the ecology of *Fusarium roseum* f. sp. *cerealis* 'Culmorum.' *Plant Soil* **32**, 131–145.

Cook, R. J., and Papendick, R. I. (1970b). Effect of soil water on microbial antagonism and nutrient availability in relation to soil-borne fungal diseases of plants. *In* "Root Diseases and Soil-Borne Pathogens" (T. A. Toussoun, R. V. Bega, and P. E. Nelson, eds.), pp. 81–88. Univ. of California Press, Berkeley.

Cook, R. J., and Papendick, R. I. (1972). Influence of water potential of soils and plants on root disease. *Annu. Rev. Phytopathol.* **10**, 349–374.

Cook, R. J., Papendick, R. I., and Griffin, D. M. (1972). Growth of two root rot fungi as affected by osmotic and matric water potentials. *Soil Sci. Soc. Am., Proc.* **36**, 78–82.

Cother, E. J., and Griffin, D. M. (1974). Chlamydospore germination in *Phytophthora drechsleri*. *Trans. Br. Mycol. Soc.* **63**, 273–279.

Curran, P. M. T. (1971). Sporulation in some members of the *Aspergillus glaucus* group in response to osmotic pressure, illumination and temperature. *Trans. Br. Mycol. Soc.* **57**, 201–211.

Davidson, D. E. (1974). The effect of salinity on a marine and a freshwater Ascomycete. *Can. J. Bot.* **52**, 553–563.

Davis, J. R., McMaster, G. M. Callihan, R. H., Garner, J. G., and McDole, R. E. (1974). The relationship of irrigation timing and soil treatments to control potato scab. *Phytopathology* **64**, 1404–1410.

Dick, M. W. (1968). Consideration of the role of water on the taxonomy and ecology of the filamentous biflagellate fungi in littoral zones. *Veröeff. Inst. Meevesforsch. Bremerhaven* **3**, 27–38.

Dubé, A. J., Dodman, R. I., and Flentje, N. T. (1971). The influence of water activity on the growth of *Rhizoctonia solani*. *Aust. J. Biol. Sci.* **24**, 57–65.

Duniway, J. M. (1975a). Formation of sporangia by *Phytophthora drechsleri* in soil at high matric potentials. *Can. J. Bot.* **53**, 1270–1275.

Duniway, J. M. (1975b). Limiting influence of low water potential on the formation of sporangia by *Phytophthora drechsleri* in soil. *Phytopathology* **65**, 1089–1093.

Duniway, J. M. (1976). Movement of zoospores of *Phytophthora cryptogea* in soils of various textures and matric potentials. *Phytopathology* **66**, 877–882.

Ekundayo, J. A., and Carlile, M. J. (1964). The germination of sporangiospores of *Rhizopus arrhizus;* spore swelling and germ-tube emergence. *J. Gen. Microbiol.* **35**, 261–269.

Flentje, N. T. (1964). Pre-emergence rotting of peas in South Australia. I. Factors associated with the soil. *Aust. J. Biol. Sci.* **17**, 651–654.

Gardner, W. R. (1968). Availability and measurement of soil water. *In* "Water Deficits and Plant Growth" (T. T. Kozlowski, ed), Vol. 1, pp. 107–135. Academic Press, New York.

Gardner, W. R., Dalton, F. N., and Harris, R. F. (1972). Thermocouple psychrometry for the study of water relations of soil micro-organisms. *In* "Psychrometry in Water Relations Research" (R. W. Brown and B. P. van Haveren, eds.), pp. 150–153. Utah Agric. Exp. Stn., Logan.

Garrett, S. D. (1970), "Pathogenic Root-Infecting Fungi." Cambridge Univ. Press, London and New York.

Gray, T. R. G. (1976). Survival of vegetative microbes in soil. *Symp. Soc. Gen. Microbiol.* **26**, 327–364.

Griffin, D. M. (1963). Soil moisture and the ecology of soil fungi. *Biol. Rev. Cambridge Philos. Soc.* **38**, 141–166.

Griffin, D. M. (1968). A theoretical study relating the concentration and diffusion of oxygen to the biology of organisms in soil. *New Phytol.* **67**, 561–577.

Griffin, D. M. (1969). Soil water in the ecology of fungi. *Annu. Rev. Phytopathol.* **7**, 289–310.

Griffin, D. M. (1970). Effect of soil mosture and aeration on fungal activity: An introduction. *In* "Root Diseases and Soil-Borne Pathogens" (T. A. Toussoun, R. V. Bega, and P. E. Nelson, eds), pp. 77–80. Univ. of California Press, Berkely.

Griffin, D. M. (1972). "Ecology of Soil Fungi." Chapman & Hall, London.

Griffin, D. M. (1977). Water potential and wood-decay fungi. *Annu. Rev. Phytopathol.* **15**, 319–329.

Griffin, D. M., and Quail, G. (1968). Movement of bacteria in moist, particulate systems. *Aust. J. Biol. Sci.* **21**, 579–582.

Grogan, R. G., and Abawi, G. W. (1975). Influence of water potential on growth and survival of *Whetzelinia sclerotiorum*. *Phytopathology* **65**, 122–128.

Hamdi, Y. A. (1971). Soil water tension and the movement of *Rhizobia*. *Soil Biol. & Biochem.* **3**, 121–126.

Hickman, C. J., and Ho, H. H. (1966). Behaviour of zoospores in plant-pathogenic phycomycetes. *Annu. Rev. Phytopathol.* **4**, 195–220.

Hill, P. (1972). The production of penicillins in soils and seeds by *Penicillium chrysogenum* and the role of penicillin β-lactamase in the ecology of soil *Bacillus*. *J. Gen. Microbiol.* **70**, 243–252.

Hirst, J. M. (1965). Dispersal of soil microorganisms. *In* "Ecology of Soil-Borne Plant Pathogens" (K. F. Baker and W. C. Snyder, eds.), pp. 69–81. Univ. of California Press, Berkeley.

Ho, H. H., and Hickman C. J. (1967). Asexual reproduction and behaviour of zoospores of *Phytophthora megasperma* var. *sojae*. *Can. J. Bot.* **45**, 1963–1981.

Hoch, H. C., and Mitchell, J. E. (1973). The effects of osmotic water potentials on *Aphanomyces euteiches* during zoosporogenesis. *Can. J. Bot.* **51**, 413–420.

Keim, R., and Webster, R. K. (1974). Effect of soil moisture and temperature on viability of sclerotia of *Sclerotium oryzae*. *Phytopathology* **64**, 1499–1502.

Kerr, A. J. (1964). The influence of soil moisture on infection of peas by *Pythium ultimum*. *Aust. J. Biol. Sci.* **17**, 676–685.

Kirst, G. O. (1975). Beziehungen zwischen Mannitkonzentration und osmotischer Belastung bei der Brackwasseralge *Platymonas subcordiformis* Hazen. *Z. Pflanzenphysiol.* **76,** 316–325.

Lapwood, D. H. (1966). The effects of soil moisture at the time potato tubers are forming on the incidence of common scab (*Streptomyces scabies*). *Ann. Appl. Biol.* **58,** 447–456.

Lapwood, D. H., and Adams, M. J. (1973). The effect of a few days of rain on the distribution of common scab (*Streptomyces scabies*) on young potato tubers. *Ann. Appl. Biol.* **73,** 277–283.

Ledingham, R. J. (1970). Survival of *Cochliobolus sativus* conidia in pure culture and in natural soil at different relative humidities. *Can. J. Bot.* **48,** 1893–1896.

Lewis, B. G. (1970). Effects of water potential on the infection of potato tubers by *Streptomyces scabies* in soil. *Ann. Appl. Biol.* **66,** 83–88.

Links, J., Rambouts, J. E., and Keulen, P. (1957). The 'bulging factor,' a fungistatic antibiotic produced by *Streptomyces* strain, with evidence of an active water-excreting mechanism in fungi. *J. Gen. Microbiol.* **17,** 596–601.

Lumsden, R. D., and Ayres, W. A. (1975). Influence of the soil environment on germinability of constitutively dormant oospores of *Pythium ultimum*. *Phytopathology* **65,** 1101–1107.

McIntosh, D. L. (1972). Effects of soil water, soil temperature, carbon and nitrogen amendments and host rootlets on survival in soil of zoospores of *Phytophthora cactorum*. *Can. J. Bot.* **50,** 269–272.

Manandhar, J. B., and Bruehl, G. W. (1973). In vitro interactions of *Fusarium* and *Verticillium* wilt fungi with water, pH and temperature. *Phytopathology* **63,** 413–418.

Marshall, K. C. (1975). Clay mineralogy in relation to survival of soil bacteria. *Annu. Rev. Phytopathol.* **13,** 357–373.

Mazur, P. (1968). Survival of fungi after freezing and desiccation. *In* "The Fungi" (G. C. Ainsworth and A. S. Sussman, eds.), Vol. 3, pp. 325–394. Academic Press, New York.

Mexal, J., and Reid, C. P. P. (1973). The growth of selected mycorrhizal fungi in response to induced water stress. *Can. J. Bot.* **51,** 1579–1588.

Mexal, J., Fisher, J. T., Osteryoung, J., and Reid, C. P. P. (1975) Oxygen availability in polyethylene glycol solutions and its implications in plant-water relations. *Plant Physiol.* **55,** 20–24.

Miller, D. E., and Burke, D. W. (1974). Influence of soil bulk density and water potential on *Fusarium* root rot of bean. *Phytopathology* **64,** 526–529.

Moustafa, A. F. (1975). Osmophilous fungi in the salt marshes of Kuwait. *Can. J. Microbiol.* **21,** 1573–1580.

Mozumder, B. K. G., and Caroselli, N. E. (1966). The influence of substrate moisture on the growth of *Verticillium albo-atrum* R. and B. *Adv. Front. Plant Sci.* **16,** 77–83.

Mozumder, B. K. G., and Caroselli, N. E. (1970). Water relations of respiration of *Verticillium albo-atrum* conidia. *Phytopathology* **60,** 915–916.

Mozumder, B. K. G., Caroselli, N. E., and Albert, L. S. (1970). Influence of water activity, temperature and their interaction on germination of *Verticillium albo-atrum* conidia. *Plant Physiol.* **46,** 347–349.

Papendick, R I., and Campbell, G. S. (1975). Water potential in the rhizosphere and plants and methods of measurement and experimental control. *In* "Biology and Control of Soil-Borne Plant Pathogens" (G. W. Bruehl, ed.), pp. 39–49. Am. Phytopathol. Soc., St. Paul, Minnesota.

Papendick, R. I., and Cook, R. J. (1974). Plant water stress and development of *Fusarium* root rot in wheat subjected to different cultural practices. *Phytopathology* **64,** 358–363.

Park, D., and Robinson, P. M. (1966). Internal pressure of hyphal tips of fungi, and its significance in morphogenesis. *Ann. Bot. (London)* [N. S.] **30,** 425–439.

Pitt, J. I., and Christian, J. H. B. (1968). Water relations of xerophilic fungi isolated from prunes. *Appl. Microbiol.* **16**, 1853–1858.

Pratt, R. G., and Mitchell, J. E. (1975). The survival and activity of *Phytophthora megasperma* in naturally infested soils. *Phytopathology* **65**, 1267–1272.

Reeves, R. J. (1975). Behaviour of *Phytophthora cinnamomi* Rands in different soils and water regimes. *Soil Biol. & Biochem.* **7**, 19–24.

Robertson, N. F. (1958). Observations of the effect of water on the hyphal apices of *Fusarium oxysporum. Ann. Bot. (London)* [N. S.] **22**, 159–173.

Robertson, N. F. (1968). The growth process in fungi. *Annu. Rev. Phytopathol.* **6**, 115–136.

Robertson, N. F., and Rizvi, S. R. H. (1968). Some observations on the water relations of the hyphae of *Neurospora crassa. Ann. Bot. (London)* [N. S.] **32**, 279–291.

Robinson, P. M. (1973). Oxygen—positive chemotropic factor for fungi? *New Phytol.* **72**, 1349–1356.

Ruddick, S. M., and Williams, S. T. (1972). Studies on the ecology of actinomycetes in soil. V. Some factors influencing the dispersal and adsorption of spores in soil. *Soil Biol. & Biochem.* **4**, 93–103.

Schneider, R. (1954). Untersuchungen über Feuchtigkeitsanspruche parasitischer Pilze. *Phytopathol. Z.* **21**, 63–78.

Schoeneweiss, D. F. (1975). Predisposition, stress and plant disease. *Ann. Rev. Phytopathol.* **13**, 193–211.

Scott, W. J. (1957). Water relations of food spoilage microorganisms. *Adv. Food Res.* **7**, 83–127.

Shameemullah, M., Parkinson, D., and Burges, A. (1971). The influence of soil moisture tension on the fungal population of a pinewood soil. *Can. J. Microbiol.* **17**, 975–986.

Short, G. E., and Lacy, M. L. (1974). Germination of *Fusarium solani* f. sp. *pisi* chlamydospores in the spermosphere of pea. *Phytopathology* **64**, 558–562.

Short, G. E., and Lacy, M. L. (1976). Factors affecting pea seed and seedling rot in soil. *Phytopathology* **66**, 188–192.

Smith, A. M. (1972a). Drying and wetting sclerotia promotes biological control of *Sclerotium rolfsii* Sacc. *Soil Biol. & Biochem.* **4**, 119–124.

Smith, A. M. (1972b). Nutrient leakage promotes biological control of dried sclerotia of *Sclerotium rolfsii* Sacc. *Soil Biol. & Biochem.* **4**, 125–130.

Smith, A. M. (1972c). Biological control of fungal sclerotia in soil. *Soil Biol. & Biochem.* **4**, 131–134.

Sneh, B., and McIntosh, D. L. (1974). Studies on the behaviour and survival of *Phytophthora cactorum* in soil. *Can. J. Bot.* **52**, 795–802.

Sommers, L. E., Harris, R. F., Dalton, F. N. and Gardner, W. R. (1970). Water potential relations of three root-infecting *Phytophthora* species. *Phytopathology* **60**, 932–934.

Stanghellini, M. E., and Burr, T. J. (1973). Effect of soil water potential on disease incidence and oospore germination of *Pythium aphanidermatum. Phytopathology* **63**, 1496–1498.

Stanghellini, M. E., and Hancock, J. G. (1971a). The sporangium of *Pythium ultimum* as a survival structure in soil. *Phytopathology* **61**, 157–164.

Stanghellini, M. E., and Hancock, J. G. (1971b). Radial extent of the bean spermosphere and its relation to the behaviour of *Pythium ultimum. Phytopathology* **61**, 165–168.

Stotzky, G. (1972). Activity, ecology and population dynamics of microorganisms in soil. *Crit. Rev. Microbiol.* **2**, 59–137.

Taylor, P. A. (1974). Ecological studies on the occurrence of *Phytophthora cinnamomi* on Black Mountain, A.C.T. Ph.D. Thesis, Australian National University, Canberra.

Taylor, S. A. (1968). Terminology in plant and soil water relations. *In* "Water Deficits and Plant Growth" (T. T. Kozlowski, ed.), Vol. 1, pp. 49–72. Academic Press, New York.

Teitell, L. (1958). Effects of relative humidity on viability of conidia of *Aspergillus*. *Am. J. Bot.* **45**, 748–753.

Tresner, H. D., and Hayes, J. A. (1971). Sodium chloride tolerance of terrestrial fungi. *Appl. Microbiol.* **22**, 210–213.

Trevethick, J., and Cooke, R. C. (1973). Water relations in sclerotia of some *Sclerotinia* and *Sclerotium* species. *Trans. Br. Mycol. Soc.* **60**, 555–558.

Trinci, A. P. J. (1969). A kinetic study of the growth of *Aspergillus nidulans* and other fungi. *J. Gen. Microbiol.* **57**, 11–24.

Ward, E. W. B. (1966). Preliminary studies on the physiology of *Sclerotinia borealis*, a highly psychrophilic fungus. *Can. J. Bot.* **44**, 237–246.

Wearing, A. H. (1976). Studies on the saprophytic behaviour of *Fusarium roseum* 'graminearum.' Ph.D Thesis, University of Sydney, Sydney, Australia.

Williams, S. T., Shameemullah, M., Watson, E. T., and Mayfield, C. I. (1972). Studies on the ecology of actinomycetes in soil. VI. The influence of moisture tension on growth and survival. *Soil Biol. & Biochem.* **4**, 215–226.

Wilson, J. M. (1973). The effect of water potential on the growth and respiration of various soil microorganisms. M.Sc. Thesis, Australian National University, Canberra.

Wilson, J. M., and Griffin, D. M. (1975a). Water potential and the respiration of microorganisms in the soil. *Soil Biol. & Biochem.* **7**, 199–204.

Wilson, J. M., and Griffin, D. M. (1975b). Respiration and radial growth of soil fungi at two osmotic potentials. *Soil Biol. & Biochem.* **7**, 269–274.

Wong, P. T. W., and Griffin, D. M. (1974). Effect of osmotic potential on streptomycete growth, antibiotic production and antagonism to fungi. *Soil Biol. & Biochem.* **6**, 319–326.

Wong, P. T. W., and Griffin, D. M. (1976a). Bacterial movement at high matric potentials. I. In artificial and natural soils. *Soil Biol. & Biochem.* **8**, 215–218.

Wong, P. T. W., and Griffin, D. M. (1976b). Bacterial movement at high matric potentials. II. In fungal colonies. *Soil Biol. & Biochem.* **8**, 219–223.

Yarwood, C. E., and Sylvester, E. S. (1959). The half-life concept of longevity of plant pathogens. *Plant Dis. Rep.* **43**, 125–128.

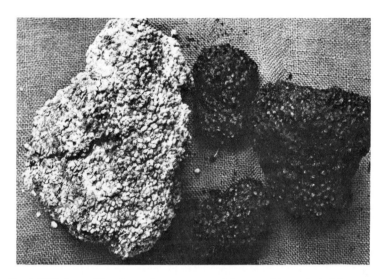

Fig. 1. "Bin-burned" soybeans from a commercial storage bin, typical of fungus-caused spoilage. In this case, the final and sudden heating, approaching near to the point of ignition, was preceded by slow and gradual rise in temperature over a period of several months, characteristic of microbiological heating.

## II. LONGEVITY OF SEEDS

Longevity of seeds is affected by many factors or agencies other than fungi, although fungi, under circumstances that permit their invasion, can slowly or rapidly reduce seed viability, Seeds of different kinds of plants, and even seeds of a single species grown under different conditions, may vary greatly in longevity. Some kinds of seeds, even when kept under conditions favorable for growth of fungi, are remarkably resistant to damaging invasion. Beals (1905) buried seeds of 23 species of plants in inverted, unstoppered bottles in moist soil out of doors in Michigan. Some of the seeds of eight species germinated after having been buried for 40 yr, and 20% of the seeds of *Verbascum blattara* (moth mullein) germinated after having been buried for 90 yr (Kivilaan and Bandurski, 1973). These must have been resistant to invasion by fungi as well as to penetration by moisture. Barton (1961) cited evidence of germination of dry-stored seeds after more than 100 yrs, and of germination of lotus seeds buried in soil for more than 1000 years. Harrington (1972) presents a recent summary of seed longevity. Unpublished evidence from my own work indicates that many kinds of hard-coated weed seeds are highly resistant to invasion by fungi, so that there are, in at least some kinds of seeds, conditions other than moisture content that determine whether seeds will or will not be invaded and damaged or decayed by fungi.

## III. FIELD FUNGI

### A. Definition

The terms "field fungi" and "storage fungi" were coined in the 1940's when intensive work first got under way on the relation of fungi to deterioration of seeds in storage. These terms are intended to describe ecological, not taxonomic, groups. Field fungi are those that invade seeds while the seeds are still developing on the plants in the field, or after the seeds have matured but before they have been threshed out. Normally the field fungi do not continue to develop in seeds after harvest because they require, for growth, a moisture content in equilibrium with relative humidities of 95–100%, and at harvest most agricultural seeds have a moisture content far below this. There are some exceptions, notably maize of high moisture content stored on the ears in cribs, but even there the fungi that invade and decay the kernels may differ somewhat from the usual field fungi. This has been recognized by designating them "advanced decay fungi" (Christensen, 1965). Since the field fungi do not normally continue to grow after seeds are harvested, the damage they cause is done by harvest time. Here again there may be exceptions: There is some evidence that wheat seeds moderately invaded by *Fusarium* at harvest may, in subsequent storage at moisture contents too low for *Fusarium* to grow, develop dark germs (in the grain trade referred to as "sick wheat").

Storage fungi are those that invade seeds during storage. Their major distinguishing characteristic is the ability to grow without free water, at moisture contents in equilibrium with relative humidities of about 68–90%. Some of them, in fact, *require* a high osmotic pressure to grow; if free water is present they grow poorly, or even die.

### B. Major Genera of Field Fungi

The predominant genera of field fungi differ to some extent according to the crop, the region or geographic location, the weather, etc. However, in the chief food and feed grains—wheat, rice, barley, oats, rye and sorghum, regardless of where in the world they are grown—the common field fungi are *Alternaria*, *Cladosporium*, *Fusarium*, and *Helminthosporium*, although in rice *Curvularia* may predominate at times. If moist and warm weather prevails after heading, the seeds of these plants may be heavily invaded by one or more of these fungi and, at times, by a multitude of other fungi. The manual on seed-borne fungi published by the International Seed Testing Association (Malone and Muskett, 1964) describes 77 species of fungi, in 56 genera, in agricultural and horticultural seeds. Close to 100% of the seeds of the crops listed above, regardless of where in the world they are grown, are, by the time they mature, invaded by *Alternaria*. Christensen and Kaufmann (1969) state, "We regularly isolate *Alternaria* from close to 100% of surface-disinfected wheat kernels after harvest. If, in fact, we obtain *Alternaria* from 90% or more of wheat kernels of a given lot that are surface disinfected and placed on an agar medium in culture dishes, but obtain no storage fungi, we consider this good evidence that the seed is newly harvested and that it has been stored under conditions that do not permit deterioration." Seeds that are produced within pods, or within fleshy fruits, if the pods or friuts are sound, may be entirely free of fungi at harvest.

## C. FORM AND LOCATION OF FIELD FUNGI IN SEEDS

### 1. Seeds of Cereals

Mycelium of fungi is commonly present inder the pericarps of wheat (Hyde, 1950; Hyde and Galleymore, 1951; Christensen, 1951) (Fig. 2). Most of this mycelium is that of *Alternaria*, but some may be of other fungi. Spores of *Alternaria* sometimes are numerous under the hulls of barley, particularly weathered barley (Kotheimer and Christensen, 1961).

### 2. Nuts and Coconuts

The relatively hard-shelled nuts such as Brazil nuts, walnuts, pecans, and almonds would appear to be resistant to invasion by fungi. Some are resistant, some are not. Some of them have a natural opening at the point of attachment, and so are readily invaded as soon as the nuts are detached from the tree. These nuts may also be damaged by insects while still on the tree, the damage being accompanied or followed by fungi. If the nuts, after they mature and fall, are allowed to lie on the ground for a time before they are gathered and dried, the danger of fungus invasion is increased. Months may elapse between the time the fruits in which Brazil nuts are borne fall from the tree and the time the nuts arrive at ports for shipment; during this time they may be heavily invaded by a great variety of fungi, as well as by *Streptomyces* and bacteria (Spencer, 1921). The author found mycelium of an unidentified fungus or fungi to be relatively abundant in the brown layer covering the meats in 100% of Brazil nuts, and large numbers of colonies of *Streptomyces*, *Cephalosporium*, and *As-*

Fig. 2. Mycelium (probably of *Alternaria*) under the outer pericarp layers of a wheat kernel.

*pergillus flavus* were recovered from the scrapings of these tissues. About 3% of the nuts were obviously decayed. All of these were U. S. Grade No. 1 Brazil nuts bought in local stores.

The thick, fibrous husk of the coconut may serve to some extent as protection again invasion of the kernel, but it also serves as a substrate for growth of many fungi. The hard shell of the coconut within the husk is composed chiefly of very dense tissues that are not readily invaded by fungi, but there is a weak portion, the "eye," through which the sprout emerges when the coconut germinates. Fungi can penetrate this fairly readily and, once inside, they are furnished with ample food and moisture by the meat and milk of the coconut. We occasionally encounter a store-bought coconut with a heavily molded interior, but we have observed many more with internal fungus infection that was not detectable by examination by the naked eye. Even if a fungus capable of causing decay penetrates into the interior of the coconut, it does not always cause decay.

## D. SURVIVAL OF FIELD FUNGI IN SEEDS

Survival of field fungi in a dormant state within seeds is favored by low moisture content and low temperature. Barley seeds moderately to rather heavily invaded by *Alternaria*, *Fusarium*, and *Helminthosporium* were stored in paper envelopes in the laboratory at "room" temperature (probably 25°–30°C). Moisture contents of the samples were not determined, but probably ranged between 9 and 12% on a wet weight basis. Periodically, kernels were surface-disinfected and plated on agar. After 20 months *Fusarium* grew from only a few kernels of seven of the 30 lots, whereas *Alternaria* survived in some kernels of all lots for 42 months, and *Helminthosporium* grew from some kernels in 18 of the 30 lots after 62 months (J. J. Christensen, 1963). Christensen (1951) recovered *Alternaria* from one kernel of a sample of wheat 8 years old.

Lutey and Christensen (1963) stored barley seed moderately to heavily invaded by *Alternaria, Fusarium,* and *Helminthosporium* at different moisture contents and temperatures, and periodically tested them for the presence of fungi. Storage at 14% moisture content and 20°C for 24 weeks resulted in death of all *Fusarium* and *Helminthosporium*, and great reduction in *Alternaria*, but with little increase in storage fungi and little reduction in germination percentage of the seed or in seedling vigor. Presumably the fungi respired sufficiently to use up some essential reserve foods, and so died. This probably is a general phenomenon: Seeds stored at moisture contents just below those necessary for germination rapidly lose viability. Lutey and Christensen suggested that by use of the right moisture content–temperature–time combination in storage it might be possible to rid seed stocks of undesirable pathogenic fungi. This still seems reasonable, but so far as I am aware it has not been tried on a commercial scale. It might be impossible to maintain the moisture content of large bulks within the narrow limits required, although with modern aeration equipment, moisture content and temperatures of large stocks of grain can be maintained fairly precisely.

## IV. STORAGE FUNGI

### A. MAJOR GENERA

The storage fungi comprise several "group" species of *Aspergillus*, a few species of *Penicillium* and, in some rather special environments, *Wallemia (Sporendonema) sebi, Chrysosporium inops,* and *Candida* or closely related forms of filamentous yeasts or yeastlike fungi. The species of *Aspergillus* and *Penicillium* involved are found in all kinds of grains and seeds and their products wherever these have been investigated; the major factor that determines which species will occur or predominate in a given situation is moisture content. *Chrysosporium inops*, on the other hand, has been found only on samples of corn taken from commercial storage in New Orleans and in San Juan, Puerto Rico, and stored in the laboratory at 15–16% moisture content in tightly closed containers where

mined that fungi were not involved in respiration of moist stored cotton-seed, because cottonseed treated with supposed fungicides (the "fungicides" that he used were not especially fungicidal under conditions of his tests, but he did not realize that, either) and then stored moist, still respired. Application of even an effective fungicide to the outside of cottonseed does not prevent fungi from growing within the seed.

D. MOISTURE CONTENT IN RELATION TO GROWTH AND SURVIVAL

By definition, storage fungi are those which grow at moisture contents in equilibrium with relative humidities of about 68–90%. Moisture contents (wet weight basis) of some common grains and seed in equilibrium with relative humidities in this range are given in Table I. These can be only approximate, because the equilibrium moisture content of a given kind of seed at a given relative humidity is affected by hysteresis, by the chemical composition of the seed, and by the relative proportion of the different tissues (in an individual kernel of corn, for example, the moisture content of the germ, which has a high oil content, is considerably lower, at a given relative humidity, than the moisture content of the endosperm, which has a low oil content).

The lower limits of relative humidity that permit growth of various storage fungi on predried strips of nutrient agar have been determined with considerable precision by Ayerst (1969). Minimum relative humidities that permit the more common fungi to develop on grains and seeds are given in Table II.

The biological complexities mentioned above make it impossible to

TABLE I

MOISTURE CONTENTS OF GRAINS AND SEEDS IN EQUILIBRIUM WITH
VARIOUS RELATIVE HUMIDITIES AT 25°–30°C

| Relative humidity (%) | Moisture content (%) | | | | | |
|---|---|---|---|---|---|---|
| | Wheat, corn, sorghum | Rice | | Soybeans | Sunflower | |
| | | Rough | Polished | | Seeds | Meats |
| 65 | 12.5–13.5 | 12.5 | 14.0 | 11.5 | 8.5 | 5.0 |
| 70 | 13.5–14.5 | 13.5 | 15.0 | 12.5 | 9.5 | 6.0 |
| 75 | 14.5–15.5 | 14.5 | 15.5 | 13.5 | 10.5 | 7.0 |
| 80 | 15.5–16.5 | 15.0 | 16.5 | 16.0 | 11.5 | 8.0 |
| 85 | 18.0–18.5 | 16.5 | 17.5 | 18.0 | 13.5 | 9.0 |

TABLE II

MINIMUM $A_w$ FOR THE GROWTH OF COMMON STORAGE FUNGI AT THEIR
OPTIMUM TEMPERATURE FOR GROWTH (26°–30°C)

| Fungus | Minimum $A_w$ for growth relative humidity |
|---|---|
| *Aspergillus halophilicus* | 0.68 |
| *A. restrictus, Wallemia (Sporendonema) sebi* | 0.70 |
| *A. glaucus* | 0.73 |
| *A. candidus, A. ochraceus* | 0.80 |
| *A. flavus* | 0.85 |
| *Penicillium* (depending on species) | 0.80–0.90 |

set very precise lower limits of moisture content for growth of these fungi in seeds. Added to this is the difficulty of precise determination of moisture content in seeds, which will be discussed below. At times, however, the fungi that develop in stored seeds provide considerable information about conditions of storage, including moisture content.

If, for example, we recover *Aspergillus halophilicus* from a sample of stored wheat, we can be certain that the wheat has been stored for at least 6 months or more at a moisture content of 13.8–14.3% and at a temperature of 20°–30°C, because only under those circumstances will *A. halophilicus* develop on wheat. We have never been able to isolate *A. halophilicus* from freshly harvested wheat, but have observed it developing on many samples of wheat taken from commercial bins and stored at the moisture content given above, and on many samples of some other kinds of seeds stored at equilvalent moisture contents. *Aspergillus halophilicus* will grow in agar media saturated with sodium chloride or sucrose, and at least some strains will not grow in agar media containing less than about 40% sucrose or 10% sodium chloride (Christensen *et al.*, 1959).

Similarly, if *A. restrictus* grows from surface-disinfected kernels of wheat or corn plated on agar (preferably, and sometimes necessarily, an agar medium containing 6–10% sodium chloride), we know that the grain has been stored for some months at a moisture content of 14–15%, regardless of the moisture content given on the warehouse records, and at a temperature of 20°–30°C, because *A. restrictus* will invade these grains only under those conditions. Similarly, if a considerable number of noninsect-injured kernels of corn from storage yield *A. flavus*, the moisture content of the grain at some time in storage must have been at least 18%, because *A. flavus* cannot grow at moisture contents below that.

E. INFLUENCE OF SMALL DIFFERENCES IN MOISTURE CONTENT ON GROWTH OF STORAGE FUNGI

In the range of moisture content between 14.0 and 15.0%, in winter wheat, a difference of only 0.2% can mean the difference between no or very light invasion by *A. restrictus*, and relatively heavy invasion with the accompanying development of brown germs (sick wheat) and reduction in quality (C. M. Christensen, 1963; Christensen and Linko, 1963). As will be seen below, differences that small cannot be detected by the procedures and meters used in grain warehouses and grain inspection offices. Representative data are given in Tables III and IV.

Concerning invasion of wheat by *A. ochraceus* Christensen (1962) stated, ''A difference of less than 1% in moisture had a great effect on the rate at which the fungus invaded the grain.'' The same probably is true of all of the storage fungi when growing near the lower limit of moisture content that permits them to develop.

## V. MEASUREMENT OF MOISTURE CONTENT

### A. METHODS

Moisture content of grains and seeds is impossible to measure precisely. Some of the water is physically or chemically bound to some of the constituents of the seeds and can be driven off only by temperatures above 100°C, and at those temperatures some of the volatile substances other than water are driven off also. In the laboratory, moisture content may be determined by a variety of means: drying in a circulating air oven, drying in a vacuum oven, drying by desiccants, distillation in toluene, or by chemical reaction as with the Karl Fischer reagent. In the grain trade, where speed is essential, moisture content is determined by an electric moisture meter, of which there are many makes and models. The one officially approved by the United States Department of Agriculture for use in grain inspection offices is the Motomco, Model No. 919. Other makes and models may give equally good results. Probably none of these meters can be expected to have an accuracy, on a given sample, of better than ±0.1%, as compared with oven-drying, and on some samples the error (if that is the correct word) may exceed ±0.5 or even ±1.0%.

Table V, from Hunt and Pixton (1974), summarizes the various ''official'' methods for determination of moisture contents of cereal grains and their products. Note that for whole kernel corn there are six different methods. These do not all give the same results, as shown in Table VI from unpublished data in our own laboratory. Moisture contents as de-

TABLE III

INFLUENCE OF SMALL DIFFERENCES IN MOISTURE CONTENT UPON THE
DEVELOPMENT OF *Aspergillus restrictus* AND *A. repens* IN HARD
RED WINTER WHEAT STORED VARIOUS LENGTHS OF TIME
AT A TEMPERATURE OF 20°–25°C[a]

| Storage time (days) | Moisture content (%) | Colonies per gram | | Surface-disinfected kernels | |
| | | *Aspergillus restrictus* (1000's) | *Aspergillus repens* (1000's) | Yielding *Aspergillus restrictus* (%) | Yielding *Aspergillus repens* (%) |
| --- | --- | --- | --- | --- | --- |
| Test No. 1 (three replicates at each moisture content) | | | | | |
| 36 | 14.4 | 16 | 0 | —[a] | — |
| | 14.9 | 223 | 0 | — | — |
| | 15.3 | 1320 | 0 | — | — |
| 68 | 14.5 | 226 | 0 | — | — |
| | 15.0 | 2330 | 0 | — | — |
| | 15.5 | 7216 | 0 | — | — |
| 170 | 14.7 | 12,930 | 0 | 96 | 4 |
| | 15.1 | 30,460 | 460 | 68 | 31 |
| | 15.5 | 20,930 | 1460 | 45 | 55 |
| Test No. 2 (two replicates at each moisture content) | | | | | |
| 76 | 14.2 | 1 | 0 | — | — |
| | 14.4 | 113 | 0 | — | — |
| | 14.6 | 457 | 0 | — | — |
| | 14.9 | 2100 | 0 | — | — |
| | 15.0 | 3150 | 0 | — | — |
| 153 | 14.3 | 1650 | 0 | 99 | 0 |
| | 14.4 | 5900 | 0 | 97 | 4 |
| | 14.5 | 9250 | 0 | 99 | 1 |
| | 14.9 | 15,250 | 0 | 73 | 26 |
| | 15.0 | 28,000 | 0 | 66 | 30 |

[a] From Christensen (1963).

termined on commercial grain samples by different meters and by oven-drying are summarized in Tables VII and VIII. From these data it should be evident that the methods of determining moisture content in grains and seeds are empirical and that the "true" moisture content is elusive. However, with wheat and corn, moisture contents as determined by drying in a circulating air oven at 103°C for 72 hr (wheat) or at 130°C for 24 hr (corn), and those determined by drying in a vacuum oven at 95°C to constant weight, and those determined by distillation in toluene will agree with ±0.1–0.2%. Different subsamples of the same sample will vary that much,

TABLE IV

Influence of Moisture Content upon Increase of Storage Fungi and
Brown Germs in Samples of Hard Red Winter Wheat Stored 16 Months
at Room Temperature[a]

| Moisture content (oven) (%) | No. of samples | Brown germs (%) | Surface-disinfected kernels yielding | | Colonies | |
|---|---|---|---|---|---|---|
| | | | *A. restrictus* (%) | Other spp. of *A. glaucus* (%) | *A. restrictus* (1000's/gm) | Other spp. of *A. glaucus* (1000's/gm) |
| 13.0–13.8 | 2 | 0 | 8 | 0 | 0 | 0 |
| 14.1–14.5 | 4 | 3 | 75 | 1 | 161 | 0.4 |
| 14.8–15.1 | 2 | 24 | 47 | 48 | 195 | 30 |
| 15.2–15.6 | 2 | 100 | 64 | 94 | 1920 | 442 |

[a] From Christensen and Linko (1963).

at least in the range of moisture contents with which we usually are concerned in storage. To insist that the moisture content of the grain in a given lot of 100,000 or 1,000,000 bushels should be or can be known within ±0.1 or 0.2% is unrealistic to the point of absurdity. Nevertheless it has been so insisted.

B. Range of Moisture Content in a Given Bulk of Seeds

Considerable effort has gone into development of sampling procedures to make sure that a representative sample is obtained from a given seed lot, for purposes of inspection or testing of different characteristics, including moisture content, that determine the commercial grade, and therefore the price, of the lot. The effort devoted to obtaining a truly representative sample is tacit recognition that the moisture content, as well as other characteristics, may vary from place to place in the bulk, a recognition that nonuniformity, not uniformity, prevails. This elementary fact, however, if it is recognized, certainly is not given the weight it deserves by many practical grain merchants and warehousemen. As noted above in the discussion of *Aspergillus restrictus*, in the range of moisture content between 14.0 and 15.0%, a difference of only 0.2% can make a considerable difference in the degree of fungus invasion in a given storage period. In the starchy cereal seeds, in general, a difference of ±0.5% in moisture content, in the range between 14.5–15.0 and 16.5–17.0%, can mean the difference between safe storage and spoilage. Yet many of those who deal in storage or transport of lots of a million bushels or more

## Official Methods of Various Countries and Technical Societies for Moisture Determination of Cereal Grains and Their Products[a,b]

| Crop | AACC[c] | AOAC[d] | AOCS[e] | ASBC[f] | USDA[g] | U.S.S.R. | Canada | England | France | ICC[h] | ISO[i] | CEE[j] |
|---|---|---|---|---|---|---|---|---|---|---|---|---|
| Barley | 9,18,27 | 9,18 | | 14,17 | 9,27 | 6,8,19 | 9,27 | 2,9,10 | 5,10,22 | 10,22 | 10,22 | 10 |
| Beans, dry, edible | 3,27 | | | | 3,27 | | 3,27 | 2,9,24,25 | | | | |
| Brewers grains | | | | 1[k] | | | | 2,9 | | | | |
| Buckwheat | | 9,18 | | 1 | | 6,8,19 | 9,27 | 2,9,10 | | | | |
| Cereal adjuncts | 18 | | | 1 | | | 1,18 | | | | | |
| Corn, whole | 3,27 | 9,18 | | | 3,27 | 6,8,19 | 3,27 | 2,9,10 | 13 | | | |
| Corn, ground | 9,18 | | | | | 6,8,19 | | | 5,12,22 | 12,22 | 12,22 | 12 |
| Corn gluten feed | | 14,17,21,25 | | | | | | 2,4,17,24,25 | 4,15 | | | 4,15 |
| Corn-oil meal | | 14,17,21,25 | | | | | | 2,4,9,17,24,25 | 4,15 | | | 4,15 |
| Feed and feeding stuffs | 18 | 14,17,21,25 | | | | | | 4,17,24,25 | 4 | | 4 | 4,15 |
| Flaxseed, whole | 2,27 | | 11 | | 3,27 | | 11,27 | 9,10 | 5,10,22 | 10,22 | 10,22 | 10 |
| Flour | 9,12,18 | 9,18 | | | 9 | 6,8,19 | | 5,23 | 5,10,22 | | | |
| Grain | 9,18 | | | | 9,27 | | 9,27 | | | | | |
| Malt | 1 | | | 1 | | | | | | | | |
| Oats, ground | 9,27 | 9,18 | | | 9,27 | 6,8,19 | 9,27 | 9,10 | 5,10,22 | | | |
| Peas and lentils | 9,27 | | | | 9,27 | 8,19 | | 2,9,24,25 | 4 | | 4 | |
| Rice | 9,27 | 9,18 | | | 9,27 | 6,8,19 | | 9,10 | 5,10,22 | 10,22 | 10,22 | 10 |
| Rye | 9,27 | 9,18 | | | 9 | 6,8,19 | | 9,10 | 5,10,22 | 10,22 | 10,22 | 10 |
| Semolina | 9,18 | 9,18 | | | 9,27 | 6,8,19 | | 9,10 | 5,10,22 | 10,22 | 10,22 | 10 |
| Soybeans | 9,27 | | 11 | | 9 | | 9,27 | 2,9,10,24,25 | 4 | | 4 | 4 |
| Safflower seed, whole | | | | | | | | 2,4,17,24,25 | | | | |
| Sunflower seed, whole | | | | | 11 | 5 | | 2,4,17,24,25 | 4 | | 4 | |
| Wheat, ground | 9,18,27 | 9,18 | | | 9,27 | 6,8,19 | 9,27 | 9,10 | 5,10,22 | 10,22 | 10,22 | 10 |

[a] From Hunt and Pixton (1974).

[b] Key to numbers in table: 1. Air oven, 103° to 106°C for 3 hr. 2. Air oven, 103°C for 4 hr. 3. Air oven, 103°C for 72 hr. 4. Air oven, 100° to 105°C for 3 hr, plus 1-hr intervals to constant weight. 5. Air oven, 102° for 17 hr, constant weight-corrective for relative humidity [CNERNA (Centre National d' Etudes et Recherches en Nutrition et Alimentation)]. 6. Air oven, 105°C for 30 min, plus 130°C for 40 min. 7. Air oven, 105°C for 30 min. 8. Air oven, 130°C for 40 min. 9. Air oven, 130°C for 1 hr. 10. Air oven, 130°C for 2 hr. 11. Air oven, 130°C for 3 hr. 12. Air oven, 130°C for 4 hr. 13. Air oven, 130°C for 38 hr. 14. Air oven, 135°C for 2 hr. 15. Vacuum oven, 70°C at 10 mm. Hg pressure-constant weight (CEE). 16. Vacuum oven, 100°C for 3 hr. 17. Vacuum oven, 95° to 100°C for 5 hr, or constant weight. 18. Vacuum oven, 98° to 100°C for 5 hr, or constant weight. 19. Vacuum oven, 105°C for 30 min, plus 130°C for 1 hr. 20. Vacuum oven, 80°C for 20 hr. 21. Vacuum desiccator, no heat, to constant weight. 22. Glass drying tube, at 10 to 20 mm. Hg pressure, temperature of 45° to 50°C, with $P_2O_5$ as desiccant (CNERNA, ISO, and ICC Basic Method). 23. Aluminum plate, 140°C for 15 min. 24. Bidwell-Sterling, benzene. 25. Bidwell-Sterling, toluene. 26. Karl Fischer titration method. 27. Model 919 moisture meter (Motomco, CAE, Halross).

[c] American Association of Cereal Chemists.
[d] Association of Official Analytical Chemists.
[e] American Oil Chemists' Society.
[f] American Society of Brewing Chemists.
[g] United States Department of Agriculture.
[h] International Association for Cereal Chemistry.
[i] International Standardization Organization.
[j] European Economic Community.
[k] Predried at 50° to 60°C.

TABLE VI
MOISTURE CONTENTS OF SAMPLES OF CORN, OF DIFFERENT BEGINNING MOISTURE
CONTENTS, AS DETERMINED BY DIFFERENT METHODS[a]

| | Original moisture content (%) | | |
|---|---|---|---|
| Method | 10.5–11.0 | 14.0–14.5 | 17.5–18.0 |
| 1. Motomco meter (whole kernel) | 10.57 | 14.25 | 17.77 |
| 2. Circulating air oven (whole kernel) | | | |
| 130°C | | | |
| 1 hr | 6.01 | 11.47 | 15.69 |
| 3 hr | 7.74 | 13.03 | 17.27 |
| 24 hr | 10.45 | 14.45 | 17.59 |
| 103°C, 72 hr | 9.73 | 15.15 | 18.94 |
| 3. Vacuum over (whole kernel) | | | |
| 85°–90°C | | | |
| 6 hr | 6.52 | 13.22 | 17.20 |
| 12 hr | 8.37 | 14.45 | 18.01 |
| 24 hr | 10.33 | 15.58 | 19.42 |
| 4. Two-stage[b] | 10.33 | 14.92 | 19.32 |

[a] M. A. Sellam and C. M. Christensen, unpublished data.
[b] Air-dried, then ground in Wiley Experimental Mill through 20-mesh sieve and replicate samples dried 1 hr at 130°C.

assume that the moisture content of the grain is uniform throughout. This is a highly unlikely circumstance and one that, if it does prevail at any one time, is not likely to do so for long. When dealing with large bulks of grain the sensible thing is to take a number of samples and determine the moisture content of each one separately. Good warehousemen do this,

TABLE VII
MOISTURE CONTENTS AS DETERMINED BY METERS AND BY OVEN-DRYING
OF HARD RED WINTER WHEAT[a]

| | Meter | | | | |
|---|---|---|---|---|---|
| Sample | Motomco (%) | Radson (%) | Steinlite (%) | Weston (%) | Oven-drying (%) |
| 1 | 14.20 | 14.20 | 13.31 | 13.51 | 14.2 |
| 2 | 13.70 | 13.40 | 13.85 | 12.70 | 14.3 |
| 3 | 14.30 | 14.00 | 13.44 | 12.76 | 14.5 |
| 4 | 13.30 | 14.60 | 16.35 | 13.97 | 14.5 |
| 5 | 14.77 | 14.50 | 15.00 | 13.52 | 15.1 |

[a] In St. Louis, Missouri, 1960 crop.

TABLE VIII

MOISTURE CONTENTS AS DETERMINED BY METERS AND BY OVEN-DRYING
OF SOFT RED WINTER WHEAT[a]

|  | Meter | | | | |
|--------|--------------|-------------|-----------------|----------------|--------------------|
| Sample | Motomco (%) | Radson (%) | Steinlite (%) | Weston (%) | Oven-drying (%) |
| 1 | 13.04 | 13.50 | 13.05 | 11.75 | 12.5 |
| 2 | 13.99 | 14.50 | 14.20 | 13.40 | 13.6 |
| 3 | 14.34 | 14.70 | 14.50 | 13.43 | 13.8 |
| 4 | 14.20 | 14.50 | 14.70 | 13.43 | 13.9 |
| 5 | 14.90 | 14.90 | 15.05 | 14.00 | 15.5 |

[a] In New Orleans, 1961 crop.

and such repeated sampling and testing are also the rule in loading of large bulks of grain for export. That is, for buying and selling of grains, it is important to know the moisture content of a representative sample, but for avoidance of loss in storage and transport it is important to know the range of moisture content throughout the bulk. As indicated above, moisture that is uniformly distributed throughout a given bulk today may not be uniformly distributed throughout the bulk tomorrow. This is discussed below.

## C. MOISTURE TRANSFER OR MIGRATION

It has long been known that moisture within bulk stored products may shift with time, but there is relatively little information on the rate and magnitude of such transfer. Such transfer depends largely on the moisture content of the material (a lot or parcel of wheat of 100,000 bushels at 15% moisture contains 90,000 lb water, whereas at 11.0% moisture it contains 66,000 lb water) and on the magnitude of temperature differences between different portions of the bulk. Sometimes this moisture transfer can be surprisingly rapid. Johnson (1957) calculated that in corn with 14.5% moisture when stored, at a temperature of 85°F (29.4°C), enough moisture could be transferred in 21 days, as the top layer cooled, to raise the moisture content of a top layer 6 in. deep to 20%. This increase in moisture content of the grain in the center of the top is the more or less "standard" picture in grain stored through the winter, because of the pattern of air circulation through the grain mass—downward along the cool sides and upward through the warm center. Holman (1950) reported that soybeans stored in November, 1942, with an average moisture content of 12–13% had, by February, 1943, a moisture content of 16–17% in the center of the upper surface of the bulk. By February, 1944, after 15 months of storage, the moisture content of the beans in the center of the upper surface of the bulk ranged from 20 to 24%. Christensen (1970) stored sorghum seeds of 14.3% moisture in a gallon glass jug and maintained a temperature of 10°–15°C higher in the grain on one side than in the grain at the opposite side. After only 3 days, the moisture content of the grain on the cool side was 1.4% higher than that of the grain on the warm side; after 6 days the difference was 2.0%. Sellam and Christensen (1977) stored corn of different moisture content in plastic boxes in the laboratory and maintained temperature differences of approximately 5° and 10°C between the warm and the cool ends of the boxes. Periodically samples were removed and tested for moisture content and fungus invasion. Representative data from these tests are given in Table IX.

D. GAIN AND LOSS OF MOISTURE

Grain exposed to air of lower or higher relative humidity than that in equilibrium with its moisture content at the time will lose or gain moisture and, of course, weight. In most practical situations this loss or gain is so small and slow that it is of relatively little importance or concern. However, when grains in commerce are transferred into and out of warehouses, as all grains in commerce are, usually at least several times, they may lose or gain a detectable amount in weight. Not uncommonly a given bulk of wheat or corn will, within a period of 6 months, or during its storage life, be transferred at monthly intervals, in addition to the transfers when it is entering or leaving the elevator. If the grain has a moisture content of 12–14% when it arrives, and is transferred several times through air of 85–90% relative humidity, which often prevails in the southern ports of the United States such as Houston, Port Allen, Port Destrehan, and New Orleans, it may absorb enough moisture to gain 0.1 or 0.2% in weight. If several hundred million bushels are involved, with a price of several dollars per bushel, a gain or loss of 0.2% can be of considerable importance.

In laboratory tests to explore this phenomenon, 500-gm portions of grain of different initial moisture contents were placed in a balance scoop 13.5 cm wide, 30 cm long, and 6 cm deep. The surface:depth ratio approximated that of grain on a warehouse transfer belt. The scoop containing the grain was put into a climate chamber in which the relative humidity was maintained at 85–90%, with a temperature of 30°C. Weight gains were determined at intervals of 20 minutes. Representative results are given in Table X. At this time, tests are under way to determine whether gains of

## TABLE IX

Moisture Content of Warm and Cool Side of Corn (Grade II) Subjected to Different Degrees of Temperature

| Time of exam. (days) | 13.5% | | | | 15.5% | | | | 17.5% | | | |
|---|---|---|---|---|---|---|---|---|---|---|---|---|
| | 5°C ± 1 | | 10°C ± 1 | | 5°C ± 1 | | 10°C ± 1 | | 5°C ± 1 | | 10°C ± 1 | |
| | Warm | Cool | Warm | Cool | Warm | Cool | Warm | Cool | Warm | Cool | Warm | Cool |
| 2 | 13.13 | 13.87 | 12.90 | 14.04 | 14.71 | 15.80 | 14.44 | 16.66 | 16.13 | 17.14 | 14.81 | 18.03 |
| 4 | 12.05 | 14.76 | 12.68 | 14.21 | 14.40 | 16.16 | 13.66 | 17.93 | 14.98 | 18.05 | 14.01 | 18.48 |
| 6 | 12.51 | 13.87 | 12.27 | 14.37 | 14.12 | 16.22 | 13.45 | 17.37 | 14.99 | 18.11 | 14.04 | 18.56 |
| 8 | 12.41 | 13.92 | 12.01 | 14.98 | 14.58 | 16.20 | 13.38 | 17.73 | 14.65 | 18.34 | 14.38 | 18.34 |
| 15 | 11.54 | 14.58 | 10.89 | 15.53 | 13.43 | 16.78 | 11.82 | 19.23 | 14.48 | 18.92 | 13.03 | 21.13 |

TABLE X

GAIN IN WEIGHT OF 500-G/SAMPLES OF GRAIN AND GRAIN PRODUCTS
EXPOSED TO A RELATIVE HUMIDITY OF 85–90% AT 30°C[a]

| Material | Beginning moisture (%) | Percentage gain according to exposure time (min) | | |
|---|---|---|---|---|
| | | 20 | 40 | 60 |
| Grade 2, yellow corn | 13.5 | 0.17 | 0.19 | 0.26 |
| White wheat | 11.7 | 0.13 | 0.20 | 0.29 |
| Hard red winter wheat | 9.70 | 0.08 | 0.23 | 0.25 |
| Soybean meal | 11.2 | 0.15 | 0.22 | 0.27 |
| Alfalfa pellets | 9.0 | 0.16 | 0.36 | 0.38 |

[a] Each figure is an average of two to three replicates.

the same magnitude occur in commercial practice, but data from these tests are not yet available. However, in some situations it seems highly probable that the weight of a given bulk of grain may fluctuate measurably from gain or loss of moisture during relatively brief exposure to air of high or low relative humidity.

REFERENCES

Ayerst, G. (1969). The effects of moisture and temperature on growth and spore germination in some fungi. *J. Stored Prod. Res.* 5, 127–141.

Barton, L. V. (1961). "Seed Preservation and Longevity." Wiley (Interscience), New York.

Beals, W. J. (1905). The viability of seeds. *Bot. Gaz. (Chicago)* 38, 140–143.

Christensen, C. M. (1951). Fungi on and in wheat seeds. *Cereal Chem.* 28, 408–415.

Christensen, C. M. (1962). Invasion of stored wheat by *Aspergillus ochraceus. Cereal Chem.* 39, 100–106.

Christensen, C. M. (1963). Influence of small differences in moisture content upon the invasion of hard red winter wheat by *Aspergillu restrictus* and *A. repens. Cereal Chem.* 40, 385–390.

Christensen, C. M. (1965). Fungi in cereal grains and their products. *In* "Mycotoxins in Foodstuffs" (G. N. Wogan, ed.) pp. 9–14. Mass. Inst. Tech. Press.

Christensen, C. M. (1970). Moisture content, moisture transfer, and invasion of stored sorghum seeds by fungi. *Phytopathology* 60, 280–283.

Christensen, C. M., and Kaufmann, H. H. (1969). "Grain Storage: The Role of Fungi in Quality Loss." Univ. of Minnesota Press, Minneapolis.

Christensen, C. M., and Linko, P. (1963). Moisture contents of hard red winter wheat as determined by meters and by oven drying, and influence of small differences in moisture content upon subsequent deterioration of the grain in storage. *Cereal Chem.* 40, 129–137.

Christensen, C. M., Olafson, J. H., and Geddes, W. F. (1949). Grain storage studies. VIII. Relation of molds in moist stored cottonseed to increased production of carbon dioxide, fatty acids, and heat. *Cereal Chem.* 26, 109–128.

Christensen, C. M., Papavizas, G. C., and Benjamin, C. R. (1959). A new halophilic species of *Eurotium*. *Mycologia* **51**, 636–640.

Christensen, C. M., Nelson, G. H., Speers, G. M., and Mirocha, C. J. (1973). Results of feeding tests with rations containing grain invaded by a mixture of naturally present fungi plus *Aspergillus flavus* NRRL 2999. *Feedstuffs* **45**, 20.

Christensen, J. J. (1963). Longevity of fungi in barley kernels. *Plant Dis. Rep.* **47**, 639–642.

Harrington, J. F. (1972). Seed storage and longevity. *In* "Seed Biology" (T. T. Kozlowski, ed), Vol. 3, pp. 145–245. Academic Press, New York.

Holman, L. H. (1950). Handling and storage of soybeans. *In* "Soybeans and Soybean Products" (K. S. Markley, ed.), pp. 455–482. Wiley (Interscience), New York.

Hunt, W. H., and Pixton, S. W. (1974). Moisture—its significance, behavior, and measurement. *In* "Storage of Cereal Grains and Their Products" (C. M. Christensen, ed.), pp. 1–55. Am. Assoc. Cereal Chem., St. Paul, Minnesota.

Hyde, M. B. (1950). The subepidermal fungi of cereal grains. 1. A survey of the world distribution of fungal mycelium in wheat. *Ann. Appl. Biol.* **37**, 179–186.

Hyde, M. B., and Galleymore, H. B. (1951). The subepidermal fungi of cereal grains. II. The nature, identity and origin of the mycelium in wheat. *Ann. Appl. Biol.* **38**, 348–356.

Johnson, H. K. (1957). Cooling stored grain by aeration. *Agric. Eng.* **38**, 238–241 and 244–246.

Kaufmann, H. H. (1959). Fungus infection of grain upon arrival at terminal elevators. *Cereal Sci. Today* **4**, 13–15.

Kivilaan, A., and Bandurski, R. S. (1973). The ninety-year period for Dr. Beal's seed viability experiment. *Am. J. Bot.* **60**, 140–145.

Kotheimer, J. B., and Christensen, C. M. (1961). Microflora of barley kernels. *Wallerstein Lab. Commun.* **24**, 21–28.

Lillehoj, E. B., Kwolek, W. F., Shannon, G. M., Shotwell, O. L., and Hesseltine, C. W. (1975). Aflatoxin occurrence in 1973 corn at harvest. I. A limited survey in the southeastern U. S. *Cereal Chem.* **52**, 603–611.

Lutey, R. W., and Christensen, C. M. (1963). Influence of moisture content, temperature, and length of storage period upon survival of fungi in barley kernels. *Phytopathology* **53**, 713–717.

Malone, J. P., and Muskett, A. E. (1964). Seed-borne fungi. Descriptions of 77 fungus species. *Proc. Int. Seed Test. Assoc.* **29**, 179–384.

Qasem, S. A., and Christensen, C. M. (1958). Influence of moisture content, temperature, and time on the deterioration of stored corn by fungi. *Phytopathology* **48**, 544–549.

Sellam, M. A., and Christensen, C. M. (1977). Temperature differences, moisture transfer, and spoilage in stored corn. (In press.)

Spencer, E. R. (1921). Decay of Brazil nuts. *Bot. Gaz. (Chicago)* **72**, 265–292.

Taubenhaus, J. J. (1920). A study of the black and yellow molds of ear corn. *Tex., Agric. Exp. Stn., Bull.* **270**.

Tuite, T. F. (1959). Low incidence of storage molds in freshly harvested seed of soft red winter wheat. *Plants Dis. Rep.* **43**, 470.

Tuite, J. F. (1961). Fungi isolated from unstored corn seed in Indiana in 1956–1958. *Plant Dis. Rep.* **45**, 212–215.

Tuite, J. F., and Christensen, C. M. (1955). Grain storage studies. XVI. Influence of storage conditions upon the fungus flora of barley seed. *Cereal Chem.* **32**, 1–11.

Tuite, J. F., and Christensen, C. M. (1957). Grain storage studies. XXIII. Time of invasion of wheat seed by various species of Aspergillus responsible for deterioration of stored grain, and source of inoculum of these fungi. *Phytopathology* **47**, 265–268.

CHAPTER 8

# MOISTURE AND DETERIORATION OF WOOD

## D. W. French and C. M. Christensen

DEPARTMENT OF PLANT PATHOLOGY, UNIVERSITY OF MINNESOTA, ST. PAUL,
MINNESOTA

# I. INTRODUCTION

This chapter summarizes information on deterioration of wood and wood products by fungi, with special emphasis on the relation of moisture to the processes involved. Under some circumstances a few kinds of wood-inhabiting bacteria may digest portions of wood sufficiently to weaken the wood, but as agents of deterioration they are of minor significance compared with fungi. Also under some circumstances, as with roofs or buildings shaded by trees in humid climates, the wood surface may be overgrown by algae, mosses, and lichens, but ordinarily these organisms are not considered to be primary causes of deterioration of wood.

# II. KINDS OF DETERIORATION

## A. DECAY

Decay is by far the most important in deterioration of wood caused by fungi, and most of the chapter will be devoted to decay of wood and wood products in use. The decay of heartwood in living trees, also caused by fungi, is outside the scope of this chapter, and will be mentioned only in passing. As the term is used here, decay means actual consumption of the wood substances by fungi. Characteristics of different types of decay are described below.

## B. SOFT ROT

This term was coined to describe a condition that develops when wood in use is continually kept wet, as in cooling towers (Savory, 1954). The surface of the wood is invaded by a large number of saprophytic fungi, mainly Fungi Imperfecti, and probably also by bacteria, and is partly digested so that it becomes mushy. Soft rot occurs in marine habitats and is caused by similar fungi.

## C. STAIN

The sapwood of newly cut logs or of freshly cut lumber of many kinds of trees is susceptible to invasion by blue stain fungi. These comprise a somewhat special group. Many fungi with pigmented mycelium or spores or those that produce a pigment in the substrate in or on which they are growing can grow on the surface of wood and wood products and cause staining. In wood in the form of lumber this staining is likely to be only superficial, but in products such as fiberboard it may extend beneath the surface. The "foxing" of books is an example of staining by fungi.

## D. WEATHERING

Wood can be discolored by weathering, a process that can camouflage some forms of degradation by microorganisms. Moisture plays a role in weathering of wood along with ultraviolet radiation and in places the abrasive action of soil particles carried by the wind can speed the weathering process. Normally weathering is a slow process affecting only the exposed surfaces.

## II. ORGANISMS RESPONSIBLE

### A. DECAY FUNGI

Most decay of wood is caused by fungi of two orders, Aphyllophorales and Agaricales, in the Basidiomycetes (Fig. 1). Some species of the Tremellales, also in the Basidiomycetes, cause decay, but mostly of wood or bark of fallen trees in the forest, not of manufactured wood products. Some ascomycetes also are adapted to a life in wood, but they seldom cause any economically significant decay of wood products in use (Fig. 2). The major wood-rotting fungi, in the Aphyllophorales and Agaricales, are specialists to the extent that they are adapted to living on

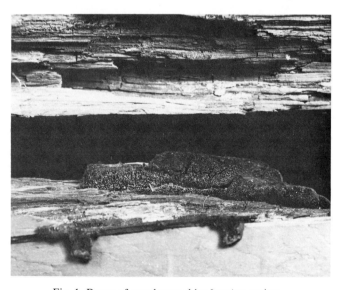

Fig. 1. Decay of wood caused by *Lenzites trabea*.

wood. Their role in the economy of nature is to consume wood, and to recycle it. Many of them are further specialized, so that they will grow on and decay wood of only certain kinds or in special situations. *Merulius lacrymans*, for example, is an economically important cause of decay in houses and other buildings in Europe and Great Britain. The decay it causes is called "dry rot"—a misnomer in that the wood undergoing decay is moist, not dry. If a house is certified to be free of dry rot, this means that it is free of rot caused by *M. lacrymans*. That houses are so certified when they are bought and sold indicates the prevalence of this kind of decay. Yet *M. lacrymans* has never been found in nature. It must, of course, occur in nature, but evidently it is so inconspicuous that it never has been detected. In buildings it is extremely conspicuous, often taking over entire homes to the point where they simply are abandoned.

Some hundreds of species of fungi cause decay in the heartwood of

Fig. 2. Sporocarps of *Daldinia concentrica*, a fungus species in the Ascomycetes which can deteriorate wood.

living trees, some of them being so specialized that they will grow in the heartwood of only a single species of tree. Few of these fungi cause decay in wood in use; they are specialists, adapted to the life they lead. Among the major genera of fungi that decay wood or wood products in use are *Poria* and *Polyporus* in the Polyporaceae, *Lentinus* (Fig. 3) and *Lenzites* in the Agaricaceae (*Lenzites* has gill-like pores, or porelike gills, and so is placed in both the Polyporaceae and Agaricaceae). *Coniophora*, in the family Thelephoraceae of the order Aphyllophorales, also is an important cause of decay of wood in use. Those five genera probably include more than 70% of the species of fungi that are of importance in decay of wood. Unfortunately, the fungi in these groups have in recent years been undergoing taxonomic rearrangements. Many of the generic and specific names that had been in use in many lands for a century or more have been changed, necessitating the learning of both the old and the new systems (Johansen and Smith, 1974; Käärik, 1976). As new wood products come into use, fungi that previously were of little significance may become major agents of deterioration.

## B. Soft Rot Fungi

Soft rot is caused mainly by Ascomycetes and Fungi Imperfecti. Common genera are *Bispora*, *Chaetomium*, *Coniothyrium*, *Phialophora*, *Stysanus*, and *Trichurus* (Savory, 1955).

Fig. 3. Sporocarp of *Lentinus lepideus* growing on the side of a log cabin.

## C. Stain Fungi

The blue stain or sap stain fungi, as indicated above, constitute a somewhat special group. Many of them are in the genus *Ceratocystis*, in the Ascomycetes, with conidial or imperfect stages in the genus *Graphium* (some of them have more than one conidial or imperfect stage). Some of these are associated with bark beetles, which carry them into the sapwood of the trees or logs they inhabit. Other blue staining fungi are adapted to the sapwood of freshly cut lumber (Fig. 4). All of them grow almost exclusively in the cell lumens of the wood they inhabit. They penetrate from cell to cell through extremely small boreholes, and so weaken the wood very little or not at all (Fig. 5). For wood in most uses their effect is mainly esthetic. For wood in some uses, as in rustic cabins, for example, blue-stained wood may be very attractive.

Many fungi with dark-colored mycelium or spores, or both, cause superficial or deep-seated staining of wood. Two very common genera are *Cladosporium* and *Aureobasidium* (also referred to as *Pullularia*) (Fig. 6).

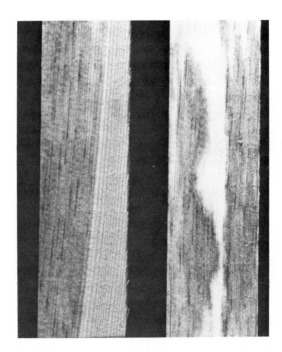

Fig. 4. Blue stain in the sapwood portion of pine boards.

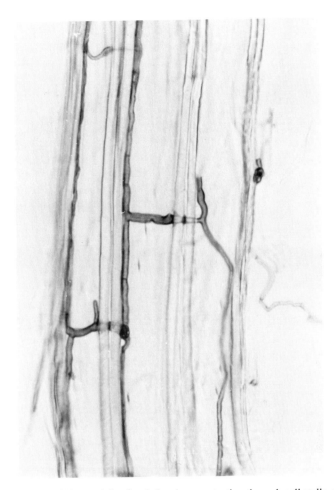

Fig. 5. Hyphae of blue-staining fungi showing penetration through cell walls. (Photo by Robert Campbell.)

Both of these have dark mycelium and spores, and can grow in a great variety of materials, including most kinds of wood and wood products. They also grow in and cause spotting of paints and coatings. *Aspergillus restrictus* can grow in materials whose moisture content is in equilibrium with relative humidities of about 70–75%. It does not require free water to grow; some strains of it, in fact, are inhibited by free water. It was found to cause spotting of coated fiberboard used in regions of periodic or regular high humidity (French and Christensen, 1958).

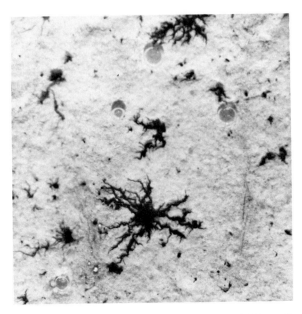

Fig. 6. Mycelium and spores of *Aureobasidium pullulans* which have discolored a paint surface.


## IV. MEASUREMENT OF MOISTURE CONTENT

Measurement of moisture content of wood, like the measurement of moisture content of most other biological materials, is to some extent arbitrary, in that different procedures and different drying schedules give somewhat different results. In general, moisture content is determined by drying in an air oven or in a vacuum oven, by distillation in a nonwater-miscible liquid such as toluene, or by electric meters. Each method has its virtues and limitations. As a general rule meters are less accurate than oven drying in determining moisture content of wood, but they have the virtues of being usable in the field, of rapidity, and of not requiring removal of samples from the materials whose moisture content is being measured.

Electric moisture meters operate on the principle that electrical resistance in wood changes with moisture content. Two major kinds of meters are in use today, the resistance and dielectric types. As a general rule, these meters provide reasonably accurate readings at moisture content below fiber saturation. At higher moisture content above 30% the meters provide only qualitative readings (James, 1975).

## V. DISTRIBUTION OF WATER IN WOOD

When the cell walls of wood have absorbed all the water they can hold, but no free water occurs in the cell cavities, the wood is said to be at the "fiber saturation point." The quantity of water present at the fiber saturation point amounts to about 30% of the dry weight of the wood (24–32%, depending on the wood). Wood with less water present than required to saturate the cell walls will not decay. Some fungi will grow in a variety of materials whose moisture contents are in equilibrium with relative humidities of about 68–80%. Some of these may cause spotting of wood products such as fiberboards (French and Christensen, 1958, 1969), but grow on the coatings of the boards, not in the wood fibers; they usually do not cause decay. Some of these species over a long period of time, however, could cause small losses in strength and weight of fiberboards (Merrill et al., 1968).

## VI. CHARACTERISTICS OF DETERIORATED WOOD

### A. DECAY

As mentioned above, wood decay fungi actually consume the wood that they inhabit. Wood is their food of choice, and furnishes them with all of the ingredients they need to grow, except water and oxygen. Some of them cause severe loss in strength of the wood before the wood is visibly altered. The wood may appear sound, but if it is used where strength is an important characteristic, as in supporting members of a structure or in a ladder or as scaffolding, it may fail, sometimes with dire consequences. Other decay fungi cause at least moderate and obvious decay before some of the strength properties are much affected. In practice, various strength tests, particularly resistance to impact breaking, are used as a measure of weakening by decay.

Traditionally, wood decays have been characterized as either white rots or brown rots, a perhaps crude but useful grouping. Wood decayed by white-rotting fungi will, at least in the early stages of decay, retain its shape and will not crack across the grain, whereas wood decayed by brown-rotting fungi will crack into somewhat cubical portions, similar to the cracks that appear in mud as it dries (Fig. 7). White-rotting fungi cause a progressive thinning of the secondary wall, from the lumen toward the middle lamella (Wilcox, 1968). The brown-rotting fungi remove primarily the carbohydrate fraction and supposedly this accounts for the brown color of the decayed wood. These fungi do not cause a gradual thinning of the cell walls but have the ability to attack unexposed surfaces. The

Fig. 7. Brown cubical rot showing the characteristics of a fungus classed as a brown rotter.

enzymes of these fungi move throughout the cell walls and their activity from cell to cell varies. The brown-rotting fungi in early stages of their attack can cause substantial amounts of strength loss before their presence is evident, based on macroscopic observations. In advanced stage the wood can be crumbled between the fingers. In wood and wood products in use, brown rots are far more common than white rots.

### B. Soft Rots

The soft rots often decay the superficial surfaces of wood substrates that are quite wet, such as the lining of cooling towers. These fungi also cause deterioration of wood in salt water such as quays, docks, and piers. The presence of the soft-rotting fungi can be identified by diamond-shaped cavities in the cell walls of affected wood. Unfortunately for those who desire distinct categories, the diamond-shaped cavities can be caused by some of the true wood-rotting fungi. The soft-rotting fungi apparently are intermediate in their capacity to destroy wood, being somewhere between the true wood-rotting fungi and the stain fungi (Fig. 8).

### C. Wood Stains

In the blue stains or sap stains caused by species of *Ceratocystis* and related fungi, the mycelium is restricted almost entirely to the cells of the ray parenchyma, and the hyphae pass from cell to cell through the natural pits in the wall. The mycelium of some of the blue stain fungi grows in the tracheids of the coniferous woods, and penetrates from cell to cell directly

Fig. 8. Soft-rotting fungi have shortened the handle of the brush stored in the block of wood.

through the cell wall, by means of very small bore holes. Fungi such as *Cladosporium* and *Aureobasidium* grow mainly on the surface of the wood, although in products such as fiberboard or paper they penetrate through the substrate.

## D. FUNGUS SUCCESSION

Decay of wood or wood products in use is likely to involve a succession of fungi, although one or a few of these may cause most of the decay. Wood substrates in living trees and to some extent in products are often invaded first by nonwood-rotting fungi in the Ascomycetes and Fungi Imperfecti. These pioneer organisms utilize some of the stored materials in wood rays, causing their partial destruction, thus providing easier access for subsequent invasion by true wood-rotting fungi. These pioneer fungi may alter the wood in subtle ways providing a better substrate for succeeding fungi. Bacteria can act in a similar manner. Some woods are by nature highly susceptible to attack by fungi. These decay-susceptible woods, such as aspen or balsam fir, contain no toxic extractives in the heartwood, and are invaded readily by a great variety of fungi, including those capable of causing decay. Other woods, such as cypress, redwood, cedars, catalpa, and white oak, to name a few, contain toxic materials in the heartwood, and so are resistant to invasion by fungi, including fungi that cause decay. They are resistant in various degrees to decay, but are not immune. Once decay-causing fungi are established in these decay-resistant woods, they may be accompanied or followed by other fungi, and also by bacteria. Malt extract agar or other mildly acid agar is used for isolation of wood rotting fungi from wood because it is somewhat unfavorable for growth of bacteria. Even so, bacteria very commonly are isolated, along with typical decay fungi, from decayed wood, even in the early stages of decay. If the bacteria are so regularly associated with the decay fungi in wood, they presumably play some role in the decay process, but the nature of this association has not been fully studied. Wood in the last stages of decay is a favorite substrate for slime molds, or Myxomycetes. Slime molds sometimes fruit in abundance on window sills or near bathtubs or toilets or sinks in homes, to the consternation of the inhabitants. Their appearance is positive evidence of high moisture contents and possibly decay in the wood from which they grew.

## VII. THE ROLE OF MOISTURE IN DETERIORATION

### A. DECAY

All wood-rotting fungi require free water to grow—water, that is, in excess of that required to reach the fiber saturation point. Wood that is kept at a moisture content below the fiber saturation point may be superficially invaded by xerophytic fungi such as the *A. restrictus* mentioned above, but it will not be decayed. Really dry wood is immune from attack by decay-causing fungi; in the absence of fire it should endure forever. Too much water also limits decay, either by excluding oxygen or by diluting out the digestive enzymes and acids so that they are inoperative. Table I lists the moisture contents inhibitory of decay in woods of different specific gravity.

### 1. Different Moisture Requirements

Although all wood-rotting fungi require free water to grow, they differ in what for a better term might be called their water economy. *Lenzites trabea*, for example, is a very common cause of wood decay in posts, poles, and lumber and wood construction items of all kinds. Once it is

TABLE I

UPPER LIMIT OF MOISTURE CONTENT AT WHICH DECAY CAN OCCUR
IN WOOD OF DIFFERENT SPECIFIC GRAVITY[a]

| Species | Specific gravity | Upper limit of optimum growth (%) | Inhibitory moisture content (%) |
|---|---|---|---|
| Sitka spruce | 0.34 | 150 | 190–200 |
| Ponderosa pine | 0.44 | 100 | 145–150 |
| Douglas fir | 0.54 | 70 | 110–120 |
| Southern pine | 0.70 | 50 | 75–80 |

[a] From Snell (1929).

established in wood it can endure drying for years and remain alive, and if the wood again becomes wet it can begin growing at once. Wood may be lightly invaded by *L. trabea* during shipment, or during storage at a lumber yard, or during construction. If such wood is used in the rafters of a house roof, and the roof is well made, in temperate climates the decay will not progress. The wood will dry out, and the fungus become dormant. But if a leak should develop in the roof some years later, the fungus will revive and decay will get under way immediately. *Poria incrassata*, on the other hand, a common cause of rot in houses in the United States, will die within a few days if the wood in which it is growing is thoroughly dried. It cannot endure desiccation (Verrall, 1954, 1968).

## 2. Sources of Water

The water necessary for decay to occur can come from different sources. The exteriors of houses and other buildings are constantly exposed to the weather, and in humid climates if such exposed wood is not of decay-resistant species or is not treated with a preservative, it may decay fairly rapidly. Framing members are exposed to the weather during construction and if they become soaked with water, decay may become established before they dry out. In mild climates most houses are constructed without basements. Usually they are raised above the ground on stone or concrete, or preservative-treated wood supports, leaving a so-called crawl space underneath. If this crawl space is enclosed, and the soil within it is not covered with moisture-proof paper or plastic, the wood above it will soon become moist enough to permit decay. In a relatively humid climate such as that in the southeastern United States, this decay may progress so rapidly that the joists and floors of the house collapse within a few months after the house is completed. Usually *Poria incrassata* is the fungus involved in such situations.

Sometimes poor construction permits water to accumulate in the roof or walls of wood buildings, in which case decay is inevitable. Many architects and builders evidently have only a very limited knowledge of the biology of wood decay fungi. Even in well-constructed houses moisture may accumulate in some places in sufficient amounts to cause decay. Some air humidification systems add large quantities of water to the air, some of which is absorbed by the wood. If there is actual condensation, as often is the case, decay almost inevitably will follow. Even in relatively dry climates, in humid periods water condenses on toilet tanks and bowls and soaks into the wood. Sinks overflow and windows are left open during rains. Even with the best of construction, intelligent vigilance on the part of those occupying the house, and in a relatively low decay-hazard region such as Minnesota, some decay eventually occurs in almost every home. With poor construction, poor maintenance, and a high decay-hazard region, extensive decay is the rule.

*3. Dry Rot*

Dry rot is so-called because when the decay is discovered by the owner or occupant, the wood appears to be dry. It was not dry when the decay occurred. *Merulius lacrymans* is the major cause of dry rot in buildings, especially homes, in northern Europe and the British Isles. This fungus prefers enclosed places such as cellars or vaults or the spaces between walls. In the cool and humid climate that prevails in much of the region where it predominates, wood in such enclosed places may become wet in a rainy period and remain wet for a long time. Once the fungus becomes established it may produce enough metabolic water to enable it to continue. The specific name *lacrymans* was given to it because of the characteristic droplets of liquid exuded by the fruit bodies (or sporocarps). Usually it is not discovered until the decay has progressed to the stage of collapse of walls or floors.

*Poria incrassata* is a common cause of dry rot in much of the United States, although a number of other fungi cause decay in buildings, and all of this type of decay is referred to as dry rot. Once *P. incrassata* is established in moist wood it can spread by means of rhizomorphs that serve as water conduits (Verrall, 1954, 1968). Water can be transported by these rhizomorphs as far as 3 m vertically and 7 m horizontally. These may grow up through the space between the walls, from the cellar to the upper floors, and in this way decay advances throughout the building. As with the dry rot caused by *M. lacrymans*, the decay caused by *P. incrassata* often goes undetected until it has progressed to the point of failure of the wood.

## 4. Soft Rot

The group of fungi referred to as soft-rotting fungi were recognized in the early 1950's. The so-called soft-rotting fungi have been responsible for deterioration for centuries but their activities were not recognized. Hyphae were observed within wood cell walls in 1863 by Schacht, and in 1913 Bailey reported cavities in the secondary walls of pine (Duncan, 1960). Findlay and Savory demonstrated that *Chaetomium globosum* and other fungi were responsible for the cavities and the deterioration of wood slats in cooling towers. The excessive deterioration of wood in cooling towers stimulated the interest that led to the understanding of these fungi. In 1954 replacing the wood in one tower cost $40,000 and some towers required complete replacement after 3 to 5 years of service (Savory, 1954).

The soft-rotting fungi attack a wide range of hardwoods and softwoods, but probably are more of a problem in the former. These fungi can tolerate greater extremes of environment than the so-called true wood-rotting fungi. The soft-rotting fungi tolerate high temperatures, high levels of pH, toxic compounds including commonly used preservatives, and low levels of oxygen. The soft-rotting fungi can deteriorate wood at moisture contents below the fiber saturation point but these may be a special group deserving a unique designation. Microfungi or soft-rotting fungi can attack wood in seawater (Jones, 1971). These marine fungi are not completely documented but number well over 100 species (Jones, 1972). In marine habitats the soft-rotting fungi have been reported at depths of over 1600 m. At greater depths deterioration is caused by bacteria (Jones, 1971).

Bacteria are almost always restricted to wet wood and are a factor in such places as mill ponds when the raw wood is stored for long periods of time. Bacteria can alter the porosity of wood (DeGroot and Scheld, 1973; Unligil and Krzyzewski, 1974). *Bacillus polymyxa* has been reported to be an important cause of this increased porosity. Bacteria apparently can reduce the strength of pilings as evidenced by loss of 20% of the original strength of piling below the mudline in the Potomac River for 62 years (Scheffer and Wilkinson, 1969). Above the mudline the loss in strength was more than 50% of the original strength.

## B. STAIN

The fungi that cause blue stain or sap stain grow only in freshly cut wood, in which the sapwood has a moisture content approaching that of living trees. The stain fungi will not survive in wood once the moisture content drops below 20% based on oven-dry weight. Most of these blue stain fungi will not grow in wood that has been dried and then rewetted. Presumably this is because they cannot compete effectively with the great variety of other fungi that colonize such wood, since they will readily grow in wood that has been dried and then rewetted if it has been autoclaved and inoculated with a pure culture. Most of the fungi that cause superficial stain require free water to grow. One exception to this is the *Aspergillus restrictus*, mentioned above as having been responsible for spotting of coated insulation board (Fig. 9). The spots caused by this fungus were especially prominent on board in which a preservative had been added to the coating (French and Christensen, 1969). The preservative used required being dissolved in water to be effective, whereas the fungus was growing at a moisture content in equilibrium with relative humidities of 70–75 or 80%, where no free water was available.

Some fungi which have the capacity for discoloring wood substrates

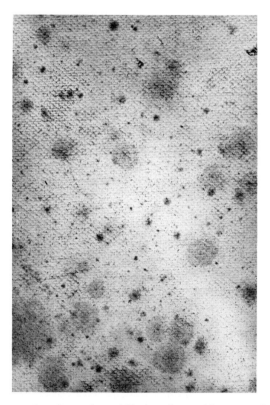

Fig. 9. Brown spotting of coated wood fiberboard caused by *Aspergillus restrictus*.

are able to grow in wood which is near saturation. The most common species in this group is *Trichoderma viride*; it is often the only fungus isolated from wood which has been stored in ponds.

## VIII. SUBSIDIARY FACTORS

A number of factors other than moisture influence the growth of fungi that cause deterioration of wood, and so determine the rate and extent or degree of deterioration that will occur. Two of the more important of these are temperature and the components of wood.

### A. TEMPERATURE

Most of the typical wood-rotting fungi have a minimum temperature for growth of 0°–5°C, an optimum of 24°–30°C, and maximum of 35°–40°C. These cardinal temperatures for growth have been determined by measuring the rate of radial growth of colonies of the fungi on agar media in petri dishes. The results probably do not apply strictly to the same fungi when growing in wood. Some wood-rotting fungi can endure a temperature of 60°C for some time, and so they probably are capable of growing and causing decay at a temperature not much below this. At least one important decay-causing fungus, *Chrysosporium pruinosum*, has an optimum temperature for growth of 37°C, and there probably are other thermotolerant wood-rotting fungi (Merrill and French, 1966). Some of the true thermophilic fungi that are responsible for microbiological heating in organic materials, as described below, grow well at 55°–60°C. These do not include what we consider to be typical wood-rotting fungi, but they may cause soft rot in heating piles of wood chips.

At temperatures close to 0°C wood-rotting fungi cease growing, as do those that cause soft rot and most but not all of those that cause staining. This is taken advantage of by felling trees in winter in regions of low winter temperature. Some strains of *Cladosporium herbarum* that grow in and cause blackening of wood and wood products in use, including painted wood siding of houses, can grow slowly at temperatures down to −5°C. Some species of *Penicillium* that cause staining in pulp stock can also grow at or near 0°C.

### B. CHEMICAL CONSTITUENTS OF WOOD

The basic structural materials of all wood are cellulose and lignin. A great variety of materials are deposited in the cell lumens and cell walls, especially as heartwood is formed. In trees with dark heartwood these include terpenoids, tropolenes, flavonoids, and stilbenes, phenolic compounds that are highly toxic to fungi. It is to these compounds that some woods owe their high resistance to decay (Scheffer and Cowling, 1966). Where wood is used in places of high deterioration risk it is of advantage to choose one of these naturally decay-resistant species. The choice must be supported by knowledge: cypress was for many years the wood of choice for use in greenhouse benches, because of its high resistance to decay (even though much of this wood was "pecky" cypress, which in itself is a product of decay). This was from old-growth cypress trees. Wood from younger new-growth cypress trees, even heartwood, is much less resistant to decay. Some of the "true" mahoganies are very highly resistant to decay, although most of them are so expensive that they are not likely to be used for materials in high deterioration-risk places. Philippine mahogany, unrelated to the true mahoganies, is not decay resistant.

## IX. PREVENTION OF DETERIORATION

In nature the normal fate of wood is to eventually be decayed by fungi or consumed by insects or fire, and in this way be recycled. Wood in use is subject to the same hazards. Various strategies have been developed to minimize the deterioration risk of wood and wood products and so extend their useful life. Some of these strategies can be applied as soon as the trees are felled in the woods. Others are applicable before, during, or after the wood has been processed into its final form, or even after it has been put into use.

### A. IN THE WOODS

In regions of low winter temperatures trees can be felled in winter, cut into logs, and left on the ground until spring, with no danger of significant damage by fungi. Some stands of timber, such as those of black spruce, in swampy areas, can be harvested only in winter because of the nature of the terrain. Logs cut in warm weather for veneer must be utilized almost immediately or else must be protected from checking and from stain. This is done by spraying the ends of the logs with a fungicide, followed by a waterproof coating of asphalt or pitch (Scheffer and Eslyn, 1976). For really effective protection these materials must be applied as soon as the logs are cut. Logs treated in this way can be held through the summer without loss in processing quality or in quality of the veneer made from them.

## B. At Processing Plants

Logs, bolts, and chips are vulnerable to rapid deterioration by both decay and stain fungi. Normally, to avoid costly shutdowns that would result from lack of supplies, these materials are accumulated in large quantities at the mills and processed over a period of weeks or months. During this time they must be protected from deterioration. The average moisture content of cottonwood bolts on their arrival at the mill ranges from about 50 to 180% (Bois and Eslyn, 1966).

### 1. Logs and Bolts

For a time severe losses in both quantity and quality were sustained in bolts of southern pine stored for processing into pulp for paper. Losses were reduced by eliminating the peeling or debarking that previously had been customary and by stacking the bolts as tightly as possible in large piles (Lindgren, 1951). Later a sprinkling system was developed to keep the bolts continuously wet, and this further reduced losses (Brodie and DeGroot, 1976; Wright *et al.*, 1959). The same or a similar sprinkling system has been used for decades in Minnesota to protect birch match-veneer stock from staining. It has the additional virtue of increasing the quantity of veneer stock over that cut from nonsprinkled bolts, and of reducing the frequency of sharpening of the veneer cutting knives.

As a result of the increased moisture content of the stored wood, its permeability is abruptly increased during the 3 months following the first month of storage (DeGroot and Scheld, 1973). Increased permeability may be advantageous if the wood is to be treated with preservative (Unligil and Krzyzewski, 1974).

Underwater storage has long been used to protect pulpwood and logs from deterioration, especially in regions where deterioration can occur rapidly (Osborne *et al.*, 1956). The moisture content of the wood should be maintained as high as possible during storage (Wilhelmsen, 1968). Deterioration occurs at an accelerated rate between about 30 and 50% moisture content.

An unusual problem developed in the processing of sugar maple logs. A fungus, *Crytostroma corticale,* inhabits the bark of sugar maple. Once the trees are felled and cut into logs, and these are held under conditions favorable to the growth of this fungus, it develops rapidly and produces tremendous numbers of spores. Some people develop disabling pneumonialike symptoms when they inhale the spores. The problem is alleviated by utilizing the logs before the fungus can sporulate heavily, and by storage of logs under continuous water spray, which prolongs the low-risk period for several months. Logs so stored can also be debarked more rapidly and cheaply than those not so stored (Ohman *et al.*, 1969).

## 2. Chips

Wood in the form of chips often can be handled much more economically than in the form of bolts or logs, and the use of chips has been increasing. Losses in wood substance and pulp quality are of the same magnitude whether stored as roundwood or as chips (Hajny, 1966). Wood chips stored in bulk, however, present some special storage problems. One of these problems is microbiological heating, because of the large surface exposed as compared with logs or bolts. Much the same sort of heating occurs in many kinds of materials—hay, straw, stored grains and seeds of all kinds, alfalfa pellets, baled cotton and wool, mushroom compost, and in other biological materials subject to moderately rapid invasion by fungi and bacteria. In some of these materials the heating at times proceeds to spontaneous combustion. The usual heating pattern in wood chips, as also in stored grains, is for the temperature to rise to 55°–65°C, remain there for a week or more, then decrease to 30°–40°C (Feist *et al.*, 1973; Shields, 1967). This heating has been studied extensively in some materials, and the microflora responsible for it have been identified. The fungi and bacteria responsible for heating of wood chips may not be the same as those that cause heating in baled wool or in moist stored soybeans, although this seems unlikely because essentially the same groups have been found in all of the materials that have been studied thoroughly. Some of the common thermotolerant fungi carry the heating up to about 55°C, where thermophilic fungi take over and carry it up to 60°–65°C. These are followed by thermophilic bacteria that may carry the heating up to 75°C, after which purely chemical oxidation may enter and raise the temperature to that necessary for ignition. Continuous water sprinkling reduces or eliminates heating in wood chip piles, but it may promote soft rot (Shields, 1967). Anaerobic storage and treatment with fungicidal chemicals are being tested as preventives of heating (Eslyn and Laundrie, 1973, 1975).

## C. After Processing

Once logs are cut into lumber the lumber must be dried immediately to a moisture content too low to permit deteriorating fungi to grow, or it must be treated with a fungicide. It is routine practice at mills in the southern United States to pass the newly sawn lumber through a vat containing a water-soluble fungicide. The passage requires 10–15 sec. Ethyl mercury phosphate was first used for this fungicidal treatment when it was begun in the 1920's. Later sodium pentachlorophenate was favored. The trend now is toward formulations of borax alone, or borax formulations in mixtures with chlorinated phenols. Copper 8-quinolinolate is used for wood to be used in contact with foodstuffs. Some mixtures are more effective than single compounds. If protection against insects is necessary, an insecticide is added. Where feasible, lumber is kiln dried as soon as possible.

## D. IN SHIPMENT

Much lumber of high moisture content is carried by rail or ship from the West to the East Coast of the United States or from British Columbia to Europe or the Orient. More than 90% of the lumber shipped from British Columbia in 1973 was not seasoned (Roff *et al.*, 1974). Shingle stock of luan wood is shipped green from the Philippine Islands to the United States. Much of this lumber is in unit lots, for ease of mechanical transfer, the boards fastened together in relatively large and tight packages, and the unit lots are stacked up so that there is little opportunity for the lumber to dry during the journey. These practices make for high deterioration hazard, and unless the lumber is protected by an effective fungicide, stains and decay may develop. Sodium pentachlorophenate at concentrations of 0.5–1.4% has provided good protection. A boron-diffusion treatment has protected lumber in storage (Roff, 1974).

Even if lumber of other wood products is dried to a safe moisture content it can become wet in transit or in lumber yards or at construction sites, with consequent deterioration.

## E. IN BUILDINGS

In the cases described above the need was for protection for a relatively short time, usually weeks or months at most. Once wood is fabricated into structures, we expect or hope that these will endure for a long time—in the case of houses, preferably for centuries. Sometimes they do, as witness the houses and barns more than two centuries old in New England, and still sound. Sometimes they do not, as in houses that decay to the point of failure or collapse within a few years after they are built, or, in extreme cases, even before they are completed. The useful life of homes and of many other structures made principally of wood is determined to a large extent by whether the wood remains free of decay. This in turn is affected by climate, by the wood chosen, by construction practices, by use or nonuse of preservatives, and by maintenance practices, to name the more important factors (Fig. 10).

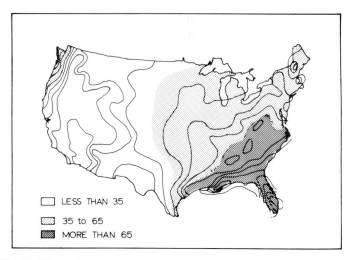

Fig. 10. Climate-index map of the United States based on the formula

$$\text{Climate index} = \frac{\sum_{\text{Jan.}}^{\text{Dec.}} [(T - 2)(D - 3)]}{16.7}$$

$T$, mean monthly temperatures (°C); $D$, the mean number of days in the month with 0.25 mm or more precipitation.

## 1. Climate

In dry or desert climates wood structures are not subject to high decay hazard. In the southwestern United States and along most of the west coast of South America the climate is so dry that very little care is necessary to make wood buildings last almost indefinitely. In those regions the major source of water is internal to the structures themselves, from use or misuse of water by the occupants. Even in those regions wood in constant contact with the soil will decay in time, because if there is enough rainfall, even at relatively long intervals, to permit flowering plants to grow, there will be enough moisture in the soil so that wood in contact with it will be sufficiently wet long enough to decay.

The decay hazard in buildings increases with increasing rainfall and increasing relative humidity. On the opposite extreme from desert climates, so far as building decay is concerned, is the southeastern United States from Louisiana into Alabama, Florida, and Georgia, and many cool and humid areas in northern Europe and Great Britain. In those high deterioration-risk areas a combination of use of naturally decay-resistant or preservative-treated wood, good construction, and good maintenance is required to keep wood structures free of decay, and even with this combination total protection from decay is difficult to achieve. Thomas Jefferson nearly 200 years ago suggested that houses in Virginia should not be constructed of wood, because of their short life. At that time there was no lack of well-seasoned, decay-resistant wood for construction, as is now the case. In the high decay-hazard region of northern Europe and Great Britain masonry construction is favored.

So far as concerns decay in buildings, most of the United States and Canada are between the two extremes mentioned above (Scheffer, 1971). With good selection or preparation of building site, use of sound, dry lumber, good construction with attention paid to the details that make for low decay hazard, use of preservatives where the decay hazard is high, and good maintenance, decay can be kept to a minimum. If any one of these is lacking, decay of greater or lesser extent is likely to occur.

## 2. Site Selection or Preparation

Good drainage of the building site is essential. Water should drain away from the building, not toward it. The drain tiles under a house in Illinois, installed to carry water from the downspouts, were cut, allowing water to accumulate beneath the house in periods of rain. Within a short time the supporting members under the house had been invaded by *Poria monticola* (now called *P. placenta*) and the house collapsed from decay, necessitating costly repairs.

Many communities are built on the floodplains of rivers. About the only procedure that can be followed to alleviate the decay hazard after flooding is to dry out the structure as quickly as possible. If water has accumulated between the inner and outer walls during flooding it may be necessary to remove the siding to allow the inner portion of the wall to dry.

## 3. Use of Dry, Sound Construction Materials

During the 1960's it was common practice in the southeastern United States to use dried Douglas fir studs from the West Coast for new home construction but green southern pine for floor joists and other members in the underside of the house. Some joists had moisture contents in excess of 100% at the time they were incorporated in the house. Wood at such high moisture contents is very susceptible to rapid decay.

Lumber and other construction materials may already be partly decayed as they come from the supplier. In one case, asphalt-impregnated sheathing board was stored outside and became water soaked. Fungi present in the wood supports under the piles reacted to the increased moisture content and quickly invaded the piles of board. Much of the board appeared to be sound, but actually was so extensively decayed that it had lost much of its strength.

Construction materials are exposed to the weather as the buildings in which they are used are being erected. Every effort should be made to keep this wetting to a minimum, and to make sure that, if any materials do become water soaked, they be allowed to dry out before they are enclosed (Fig. 11). Plaster is used much less in construction than it once was, but where it is used, it can add a very large quantity of moisture to a building. If the building is kept tightly closed after it is plastered, decay may develop.

In one newly constructed building, multi-ply plasterboard walls covered with a plaster finish were quickly covered with tremendous masses of pinkish mycelium. The fungus was *Pyronema confluens* which grows rapidly in a wet substrate where few competing organisms are present. The solution was to install and start the heating system which dried out the walls and stopped the fungus.

### 4. Construction

For decay prevention good construction means prevention of moisture accumulation in any part or portion of the wood (Scheffer and Verrall, 1973). The sills that support the studdings should be on a foundation or on supports high enough to prevent splashing rain from wetting the lower portion of the siding. If the house has no basement, the ground beneath the house should be covered with waterproof plastic or paper (Amburgey and French, 1971). If the crawl space is enclosed, it should have vents to provide enough aeration to prevent uptake of water by the

Fig. 11. Apothecia of *Peziza domiciliana* on wet insulating board which had been installed in a wet basement.

floor joists and subfloors. Flashings should be installed over doors and windows. Roof valleys should have metal flashings. Eaves should project far enough to prevent wetting of the upper part of the wall during rains, and gutters and downspouts should be installed to carry water away from the house. Siding should be back dressed to eliminate the small capillaries that allow moisture to move up behind the siding and from there into the interior wall. Proper design of corners is important in avoiding openings for entry of moisture (Verrall, 1966). Bathrooms and other rooms where moisture accumulates should be vented. If insulation is installed, there should be a good vapor barrier on the inside surface of the insulation.

A common point of entry of water is under roofing as a result of ice dams which form on roofs improperly constructed (Grange and Hendricks, 1976). Warm air in the attic causes the snow on the roof to melt and as the melted snow moves down to the edge of roof it freezes, thus forming a dam and trapping the water behind it. The water then backs up under the roofing and into the house walls. Heating tapes or removing the snow are not solutions. A cold deck which prevents the snow from melting is far more effective (Grange and Hendricks, 1976).

All of these precautions seem so elementary as to be scarcely worth mentioning, yet through ignorance, indifference, carelessness, or for reasons of economy, not all of these sound practices are followed, and decay occurs. A few examples, all of them from recent personal experience, will illustrate what may happen.

A basementless house was constructed in Wisconsin, in a region or location of only moderate decay hazard. The crawl space beneath it was enclosed, and had no vents, although the contractor promised to return the following year and install them. By the following year the floor joists had decayed to the point of partial failure.

A roof was built over a terrace of a house some years old. The terrace roof was connected with the roof of the porch that already was in place, in such a way that rainwater drained into the interior of the house wall. In this case the contractor maintained that it was not his responsibility to join the new roof to the older roof in such a way as to provide good drainage of rain and snow water.

Sliding glass doors in a new house in St. Paul were not provided with adequate flashing beneath them, allowing water to seep into the subfloor adjacent to the doors. This was detected only after decay had developed sufficiently to become obvious.

A bank building was designed with uncovered projecting roof beams, giving it a pleasing modern look. The beams were not protected with a preservative, and cross members were joined to the beams in such a way as to trap water. Within a short time fairly extensive decay developed in the beams. Fortunately this was detected before it had progressed to the point where the roof might have collapsed. Presumably the architect who designed the building was not aware of the decay hazard that he designed into it.

How common is this sort of decay? In a survey of newly constructed homes in North Carolina in 1965, decay was found in almost all of them, some of the homes so new that they were not yet occupied by the owners.

### 5. Use of Preservative

Wood exposed to intermittent wetting should be treated with a preservative such as pentachlorophenol in a solvent that permits the wood to be painted where this is desirable. At least some window sash and door frames are preservative treated at the assembly plants. In many cities there are custom treating plants where wood construction components can be treated. Hardware and paint stores and lumber yards carry pentachlorophenol in 5% solution, ready to apply. Brush treatments of items such as porch and deck boards and railings and fence boards will add greatly to their useful life. Pressure-treated wood fence posts are readily available. All wood railway ties now are pressure treated with preservative; eventually they fail from mechanical wear, but not from decay. Poles and piling must be treated in relation to the decay hazard to which they are to be subjected. Transmission-line poles are so expensive to replace that it is now customary to treat these structures after they have been in service for many years.

An alternative to the use of chemical preservatives is to take advantage of the fact that the wood-deteriorating fungi need both a minimum amount of water and an adequate supply of air. Wood substrates have been modified in a variety of ways to prevent the cell walls from absorbing the moisture needed for decay or by removing or blocking the hydroxyl groups so that decay enzymes are unable to dissolve the wood components (Stamm and Baechler, 1960). There are five types of modified wood: (1) impregnated with a fiber-penetrating phenolic resin, (2) hydroxyl groups replaced with acetyl groups, (3) thermally modified, (4) crosslinked with formaldehyde, and (5) bulked with polyethylene glycol.

### 6. Maintenance

Good maintenance is an integral part of decay prevention in all wood buildings. Caulking around bathtubs, shower stalls, toilets, and sinks should be inspected regularly, and replaced if it is faulty. Roof leaks should be repaired before decay has a chance to become established in the wood beneath.

## 7. High Decay-Hazard Buildings

Some buildings because of the conditions maintained within them have a high risk of deterioration by fungi. These include cold-storage rooms, breweries, many food-processing plants, indoor swimming pools, and skating rinks (Oviatt, 1972, 1973). These have a constant high humidity combined with temperatures favorable to the growth of stain and decay fungi. Under these favorable conditions a fungus such as *Aureobasidium pullulans* can blacken the surface of unpainted or painted wood within a few days to a few weeks, contributing a look of shabby decadence to the still new building. A combination of good design to control condensation by ventilation, insulation, dehumidification, and heating should be considered. The use of preservative in the wood or in the coating, or both, will reduce the hazard.

Humidifying systems which provide excess moisture beyond reasonable levels and inadequate vapor barriers can provide high hazard conditions. The tabluation below from Prof. F. B. Rowley of the Institute of Technology, University of Minnesota, suggests maximum indoor relative humidities in relation to outside temperatures

| Outdoor temperature (°F) | Maximum indoor relative humidity |
|---|---|
| Below −20 | 15 |
| −20 to −10 | 20 |
| −10 to 0 | 25 |
| 0 to +10 | 35 |
| Above +10 | 40 |

Today's modern home needs a vapor barrier installed on the backside of the inside walls and ceilings to keep the vapor within the house and prevent it from reaching a cold surface where condensation will result.

## X. EVALUATION OF RESISTANCE TO DETERIORATION

New wood products are continually being developed and marketed, many of them for use in environments of high deterioration risk. These must be evaluated for susceptibility to stain and decay. The efficacy of preservatives available for use in these products must also be determined (Fig. 12). Ideally the tests should accurately predict the service life of the product under the most severe conditions to which it is likely to be exposed. If, for example, the product is intended for use as a covering for

Fig. 12. Samples of insulating boards, with a range of retention of preservatives, which have been evaluated for decay resistance by being buried in a bed of soil.

house roofs, it will, in actual service in the northern regions of the United States and in Canada, be subjected to periodic heavy deposits of snow and ice. In the summer it will be exposed to searing heat and, at times, mechanical damage from hail. In some regions it will be exposed to constant high humidity and may be overgrown by algae, lichens, and mosses as well as by fungi. To compete with established roof coverings such as cedar or composition shingles, it should have a service life of at least 15 to 20 years.

It is not possible to design a laboratory test that will quickly and reliably predict the behavior of a new product under the multitude of known and unknown conditions to which it might be exposed over a wide range of climates over a period of decades. To determine the resistance of wood coatings to stain fungi, panels of wood with the coatings are exposed in different locations throughout the country. This approximates the conditions likely to be encountered in use. The soil-block test is widely used to evaluate the resistance or susceptibility of untreated and preservative-treated wood to decay. Blocks of wood of standard size are inoculated with specified wood-rotting fungi and exposed for a specified time at a specified temperature on soil. The loss in weight of the test specimens is used as a measure of decay.

This is a relatively severe test, and is useful in eliminating rather quickly those woods or those treatments that are ineffective under many conditions. Judgment is needed, however, in interpreting the results. Two examples will illustrate this: one wood product was subjected to soil-block tests using the standard array of decay fungi, and passed with flying colors. It was about to be marketed, with a national promotional program. Fortunately, it was tested further, using an additional species of fungus that commonly causes decay in wood in situations where this product was to be used. This fungus caused severe decay within a short time. A formulation containing 0.2% pentachlorophenol-protected wood from decay when the wood was exposed in soil of low moisture content, but gave no protection at all when the blocks were exposed in soil of higher moisture content (Merrill and French, 1963). Other factors, known and unknown, may also affect the results of these tests, and different laboratories, although using the same soil and supposedly the same strains of the same species of fungi, may get widely different results (Sakornbut and Balske, 1957). This suggests that a certain amount of informed skepticism is desirable in those responsible for interpreting the results of these tests. In soil-block tests involving *Peniophora gigantea* the amount of decay was reduced at high light intensity but not when the moisture content of wood was increased (DeGroot, 1975). Wood moisture content appeared to influence brown-rotting fungi less than white-rotting fungi.

Laboratory methods of evaluating resistance of preservative treatments to staining and soft-rotting fungi differ from those used for the wood-rotting fungi. Coatings, paint surfaces, adhesives, and wood surfaces themselves can be evaluated by placing test samples in various chambers where the moisture content and temperature are favorable for the fungi used (Fig. 13). A very simple but effective test consists of hang-

Fig. 13. Surface molds developing on wood fiber products incubated over a saturated salt solution (left) and distilled water (right). This system has been used for evaluating fungus resistance of various products at a range of relative humidities.

ing the test sample from the cover of a jar in which water or a salt solution provides the desired relative humidity. Soft rot is evaluated by the serial-block apparatus where a nutrient solution provides for variable moisture conditions in the series of blocks (Eslyn, 1969).

## REFERENCES

Amburgey, T. L., and French, D. W. (1971). Plastic soil covers reduce the moisture content in basementless homes. *For. Prod. J.* **21**, 43–44.

Bois, P. J., and Eslyn, W. E. (1966). Deterioration rates of willow and cottonwood during storage in Georgia. *For. Prod. J.* **16**, 17–22.

Brodie, J. E., and DeGroot, R. C. (1976). Water-spray storage: A way to salvage beetle-endangered trees and reduce logging costs. *South. Lumberman* **232**, 13–15.

DeGroot, R. C. (1975). Decay fungus, light, moisture interactions. *Wood. Sci.* **7**, 219–222.

DeGroot, R. C., and Scheld, H. W. (1973). Permeability of sapwood in longleaf pine logs stored under continuous water spray. *For. Prod. J.* **23**, 43–46.

Duncan, C. G. (1960). Wood-attacking capacities and physiology of soft-rot fungi. *U. S., For. Prod. Lab., Rep.* 2173, 1–70.

Eslyn, W. E. (1969). A new method for appraising decay capabilities of microorganisms from wood chip piles. *U. S., For. Prod. Lab. Res., Pap.* **107**, 1–8.

Eslyn, W. E., and Laundrie, J. F. (1973). How anaerobic storage affects quality of Douglas-fir pulpwood chips. *Tappi* **56**, 129–131.

Eslyn, W. E., and Laundrie, J. F. (1975). Effect of anaerobic storage upon quality of aspen pulpwood chips. *Tappi* **58**, 109–110.

Feist, W. C., Hajny, G. J., and Springer, E. L. (1973). Effect of storing green wood chips at elevated temperatures. *Tappi* **56**, 91–95.

French, D. W., and Christensen, C. M. (1958). Nature and cause of spots on coated insulating boards. *Tappi* **41**, 309–312.

French, D. W., and Christensen, C. M. (1969). Influence of coating additives and relative humidity upon spotting by fungi. *For. Prod. J.* **19**, 108–110.

Grange, H. L., and Hendricks, L. T. (1976). Roof-snow behavior and ice-dam prevention in residential housing. *Minn., Univ., Agric. Ext. Serv., Ext. Bull.* **399**, 1–20.

Hajny, G. J. (1966). Outside storage of pulpwood chips. *Tappi* **49**, 97A–105A.

James, W. L. (1975). Electric moisture meters for wood. *U. S., For. Prod. Lab., Gen. Tech. Rep.* **6**, 1–27.

Johansen, C. B., and Smith, R. S. (1974). Culture collection of wood-inhabiting fungi. *Can., Dep. Environ., For. Serv. Inf. Rep.* VP-X-140, 1–45 p.

Jones, E. B. G. (1971). Chapter 12. The ecology and rotting ability of marine fungi. Ex Marine borers, fungi and fouling organisms of wood. Organization for Economic Co-operation and Development, *OECD*, pp. 237–258.

Jones, E. B. G. (1972). The decay of timber in aquatic environments. *Br. Wood Preserv. Assoc., Ann. Conv.* pp. 1–18.

Käärik, A. (1976). Recent names for some common decay fungi. The International Research Group on Wood Preservations. No. 143.

Lindgren, R. M. (1951). Deterioration of southern pine pulpwood during storage. *For. Prod. J.* **5**, 169–81.

Merrill, W., and French, D. W. (1963). Evaluating preservative treatment of rigid insulating materials. *Tappi* **46**, 449–452.

Merrill, W., and French, D. W. (1966). Decay in wood and wood fiber products by Sporotrichum pruinosum. *Mycologia* **58**, 592–596.

Merrill, W., French, D. W., and Hossfeld, R. L. (1965). Effects of common molds on physical and chemical properties of wood fiberboard. *Tappi* **48**, 470–474.

Ohman, J. H., Kessler, K. J., and Meyer, G. C. (1969). Control of Cryptostroma corticale on stored sugar maple pulpwood. *Phytopathology* **59**, 871–873.

Osborne, M. J., Chesley, K. G., Wilson, F. G., Morgan, J. W., Conner, R. O., and St. Laurent, W. (1956). Underwater storage of pulpwood. *Tappi* **39**, 129–139.

Oviatt, A. E. (1972). Moisture content in wood structures for ice skating. *For. Prod. J.* **22**, 50–54.

Oviatt, A. E. (1973). Wood is ideal in pool natatorium roof construction. *Swim. Pool Week.* pp. 1–6.

Roff, J. W. (1974). Boron-diffusion treatment of packaged utility-grade lumber arrests decay in storage. *Can., Dep. Environ., For. Serv. Inf. Rep.* **V6T 1 X 2**, 1–14.

Roff, J. W., Cserjesi, A. J., and Swann, G. W. (1974). Prevention of sap stain and mold in packaged lumber. *Can. For. Serv., Publ.* **1325**, 1–43.

Sakornbut, S. S., and Balske, R. J. (1957). The effects of soil, moisture, and aeration variables in bioassay by the soil-block culture method. *For. Prod. J.* **7**, 404–407.

Savory, J. G. (1954). Breakdown of timber by Ascomycetes and Fungi Imperfecti. *Ann. Appl. Biol.* **41**, 336–347.

Savory, J. G. (1955). The role of microfungi in the decomposition of wood. *Br. Wood Preserv. Assoc. Annu. Conv.* p. 1–17.

Scheffer, T. C. (1971). A climate index for estimating potential for decay in wood structures above ground. *For. Prod. J.* **21**, 25–31.

Scheffer, T. C., and Cowling, E. B. (1966). Natural resistance of wood to microbial deterioration. *Annu. Rev. Phytopathol.* **4**, 147–170.

Scheffer, T. C., and Eslyn, W. E. (1976). Winter treatments protect birch roundwood during storage. *For. Prod. J.* **26**, 27–31.

Scheffer, T. C., and Verrall, A. F. (1973). Principles for protecting wood buildings from decay. *U. S., For. Prod. Lab., Rev. Pap.* **190**, 1–56.

Scheffer, T. C., Duncan, C. G., and Wilkinson, T. (1969). Condition of pine piling submerged 62 years in river water. *Wood Preserv. (Chicago)* **47**, 22–24.

Shields, J. K. (1967). Microbiological deterioration in the wood chip pile. *Can., For. Branch, Dep. Publ.* **1191**, 1–29.

Snell, W. H. (1929). The relation of the moisture contents of wood to its decay. III. *Am. J. Bot.* **16**, 543–546.

Stamm, A. J., and Baechler, R. H. (1960). Decay resistance and dimensional stability of five modified woods. *For. Prod. J.* **10**, 22–26.

Unligil, H. H., and Krzyzewski, J. (1974). Permeability of white spruce wet stored in Labrador. *For. Prod. J.* **24**, 33–37.

Verrall, A. F. (1954). Preventing and controlling water-conducting rot in buildings. *U. S., For. Serv., Southeast. For. Exp. Stn., Occ. Pap.* **133**, 1–14.

Verrall, A. F. (1966). Building decay associated with rain seepage. *U. S., For. Serv., Tech. Bull.* **1356**, 1–58.

Verrall, A. F. (1968). Poria incrassata rot: Prevention and control in buildings. *U. S., For. Serv., Tech. Bull.* **1385**, 1–27.

Wilcox, W. W. (1968). Changes in wood microstructure through progressive stages of decay. *U. S., For. Prod. Lab., Res. Pap.* **70**, 1–46.

Wilhelmsen, G. (1968). Water storage of unbarked mechanical pulpwood of Norway spruce. *Nor. Skogind.* **22**, 207–211.

Wright, E., Knauss, A. C., and Lindgren, R. M. (1959). Sprinkling to prevent decay in decked western hemlock logs. *U. S., For. Serv., Pac. Southwest For. Range Exp. Stn., Res. Note* **177**, 1–10.

CHAPTER 9

# MOISTURE AS A FACTOR IN EPIDEMIOLOGY AND FORECASTING

## A F. Van der Wal

FOUNDATION FOR AGRICULTURAL PLANT BREEDING,
LABORATORY DE HAAFF, WAGENINGEN, THE NETHERLANDS

## I. INTRODUCTION

Moisture as a primary determinant of growth and development of microorganisms, such as fungi, is well known by the great majority of people by virtue of the daily routine of housekeeping. Moisture in storage rooms, cellars and cupboards is associated with the typical musty smell. Moist conditions in lawns, meadows, and forests stimulate the sometimes luxurious development of toadstools, which are a joy for those who appreciate them and a source of distress for others, such as foresters, who recognize that development of pathogenic fungi on trees means decay and financial loss.

Less common is the knowledge that plant pathogens, predominantly fungi, attack crops during the growing season, causing serious plant diseases. Some of the diseases show up early in spring, when the snow cover is vanishing, and white or pink mycelium mats become visible on the soil surface near seedlings of winter crops. The moist conditions under the snow and the long period of coverage made it possible for the fungus to grow considerably, even at relatively low temperatures.

The possible role of rain on disease incidence in crops has been appreciated for a long time. However, it was not until after World War II that attention was given to dew as a major factor in disease outbreaks. In areas with frequent rains, dew may have attracted little attention as one of the major sources of free water on plant surfaces, probably because it "touches" the plants noiselessly near darkening, and thus usually after working hours. Evidence of the importance of dew in epidemics of plant disease has been accumulated, especially by researchers in semiarid areas, where rain is of little significance in at least part of the growing season and dew is the sole possible source of free water on plant surfaces.

The importance of air humidity on plant diseases has been recognized for many years. Although certain air humidities favor the development and growth of some fungi, air humidity has been used primarily because its measurement is routine in meteorology, and not because its direct relationship with infection or sporulation, and dispersal has been proved, as will be shown later. Plant epidemiologists have shown less interest in soil moisture than in air moisture, and only a few examples of prediction of disease based on soil moisture can be given. Some interest has also been shown in effects of moisture in the plant on growth of fungi living in or on them.

The transition from study of air humidity only to free water on the plant surface, and to plant and soil water potentials as well, has been useful in progressive quantification of plant-disease relations.

Whereas epidemiology may be concerned with a variety of epidemics in an attempt to understand controlling processes and the effects of various factors on an epidemic, forecasting probably is justifiable only in cases where the disease causes severe crop loss, and has economic importance because of reduction in the amount or quality of food, fiber, etc.

Expending effort and funds on forecasting of epidemics affecting "insignificant" plant communities is not justifiable.

Epidemiology serves in three areas of agriculture. First, as will be emphasized in this chapter, it provides a basis for forecasting. Knowledge of the relations between environment, especially weather, and the sequence of events that lead to successive infections can be used in combination with weather forecasts in predicting disease outbreaks. Second, and this aspect is of increasing importance in methodology of resistance breeding and in reflections on the dynamics of populations of pathogens, accurate knowledge of the ecophysiology of the causal agent of a disease and of the diseased plant makes it possible to detect differences in rates of pathogen multiplication and to define various levels of plant resistance quantitatively, rather than in plus or minus reactions. Explicit description of the pathogen's requirements from the environment as regards light, moisture, temperature, etc., and their effect on rates of germination infection, sporulation, and dispersal are needed. This development is gaining more and more attention, supported by the belief that slow selection in the pathogen population reduces the rate of predominant appearance of new races, and thereby increases the period in which use can be made of certain resistance sources. This development fits into the general trend in trade and industry in that fine control in management is necessary in a well-balanced system and that small and gradual adjustments in response to the market at the right time and place are inherent in such systems.

The third important application of epidemiological knowledge lies in the adjustment of methods of growing the pathogen for medical purposes. *Claviceps purpurea* is grown on rye in some areas in the world to obtain alkaloids used in many pharmaceutical preparations. Proper timing of the inoculation and the knowledge of amount of inoculum necessary to provoke a wanted level of infection have been derived from knowledge of the way disease and environment are related.

This chapter deals mainly with moisture as the key factor in many systems of prediction of epidemics. Sources of moisture are discussed with respect to their impacts on epidemics and the possibilities of forecasting their appearance. Methods of determining the importance of moisture to the epidemic as well as its measurement are treated briefly. Except for a few selected cases, this chapter is meant to elucidate an approach covering the range "from the shovel to the computer," rather than serving as a compendium of accumulated and detailed knowledge. Reference will be made in the appropriate places to excellent reviews on various aspects of epidemiology and forecasting that have been published recently.

## II. FORMS OF MOISTURE: SIGNIFICANCE AND MEASUREMENT

### A. Air Humidity

Air humidity is commonly expressed as relative humidity, i.e., the actual water vapor pressure as a percentage of the maximum vapor pressure at a given temperature, as water vapor deficit, or as water potential. Air humidity usually is not directly related to spore germination. Perhaps the only confirmed exceptions are spores of powdery mildews (Erysiphales) which germinate in air humidities well under 80% relative humidity (RH). These fungi have spores with a relatively high water content and possibly they carry most of the water needed for germination. The high water content may make them more exposed to adverse environmental conditions but they are at an advantage over other spores in that free water is not required for their germination (see Chapter 5, this volume).

Diurnal changes in air humidity have been correlated many times with spore release patterns. Suzuki (1975) reported that maturing conidiophores of *Pyricularia oryzae* stopped growing during the less humid parts of the day, and they resumed growing during the highly humid conditions after sunset (Fig. 1.). Conidia, which matured fully during the night, were released and formed the morning sporulation peak at 4–5 A.M. Saturation automatically caused release of the conidia. But even when RH was 100%, free water was essential for germination.

Apart from possible direct effects of air humidity on ripening of conidiophores and conidia on top of them, water potentials of plant tissues near the boundary increase, especially when RH exceeds 98%. High water potentials favor growth and possible development of pathogenic fungi in plant tissues, and cause pycnidia to swell and release pycnospores of *Leptosphaeria nodorum* (Van der Wal and Luuring, 1976), and may enhance sporulation of the lesions of *P. oryzae*. In the latter case it was shown that when water was added to the lesion, spore release continued despite lower humidities in the atmosphere (Suzuki, 1975).

In summary, if air humidity in the field does not exceed approximately 80–90% and air humidity is the only source of moisture, there will probably be no epidemic of any of the existing airborne pathogenic fungi, except for the mildews. This information can be used in research work in controlled environments, where heavily sporulating and healthy plants can be grown next to each other for months without infection, provided that infection of the diseased plants is performed in a separate room, that no mildews are used (cf. Van der Wal and Cowan, 1974), and air humidity is set at the range mentioned.

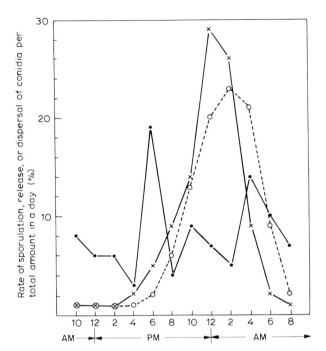

Fig. 1. Diurnal changes in rates of sporulation (●——●), release (×——×), and disper-
sal (o - - - - o) of conidia of *Pyricularia oryzae*, the cause of rice blast. [From Suzuki (1975).
Reproduced with permission, from *Ann. Rev. Phytopath*. **13**. Copyright © 1975 by Annual
Reviews, Inc. All rights reserved.]

However, air humidity figures have been used in relating weather to
disease incidence for over 40 years, before epidemiology became a recognized branch of plant pathology. The impression may have been established that air humidity controls the epidemic. But inoculation of plants,
followed by incubation, at a RH of 90% or higher, will not lead to infection, unless free water, either from guttation, dew, or mist of rain droplets,
is present for several hours following inoculation.

The 90% value is used in combination with synoptic weather maps
(Table I: Wallin late blight forecasting system; Krause and Massie, 1975).
Certain periods with 90% RH are likely to provide the kind of weather
suitable for infection with *Phytophthora infestans*. Such weather apparently meets the requirements of the fungus, viz., the period of several
hours of leaf wetness. Regional adjustments of the "rules" for potato
blight forecasting are necessary (Tarr, 1972). These adjustments are likely
to agree in that different weather spells may lead to the same environmental conditions as experienced by the pathogen, because of differences in

TABLE I

RELATIONSHIP OF TEMPERATURE AND RH PERIODS AS USED IN THE WALLIN
LATE BLIGHT FORECASTING SYSTEM[a]

| Average temperature range[b] (°C) | Severity numbers and hours of 90%, or above, RH | | | | |
|---|---|---|---|---|---|
| | 0 | 1 | 2 | 3 | 4 |
| 7.2–11.6 | 15 | 16–18 | 19–21 | 22–24 | 25+ |
| 11.7–15.0 | 12 | 13–15 | 16–18 | 19–21 | 22+ |
| 15.1–26.6 | 9 | 10–12 | 13–15 | 16–18 | 19+ |

[a] From Krause and Massie (1975).
[b] Average temperature of period when the relative humidity equals or exceeds 90%.

edaphic factors. By analyzing the components of the infection process, the environmental requirements of the pathogen can be accurately determined.

## B. MEASUREMENT OF AIR HUMIDITY

Humidity measurements still are among the most difficult in meteorology. In sciences like epidemiology, where they are important but have to be combined with other even more time-consuming measurements of pathogen and host, humidity measurements, especially those above 95% and below 30%, may be tricky.

According to Monteith (1972), air humidity meters are divided into two groups: psychrometers and hygrometers. A psychrometer is any instrument that is based on measurement of temperature differences between a wet and a dry bulb thermometer. Such sets may be thermometers with a liquid (mercury, alcohol) in glass, thermocouples, resistors, etc. Psychrometers are likely to show good response at high humidities. The ventilated types are the most accurate. The difference in temperature between the two thermometers is uniquely related to air humidity at a constant and known air movement along the wet bulb. Beyond doubt the classic instrument is the Assmann Psychrometer, often used as a standard for calibration of other types. Psychrometers with an electrical signal as their output have the advantage that they can be used to measure and record data continuously. For additional information on psychrometers the reader is referred to Monteith (1972). Hygrometers are also used for measuring the water vapor content of air. They use various principles, including contraction of hairs or strings in less humid air, condensation of vapor on a surface that is gradually cooled with the possibility of measuring the temperature of the cooled surface when condensation occurs (dew-point meters), equilibrium temperature of lithium chloride in the solid and liquid phases, capacity of aluminum oxide layers, and change in electrical resistance of a sulfonated polystyrene surface.

Hygrometers all exhibit hysteresis, and several types become permanently damaged when they are exposed to saturated air or to free water. The lithium chloride cell is the type most commonly used in micrometerological studies (Monteith, 1972).

Thermohygrographs are useful for epidemiological work. When properly calibrated and shielded from radiation in the field, they provide useful data both indoors and outdoors. Weekly cleaning with distilled water of the fibers is recommended, as is twice-yearly calibration when a meter is used continuously. The thermohygrographs do not require electric power and their maintenance is relatively inexpensive. Recording is on a paper sheet, and it takes comparatively a great deal of time to analyze the recordings as compared with instruments having an electrical output that can be transformed into a binary code and therefore are adaptable to automatic data acquisition. Another disadvantage of thermohygrographs, particularly in epidemiological work, is that the range of accurate and reproducible measurement is limited to between ca. 30 and 95% RH. Lack of precision at high humidities makes these instruments unsuitable for work on germination studies.

Recently, sensors have become available that measure humidity as a function of a change of electrical resistance over a sulfonated surface of polysterene or related materials. Such hygrometers are potentially promising, but a dearth of experience with the sensors in long-term field experiments makes it too early to evaluate their usefulness for epidemiological work. The most accurate instrument that measures very high humidities in the range of 98 to 100% is a sensor based on the Peltier effect. The Peltier effect is the counterpart of the thermocouple effect. Whereas in a thermocouple connection a voltage induced in the circuit by the difference in temperature of the two junctures is measured, in a Peltier circuit a current is passed through the thermocouple, causing the cooling of one juncture and heating of the other. The current is passed until the temperature of the cooling juncture does not drop further, because of water condensing on the juncture surface. The dewpoint temperature and the actual temperature permit determination of air humidity. The tiny thermocouples used in the Peltier circuit are fragile, and proper operation of the circuit requires some skill. This may make use of the Peltier effect in plant pathology inadvisable, even though its accuracy is very high, and its resolution is in some cases better than 0.01% RH. Soil physicists have made good use of the Peltier effect in studies of availability of soil moisture to plants. RH of 100–98.5% corresponds approximately to soil moisture suctions of 0 to $-20$ bars (Box, 1965).

Many instruments for measuring air humidity are commercially available, but good ones for measuring very high humidities in the field are not. Before funds are invested in development of high-humidity range sensors that can be used in epidemiological routine, proper examination of their possible use is required. Rapilly and Skajennikoff (1974) studied germination of *Leptosphaeria nodorum* pycnospores in the humidity range from 98 to 100% and found very slow germination at these high humidities in the absence of free water, probably far too slow to be important in the epidemic. Rapid spore germination occurred in free water. If this pattern of no germination, or at very low rates in the range at very high humidities, is more or less general, the development of methods for adequate measurement does not appear to be based on necessity. Furthermore, when free water is formed on plant surfaces, as by dew, the period of very high humidities near the plant surface is relatively short and, compared with the insignificance of those periods for spore germination, their measurement is not needed. It is much more important to concentrate on measurement of the period of free water on plant surfaces. It is unfortunate that plant pathologists have not been able to convince meteorologists of the importance of keeping accurate records of plant wetness periods. Availability of such long-term records would greatly help the progress of epidemiology.

The problem of measuring air humidity can be solved by devising reliable hygrometers or psychrometers for field use in long-term experiments (for months) so as to accurately measure RH in the range of ca. 30–95%. Such measurements should be combined with accurate determinations of periods of plant wetness. The time between transition of 95%RH toward 100% RH and formation of free water, if it forms at all, does not seem important. It is more important to know whether and when free water is present on the surface of the plant.

## C. Free Water

The presence of free water on plant surfaces is crucial to pathogens during germination, formation of appressoria, and penetration. Furthermore, the time during which free water is present on leaf surfaces is directly or indirectly (via leaf water potentials) related to the length of the latent period (Shearer and Zadoks, 1972, 1974) and probably also to the duration and rate of sporulation. Many species of plant pathogenic spores resorb moisture from the atmosphere in air humidity ranges of 98–100% (note that 98.5% RH equals ca. $-20$ bars). The process is called hydration, but the presence of free water for a few hours usually is necessary for development of the germ tubes, mildews being a notorious exception.

In the field, the length of the period of surface wetness is a given quantity that can be influenced by irrigation regimes. The infection process from germination to penetration must, therefore, be completed within the wet period, because the spores may die when the infection process is interrupted by dry periods. If interruption is not lethal, the infection process may temporarily cease, only to be completed during the next wet period. The time needed for infection is a function of temperature and the fungus species. Various species of fungi appear to differ in their infection rates, as well as in their ability to survive interruption during the wet period, and both types of "adaptation" to relatively short, wet periods occur (Bashi and Rotem, 1974). Various sources of free water differ in their roles in epidemics according to the nature of their formation.

## D. DEW

Dew forms as a result of radiative loss of heat from the plant surface and transfer of water vapor to it. Net loss by radiation is accompanied by transfer of sensible heat from the warmer air to the cooler surface of the plant. As dew condenses by release of latent heat of condensation, the transfer of water to the surface is essentially a function of the vapor pressure gradient and the transfer coefficient involving molecular diffusion near the surface of condensation and turbulent transfer from greater distances, i.e., from the soil (distillation) or from the atmosphere (dew fall) (Monteith, 1957). The fluxes of water vapor involved are therefore governed by radiation, temperature, humidity, and wind speed. With respect to disease development, the duration of dew within the different layers of the canopy is likely to be much more important than the rate of its deposition or total amount deposited (Burrage, 1972).

Conditions necessary for dew formation include plant surface temperature below the dew point of surrounding air. This is frequently realized through the loss of heat by infrared radiation from the plant parts at night which depressed the surface temperature a few tenths of a degree below that of ambient air. Clear skies and restricted air turbulence facilitate such temperature decrease (Smith, 1970). Noffsinger (1965) added fog interception by plants and guttation to the two types of dew already mentioned. It seems preferable to classify the latter two as additional source of free water on the plant surface rather than as dew. The obvious importance of rain, snow, and fog in supplying moisture to microorganisms commonly leads to underestimation of the significance of condensed water or dew (Yarwood, 1959).

After World War II, plant pathologists became aware of the importance of leaf wetness duration caused by dew (Rotem and Reichert, 1964). The pioneering work of Duvdevani *et al.* (1946) stimulated interest in dew in semiarid areas and elsewhere. Even in northwest Europe, where the weather is "rainy" with ca. 700–900 mm of precipitation yearly, up to half the periods of leaf wetness in some growing seasons can be attributed to dew (A. F. Van der Wal, unpublished). Dew appears to be of little importance in influencing water balance in plants in arid regions, because the amount is an order of magnitude smaller than the potential evaporation. Far more important to plants than the actual amount of dew is the duration of wetness by dew. In many cases this is 8–12 hr or more, beginning in the evening and ending early in the morning, usually a few hours after sunrise.

Dew is not a factor in transport of spores, except for the run-off of spores with dew droplets after heavy dew. Dew originates in the boundary of a plant and air and disappears by evaporation. However, dew is a source of free water, and fruiting bodies, like pycnidia of *Septoria* spp., protrude their cirrus containing the spores as a slimy horn via the stomata on the leaf surface (Eyal, 1971). Probably by resorption of water by the plant, and by the fungus in it, the pycnidial contents swell more than the wall of the pycnidium, and the cirrus containing the spores, having a high osmotic value (Fournet, 1969), is pushed out of the ostiole along the guard cells to the leaf surface. However, spore germination in the cirrus is negligible, but once the cirrus content is diluted with dew, the germination rate increases; in fact infection can take place in a single period of leaf wetness of ca. 10 hr at 10°C (cf. Rapilly and Skajennikoff, 1974).

### E. Measuring Dew

Since the factors influencing evaporation also affect condensation, theoretical approaches using the vapor flux technique or the energy balance technique make it possible to compute dew deposit, as discussed by Noffsinger (1965). Although the measurements needed for the computations are standard in micrometeorological studies, none are made in most epidemiological work, and a treatment of the computations is therefore omitted here. The following survey of instruments includes those used in epidemiological research. For a more detailed discussion the reader is referred to Monteith (1972) and Noffsinger (1965).

Duvdevani *et al.* (1946) used a dew gage, a standardized painted block of wood that was exposed at sunset. At sunrise the form and distribution of the dew deposit were observed. The appearance of the dew deposits was matched with a set of photographs, showing various types and quantities of deposits. This dew gage is simple to use and inexpensive per unit, but careful attention is required to time of exposure and the time of making observations. The painted surfaces deteriorate rapidly and require frequent replacement by new blocks.

Dew balances are based on measuring change in weight resulting from dew deposits on the receptacle. These changes are mechanically transmitted to the recording beam. The Hirst dew balance consists of a polystyrene cylindrical block acting as the wetness receiver, placed at the top of the crop, and raised periodically. The ink pen at the end of the arm of the balance is protected from wind, and writes on a daily chart that rotates on a drum. Other receptacles are a blackened aluminum plate bent into a flattened conical shape with its aperture directed upward, or a plate of equal size and shape made of some thermoplastic material. The latter is reported to have a closer temperature correspondence to that of the vegetation than the aluminum plate. Wind may cause the sensing element to oscillate and produce a trace that is difficult to interpret. In general, such balances show reasonable agreement in recording duration of leaf wetness. Dew recorders based on change in length of the sensing element are used in various modifications. In the Wallin-Pathemus type the sensing element consists of a water-sensitive animal membrane (lamb gut) about 20 μm thick. When dew is deposited on the membrane, it expands and allows a spring to move a stylus to a rotating recording chart. The stylus maintains its position as long as dew is present. When the dew evaporates the element dries and contracts, withdrawing the stylus from the recording chart.

In the Fuess' hemp thread recorder and the De Wit leaf wetness recorder (Post, 1959), a contracting thread serves as a sensing element. A pen driven by a clock mechanism records on a chart the amount of contraction which is then translated mechanically into a curve. Instruments based on change in length of the sensing element are known to respond at high humidities, and before dew is actually deposited. These instruments can be calibrated with a psychrometer, a flashlight, and a lens. Air humidity near the dew sensor can be measured, and the initial deposition of dew can be observed in the dark by holding the flashlight so that the shaft of light skims over the leaf surface in the direction of the observer, and by looking through the lens at the appropriate location on the leaf. Very small drops can be seen glittering in this pencil of light, and their increase in size and number can be followed periodically. By comparing such data with those in curves of the wetness recorder, calibration can be made. Rain and dew can be easily assessed from the rate of change of the curve at the beginning of the wet period. Additional problems arise when there is an overlap between the end of the wet period caused by rain and the beginning of a period caused by dew. Generally, however, such problems are not serious.

Taylor (1956) devised a dew recorder in which the sensing element was a glass disk rotated by a clock mechanism. A pencil with a stylus containing indelible ink rested on the disk and made a clear line only when the glass was wet. He stated that the surface of the disk had an emissive value of approx. 0.9 in the infrared, and concluded that the disk was useful in dew measurements because its emissive value corresponded to that of canopy surfaces.

Dew sensors based on changes in electrical resistance consist of a horizontal grid. An electrical potential difference is caused across the two ends of the grid, which is embedded in nonconductive material. Moisture deposited on the grid completes the circuit and a flow of electrical current results. After the moisture evaporates, the electrical contact is broken. The duration of the contact provides an index of the duration of wetness. The Schurer–Van der Wal leaf wetness recorder (1972) is based on changes in electrical resistance between electrodes on the leaf surface. Instead of two electrodes, a grid can be used. Sets of sensors connected to form a "line" measurement (rather than a "point" measurement) can provide very accurate determination of the duration of wetness in various layers of the crop. Using leaves instead of other surfaces the best approximation is obtained for leaf wetness duration. The change in electrical resistance, transformed into a change in mV d.c., and, like other sensors that produce an electrical signal, automatic and remote measurment and data acquisition are possible. Dew and rain can be detected from the shapes of the curves, especially the rate of increase in the output (Fig. 2).

In summary, the ideal instrument should record dew formation on

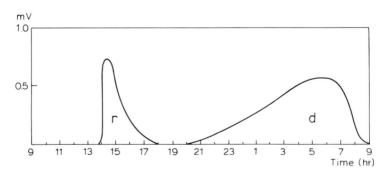

Fig. 2. A reproduced record of the output of the Schurer–Van der Wal electronic leaf wetness recorder, covering a period of 24 hours with rain (r) from approx. 14.00 to 18.00 hr, and dew (d) from 20.00 to 09.00 hr. During rain the output reaches its maximum within ca. 10 min. Dew causes a gradual increase in output, and a maximum at ca. 06.00 hr.

leaves (Schurer and Van der Wal, 1972), and modern techniques make it possible to overcome the physical difficulties inherent in the use of leaf surfaces in the sensor as mentioned by Taylor (1956). The placement of all dew meters, or at least the sensing element, is very important. The dew point is reached most rapidly in the upper canopy of a crop, and dew forms faster among plants than on bare ground. The sensor should be placed so that it is exposed to permit radiation in a position comparable to that of the plant parts to which the measurements refer. When leaves are used in the sensing element, dew forms first on those that are exposed to the open sky. When leaf positions at different canopy heights differ in exposure, a "line" measurement will provide an averaged figure for each level in the crop profile.

A general picture of dew formation shows that it is formed during clear nights preceded by restricted turbulence at sunset, usually just before or just after darkness, providing the air temperature approaches the dew point. The quantity of dew deposited gradually increases during such nights to a maximum in early summer in Northwest Europe just after sunrise (05.00–06.00 A.M.). On sunny mornings dew evaporates rapidly within 30 min, approximately between 07.00 and 07.30 A.M., beginning in the upper layers of the crop canopy.

At present, epidemiological research does not require dew measurements more often than for ca. 30 min of the actual period, because information about the duration of wet periods in relation to infection processes is not determined with time intervals shorter than 30 min.

Aside from desired accuracy, the choice of instrument is mainly determined by cost, power requirements, labor in interpreting records, and maintainance of the instrument, as well as need for automatic data acquisition. It is advisable to check performance of the instrument not only in the laboratory but also outdoors, where it should be exposed to field conditions for several weeks.

F. RAIN

Rain differs from dew in three aspects, although both are sources of free water at the plant surface. First, raindrops do not originate near the boundary of plant and air but rather in clouds of considerable thickness and usually at altitudes of a few hundred meters. The trajectory of systems carrying clouds that are likely to produce rain can be made visible by satellites, and synoptic weather maps can be used to forecast their occurrence. Second, because a system that carries rain clouds has a trajectory, fungus spores can be picked up by the system when it passes an area where infected and sporulating crops have been growing, and raindrops formed in such a system can wash out a spore load in the system and in the air masses under the clouds (Gregory, 1973). Third, raindrops have impulse when they hit a crop or the soil. Such forces may be necessary to liberate spores from their fruiting bodies (Rapilly *et al.*, 1973; Hunter and Kunimoto, 1974).

When the air masses with a cloud system pass over areas where sporulating crops were grown, spores are likely to be lifted off by strong winds and convection and mixed with air, especially if the areas are well above sea level, e.g., hills, high plains, and lower parts of mountains. Spores lifted off from fields at elevations of 2000 m can easily be carried off to altitudes of ca. 3000 m by convection and vertical mixing of air masses, enabling spores to be carried off hundreds of kilometers before being gradually deposited. To a large extent wind speed influences the length of the trajectory over which the spores can be transported.

Using satellite photographs of clouds in tracing a spore source in hills somewhere at distances over 1000 km from the spot they were deposited in the wheat crop, Nagarajan *et al.* (1976) described long-distance transport and deposition of spores. Spores of *Puccinia graminis* (stem rust of wheat) were transported a distance of ca. 500 km per day, and within a few days a subcontinent was passed by spores in the cloud system (Fig. 3). Raindrops falling through the cloud system washed out the spores, which landed on an uninfected crop as shown by spore traps designed to pick up rain water. Prediction of the first sporulation was based on the spore catch and occurred 10 days after arrival of the cloud system that contained the spores.

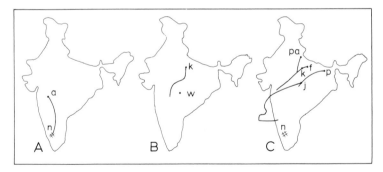

Fig. 3. Transport of *Puccinia graminis tritici* uredospores in India. (A) The 700-mb level air trajectory for the probable airborne inoculum of Ahmednagar (a) originated over the Nilgiris hills (n) and traversed 900 km in 36 hr. (B) In 48 hr the 850-mb trajectory of Kanpur had moved 700 km north from over the infected fields of Wardha (w). (C) The 700- and 850-mb combination trajectory for Jabbalpore (j) after 150 hr led southward. The 700-mb trajectory for Pusa (p) ended over Jabbalpore, and the Pantnagar (pa), Faizabad (f), and Kanpur (k) trajectory for 700 mb shows that they are off-shoots of the Jabbalpore trajectory. (From Nagarajan *et al.*, 1976.)

If the air mass in which raindrops originate does not contain spores, raindrops laden with spores, picked up on their way down to the crop surface may still be found. Spores washed out in this way usually originate less than a few hundred meters from the place of deposition. Long-distance transport of airborne spores is important when rain showers causing their deposition occur early in the season and start regularly occurring epidemic in one area or introduce new diseases in others (Meredith, 1973). If these "spore showers" occur later in an already infected crop, their contribution to the increase of the level of disease is considered unimportant, because of the relatively small quantity of spore influx as compared to the amount released by the infected crop.

Falling raindrops increase their kinetic energy. When they hit a plant or fruiting bodies of fungi on a leaf, this energy can be of use in liberating the spores, which may be airborne in tiny droplets. The latter evaporate readily and the spores behave as if they were wind borne.

## 1. Rain Puff

When a large drop falling at terminal velocity hits a *dry*, inelastic surface, the liquid in the drop spreads radially over the surface, at an initial speed of over 70 ms$^{-1}$. The speed decreases as the liquid spreads out over a radius of over 2 cm, with the whole process completed in about 2 msec. The rapid movement of the liquid disturbs the air within the laminar boundary layer. The resulting "puff" transfers spores into the turbulent layer, where they can be dispersed by eddy diffusion (Gregory, 1973).

## 2. Rain Splash

When a raindrop falls into a liquid or strikes a water film on a wet surface, "rain splash" is observed. The various events during splash are described by Gregory (1973), with respect to depth of the liquid, velocity and size of the droplet, angle of impact, etc. When a droplet falls on a surface covered with a thin liquid film, the surface film is pushed outward by the descending drop. The resulting radially moving mass of liquid is then deflected upward where the surface tension molds it into a crater that breaks up into jets and rays of droplets. Contamination of an area of the order of 1 m$^2$ is possible when a single drop hits a sporulating leaf.

With an average rainfall of 1000 mm per annum each square meter of ground would receive something of the order of a thousand million raindrops, enough to produce a real splash. When rain washing of the air occurs rapidly, it ends in the process of spore dispersal and seems to be effective even with small spores, which are only slowly and ineffectively deposited by other processes (Gregory, 1973).

In addition to washing spores out of the air, raindrops may transport spores from one atmospheric level to another by acquiring spores on their way for a time during their fall, and by evaporation of the drops before reaching the ground level. Drops smaller than 0.2 mm in diameter seldom reach the ground level. In general, raindrops reaching the ground are between 0.2 and 5 mm in diameter, with small drops more numerous than large ones. The particles captured are then in suspension in the air lower down. If carried up again in a thermal, spores can become concentrated at the base of a cumulus cloud (Gregory, 1973).

Rain can also limit spore dispersal. On papaya (*Carica papaya*) fruits with mycelium mats of *Phytophtora palmivora*, raindrops of relatively large diameter liberate sporangia on impact, but constant heavy rain cleans the air immediately from the spore-laden small raindrops, and in fact little infection is found after prolonged heavy showers. But intermittent showers and wind are ideal for liberation and dissemination of the spore-laden small drops. They are carried by the wind stream and settle on surfaces of neighboring plants. Such spores are not caught by traps that take air samples only. Special devices are needed to detect such dispersal of spores in the air by wind-blown rain (Hunter and Kunimoto, 1974).

## G. MEASURING RAIN

Meteorologists have developed standard instruments for recording rainfall in various forms. All collect water and the recording rain gages transfer the weight of rainwater kept in a container to an arm with a pen on a rotating drum. Heating and defrosting modifications are available by which amounts of snow and hail can be measured as well. I have found no reference to instruments in operation in plant pathology that measure the impulse of raindrops near the crop level. Distribution of drop sizes has been measured but such measurements are not routine.

## H. SOIL MOISTURE

Disease control by adding moisture to soil is known for scab in potatoes (Lapwood *et al.*, 1970; Lapwood and Adams, 1973). Cook and Christen (1976) report on temperature–water potential interactions for the cereal root-rot fungi. Soil moisture is not yet a common factor in epidemiology of plant diseases in the field and examples of descriptions of epidemics and forecasts based on soil moisture data are rare.

To a large extent soil moisture determines the plant water potentials to which pathogens in leaves, stems, and fruits are subjected. However, soil water and its relation to leaf water potential and airborne pathogens

example of the value of proper observation. He stated that as a result of preliminary observations and studies in the Philippines, it was established that abundant rainfall during the flowering cycles of pineapple (*Ananas comosus*) was associated with high incidence of pink disease, said to be caused by strains of acetic acid bacteria (*Acetomonas* sp.), and that small amounts of precipitation during blooming resulted in low incidence of pink disease. (Fig. 4). It also appeared from data on weather and plant development that, with abundant precipitation during flowering, increased incidence of pink desease was not always found, indicating some other determining factor. Hine concluded that periods of rainfall in combination with periods of moisture stress prior to flowering determined the appearance of disease symptoms.

Observing epidemics requires sound epidemiological and ecological knowledge, common sense, and a critical but open mind. Initial important observations are those on temperature and moisture relations during the course of an epidemic. Moisture may act directly on disease, or indirectly by influencing water potentials of plants (e.g., Hine, 1976) and soil (Lapwood and Lewis, 1967; Cook and Christen, 1976). The starting time of an epidemic is obviously of interest to the epidemiologist. Close obser-

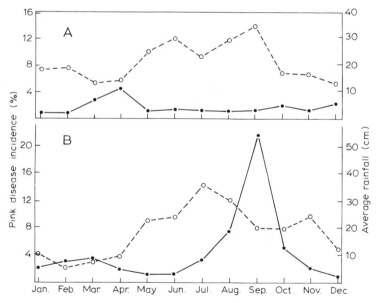

Fig. 4. Distribution of rainfall (o - - - o) and incidence of pink disease (●——●) of pineapple fruit at Del Monte, The Philippines. Data are the average monthly rainfall and incidence of disease during: (A) 1962, 1963, 1964, and 1965 (low disease incidence), and (B) 1957, 1958, 1961, and 1966 (high disease incidence). (From Hine, 1976.)

vation of distribution of the disease in the field can inform the observer much about the state of the epidemic and the possible source of inoculum. When foci stand out clearly in the crop, the inoculum is restricted to small areas.

Foci disappear when the infection cycle is repeated and spores are disseminated. For airborne diseases, a regular spread of the disease in the field indicates a rather advanced state of the epidemic, with initial infections having occurred months before, e.g., as early as the previous season. Disease progress during winter is slow, but compared with the growing season, the winter period is long and moist conditions usually do not limit pathogen development.

Abrupt increases in amount of disease indicate waves of infections following a period during which the weather was adverse to infection, but the inoculum accumulated in the crop or was transported into it in masses. The increase in percentage of infection multiplied by the increase in infected area provides a reasonable estimate of disease progress. If the relation between amount of inoculum and the resulting amount of disease is known, an estimate can be made of the amount of effective inoculum during the disease. The possibility that the infected area may increase by growth of the pathogen in the plant without reinfection has been taken into account (Van der Wal and Cowan, 1974). Once the observation leads to hypotheses on relations between weather and disease, mathematical tools can be used to quantify the relationships. Multiple regression analysis (MRA) is suitable to detect direct relations between disease and weather factors.

By combining information about the infection process with the actual situation and with some help of mathematics, reasonable conclusions about the weather–disease relations can be reached. Sometimes this is enough, but results of observations may be used as the basis for selection of situational factors and variables in an experiment that should lead to further insight into the processes involved. Tu and Hendrix (1970) combined observation and experiment in an attempt to understand the summer biology of *Puccinia striiformis*, as did Bashi and Rotem (1974), when they detected possibilities of adapting various pathogens in semiarid areas to short periods of leaf wetness.

## B. EXPERIMENTATION

Observations may lead to correlation and sometimes dissension. The relations deduced between weather and disease may have validity in certain locations but not others where the range of weather factors involved or the relation between standard meteorological observations and microclimates in crops differs from those studied. Experiments in epidemiology are performed to elucidate the relations between weather factors and sequential processes of an epidemic. Selection of situational factors, independent variables and their levels, and dependent variables which reflect responses of the variation imposed are especially important during planning an experiment.

*1. Situational Factors*

Theoretically, most factors in an experiment can be varied, but for technical reasons only a few are. Many possible variables, for example, the situational factors, are maintained at one level. However, these factors can be varied in other experiments as independent variables. The responses of plant, pathogen, or disease usually are treated as dependent variables.

The levels of the situational factors usually contribute largely to the responses. The choice of their levels is therefore critical. Suppose an experiment is conducted at 20°C (the situational factor) and the influence of the length of the period of leaf wetness (the independent variable) on the spore germination (the dependent variable) is studied. If night temperature is 10°C in the area where the disease occurs, and the aim of the experiment is to study the variables in a situation comparable to the field situation, the "scene" has been incorrectly chosen. The same holds for studies on relative growth rates of pathogens in media with water potentials of ca. 0 to -2 bars, when in diseased leaves these fungi grew at much lower water potentials (Cook and Christen, 1976; Cook and Papendick, 1972). The choice of the levels of the situational factors is based on experience, field observations, and data in the literature.

*2. Variables*

The choice of dependent and independent variables is made after a problem has been clearly defined. The usual independent variables in epidemiological studies are moisture, temperature, and, in many cases, light intensity (Politowski and Browning, 1975). Spores of some fungi do not germinate or do so slowly in the light, but they germinate rapidly in the dark. Threshold values are of interest when scheduling, for example, the germination rate in the diurnal cycle as a function of light intensity. A minimum of six levels of each independent variable is advisable in the range chosen. Six points are a minimum in the statistical analysis of curves, e.g., tests of linearity. The choice of the dependent variables may be made according to systems analysis. If the experiment is part of a series of experiments to investigate environmental influences on the epidemic of a certain disease, the entire problem can be broken down into smaller problems that can be solved, one at a time, by successive experiments. A subdivision into individual experiments should be so made that the input of the current experiment is the output of the previous experiment and that results of the second experiment can be used as an input for the third experiment, etc.

*a. Independent Variables. i. Moisture.* The possible effects of various forms of moisture on infection cycles have already been mentioned. It should be ascertained that free water is present on the surface of the plant; "100% RH" is not always sufficient. The duration of leaf wetness, expressed in hours, should be measured at least at half-hour intervals, if continuous measurement is not possible. If it appears from these studies that some spores need, for example, 14 hours of leaf wetness for germination and penetration, that value should be compared with duration of wetness occurring daily in the field. If in the field only 8 to 9 hours of leaf wetness occur (the spores needing 14 hours), a test should be made as to whether the germinating spores on the leaf surfaces survive after intermittent dry periods. Patterns of moisture duration are then introduced as independent variables. To avoid testing of unnecessary patterns, careful attention is needed in selecting these patterns. They should be based on field observations.

*ii. Temperature.* Temperature determines the rate of processes such as germination, penetration, sporulation; 6 points at an interval of 5°C will be sufficient in a first approach. If some positions of this range appear critical, smaller intervals can be used.

*iii. Light.* Both light intensity and quality (wavelength) can influence the infection cycle, particularly spore germination, formation of fruiting bodies, and sporulation (Van der Wal and Luuring, 1976; Van der Wal and Zadoks, 1976). For perthotrophic pathogens it seems that light intensities from dark to shade (0–15 kLux) are critical for spore germination and formation of fruiting bodies. For the biotrophic organisms such as rusts, it seems that their sporulation is enhanced by higher light intensities (20–40 kLux), possibly related to the rate of photosynthesis of the infected plant.

Cultivar effects on disease increase are known and used in plant breeding. Effects on germination are possible but usually the cultivars manifest their influence later in the infection process (colonization, latent period, sporulation, and injury to the plant).

*b. Dependent Variables. i. Infection.* Attention should first be given to infection in relation to the previously mentioned three primary controlling factors of the environment: duration of leaf wetness, temperature, and light. Although this set of relations may be considered classic for study by plant pathologists, amazingly little *systematic* research has been performed on these relations. Hogg *et al*. (1969) surveyed the literature on the influence of weather factors on *Puccinia graminis*, *P. striiformis*, and *P. recondita*. The infection process was divided into a series of successive subprocesses including germination, elongation of the germ tube, appressoria formation, substomatal vesicle formation, extent of the colonizing mycelium, and latent period. Little information was found for some of these important subprocesses.

*ii. Sporulation.* A 24-hr day is the time base used in sporulation studies. Sporulation rates are expressed as numbers of spores per day. In epidemiological studies, sporulation rates during the first 10–14 days after sporulation begins are important. The peak sporulation rate may occur 3–5 days after sporulation begins. Light intensity may influence sporulation. For perthotrophic fungi such as *Leptosphaeria nodorum*, the duration of leaf wetness influences the latent period (Shearer and Zadoks, 1972, 1974) and light quality influences formation of pycnidia (Van der Wal and Luuring, 1976).

*iii. Dispersal and spore landing.* The mechanisms of spore dispersal have been qualitatively characterized. However, it is difficult to establish quantitatively the relations between spore production, spore liberation and dispersal, spore density on the surface of uninfected plant parts, and the resulting disease incidence. One approach involves relating the number of spores caught in or near the crop with a spore trap to increase of infection (Kerr and Rodrigo, 1967a).

A more physiological approach is to relate spore production to disease incidence. Suppose an energy conversion factor for plant dry matter into rust dry matter (*P. recondita*) is of the order of 1:10, and daily dry matter production of a wheat crop during grain filling is 200 kg per hectare per day. Spore production of 4 kg of rust spores ($= 10^{12}$ spores, weight 4 mg per $10^6$ spores) consumes 40 kg plant dry matter per day, or 25% of the daily dry matter production. Injury of this magnitude is serious but tolerable for some time.

If in this 4 kg of spores all were effective, one day of spore production would produce enough spores to kill the crop after one night with dew during the successive latent period. Van der Wal and Cowan (1974) showed that 20 spores per $mm^2$ leaf surface of *P. recondita* (similar to 0.6 $\times 10^{12}$ spores per hectare with leaf area index of 3) caused an initial infection of 10% of the leaf area on a moderately susceptible cultivar a few days after the first open pustule was found. The infected area increased to

75% after a few weeks as a result of growth of the pathogen (newly formed pustules). Reinfection was excluded.

Approximately 40 spores per mm$^2$, one for each stoma in the upper surface of the wheat leaf (= $10^{12}$ per hectare) would have caused an initial infection of at least 20%, leading to rapid decay of the infected leaf. Not all spores that germinate on the leaf surface, even of a susceptible cultivar, cause formation of a pustule. The maximum found in our studies was 0.3 pustule per spore, an efficiency of 30%. This may partly explain why less than 100% infection occurred with 40 spores per mm$^2$, because a pustule is usually larger than the area that surrounds one stoma. The ratio of the number of infective spores to the resulting number of pustules reflects resistance of the host. The production of spores for a set of environmental conditions and a specific cultivar/race combination can be determined as well as the relation between spore density on the surface and resulting extent of the disease. Note that the density on the leaf surface is given and definitely not the density in a suspension.

Once these two relations are established, the bridge to dispersal can be made by an appropriate field experiment. Spore production and the relation between degree of disease and spore density are highly race–cultivar dependent, but spore dispersal is not. With one race–cultivar combination it is therefore possible to detect a widely applicable figure for dispersal for a pathogen species or group of species.

A variety of spore traps are commercially available. One of the first was the Hirst volumetric spore trap (Hirst *et al.*, 1967), which traps spores in samples of dry air. Other types trap spores in raindrops or splash spores (May, 1967; Hunter and Kunimoto, 1974).

Spore traps are useful in epidemiological research in two ways. First, they can be used to determine the initial arrival of immigrant spores and therefore are a help in timing the first infection (Nagarajan *et al.*, 1976). They also are useful in quantifying the relation between spores produced, dispersed, and deposed in the field. At present instruments for trapping spores that are liberated in tiny wind-blown water droplets need more attention and further development.

Once the relations between environmental factors and the various processes in an epidemic have been established experimentally, the data must be integrated in order to construct the pattern of an epidemic. Measurements or records of free water together with temperature, knowledge about the presence of inoculum, and an estimate of its amount may be sufficient as a start. Even with only a programmable table calculator an approximation of what is occurring can be evaluated quantitatively.

For experimentalists, well-chosen, well-planned, and technically well-conducted experiments are an ideal. Experiments should be planned long before they begin. Selections of situational factors, independent and dependent variables, and schemes for data acquisition must be ready when the experimentation starts. Experiments have their limitations, set by available funds, manpower, equipment, and skills. Modern electronics are of help in evaluating the output of experiments. Models can be useful to direct observation and to improve planning of experiments.

## C. MODELING

During the last fifteen years epidemiologists have become increasingly interested in modeling as a research tool. Van der Plank's (1963) description of the proportionality between time and the relative change in disease has been of particular interest. The mathematical description of epidemics has a long history. Baily (1957) reviewed much literature on medical epidemiology, and Van der Plank (1963) synthesized the literature on field crop diseases. He adopted simple exponential and logistic models to describe the development of epidemics. More explicit treatments of the influence of weather factors on the increase of epidemics are found in more recent models (De Weille, 1969; Bourke, 1970; Waggoner, 1974).

Models of epidemics should be constructed to reflect the underlying natural processes insofar as this is possible and useful, and they should not be exercises in curve fitting (Jowett *et al.*, 1974). Models merely provide a way of interpreting data. They are, however, important tools in epidemiological research. When combined with computing facilities, they serve as quick references, and mirrors for testing hypotheses, and they also facilitate the choice of situational factors, as well as independent and dependent variables. It should be emphasized that models reflect the knowledge and skill of their designers. If well-chosen and designed, they can be very useful.

According to Waggoner (1974), three main types of models are found in epidemiology: differential equations, multiple regression, and simulation.

### 1. Differential Equations

Much ecological modeling involves modification of Verhulst's basic equation $dy/dt = ay(b - y)$, where $b$ represents the upper limit of growth, the sum total of what the environment will support. Verhulst incorporated feedback into the model of exponential growth. As the population increases, the information that it is approaching its limit is fed back into the system, slowing down growth and ultimately inducing a steady state (Jowett *et al.*, 1974). The resulting equation $y$ to $t$ is the logistic:

$$y = \frac{k}{1 + be^{-at}}$$

where $y$ = size of the population and $t$ = time.* For more details the reader is referred to Pielou (1969).

The exponential and logistic models are nonlinear. The exponential model $y = ae^{bx}$ can be linearized by: $\ln y = \ln a + bx$. The logistic model cannot be directly linearized without additional assumptions, and it is therefore fitted iteratively. Such fitting requires computing facilities and will only be successful if the initial assumptions hold (convergence) (Kempthorne et al., 1954).

The major advantage of the method used by Van der Plank (1963) is that because the maximum of the disease is the known value of 100%, logistics are linearizable. Note that, paradoxically, fitting by direct least squares and by least squares after linearization does not produce the same parameter values (Jowett et al., 1974).

Exponential and logistic models are not suited to all situations. In fact, they fail to reflect underlying processes. This is the main point of Bourke's (1970) criticism of Van der Plank's book (1963). The influence of weather factors is not explicitly detectable in the model. All effects are amalgamated in the parameter that controls the rate of increase of the disease. Furthermore, the disease progress curves do not appear to be truly logistic, although the curves obtained in controlled environments with temperature constancy show good fit. But for field conditions the logistic does not reflect reality (Jowett et al., 1974). Differential equations serve best in areas where the weather shows a fairly constant pattern for a long time. They can also be used to analyze results of climate–chamber experiments in which growth rates of plant and pathogen are chosen and reproduced by controlling environmental variables.

Nevertheless, the pedagogical significance of the logistic and related functions should be emphasized. I fully agree with Jowett et al. (1974) that students in the early stages of their training should be confronted intensively with these functions. Differential equations and related more complex models seem useful when comparing epidemics in different cultivars, and in studying effects of control measures. They are therefore considered useful tools in plant breeding. The cultivars in a treatment are all exposed to the same, usually not quantified, set of situational factors in the field. To a large extent the weather during the tests influences the rate of progress of the disease. And although the absolute level of disease can be decisive (dosage effect, "disease pressure," inoculum density, and resulting disease incidence) for a resistance test, cultivar reactions are compared for differences and not in absolute terms. Tolerance tests should be carried out carefully, because plant injury and loss are closely related to

* $k$, $b$, and $a$ being constants, e the base of natural logarithm.

the environment (the situation) and to time and amount of infection (Van der Wal *et al.*, 1975).

In summary, differential equations are unsuitable for studying processes of an epidemic. However, they are very useful for comparing epidemics, as is needed, for example, in resistance breeding, and effects of chemical control of diseases.

## 2. Regression

The patterns of epidemics can also be characterized by regression equations, including the MRA technique (multiple regression analysis). The general equation is of the type:

$$\text{disease} = a + b_1X_1 + b_2X_2 + b_3X_3 + \cdots + b_nX_n$$

where disease = the estimate of disease incidence, $a$ = the intercept, $b_1 \ldots b_n$ = the partial regression coefficients, and $X_1 \ldots X_n$ = the independent variables.

Analytis (1973) concluded, that normal distribution of variables, discarding of aberrant values, linear relationships, and objective base for the formulation of variables are prerequisites for reliable application of multivariate regression analysis in studying phytopathological problems. It should be emphasized that application of statistical tools requires the fulfillment of certain prerequisites, an elementary consideration that is too frequently overlooked. I, for instance, have never encountered a normal distribution in phytopathology, except in handbooks on statistics. However, MRA should not be excluded as a technique when deviations from the normal distribution fall within certain limits (cf. Analytis, 1973). Although it is doubtful that all prerequisites are available in epidemiological studies, MRA can be useful. It is better suited than differential equations for studying epidemics because at least the various factors appear separately, in fact in a linear combination in the equation. The technique is described in various handbooks, such as that of Snedecor and Cochran (1972).

Once applied to sufficient data, estimates of the partial regression coefficients $b_1 \ldots b_n$ are obtained. Clarification is needed of the meaning of the values of the coefficients, as well as the basis of the assumption of an additive model to describe relations between the plant and pathogen. It is important to emphasize that evidence of a statistical dependence of one variable on another does not prove cause and effect relationship. In MRA the independent X-variables are combined in accordance with their direct relationship to the dependent variables. The possibility of measuring the net effect of each independent variable makes MRA valuable in epidemiology. However, some key factors may be omitted in the selection of variables that constitute to the equation (Butt and Royle, 1974). This type of omission is indicated when the accuracy of prediction by the equation varies with the source of data. The equation may not predict accurately when values of the independent variables lie outside the range of values used in constructing the equation. A most difficult task is to disentangle limitations that are imposed by intercorrelation between the X-variables. In field experiments a multitude of uncontrollable factors act and interact simultaneously. Problems caused by intercorrelation can be reduced by selecting, with the aid of a correlation matrix, candidate X-variables which are not highly correlated (Butt and Royle, 1974). This preselection technique was adopted by Analytis (1973) before he employed MRA to study factors affecting apple scab epidemics.

   In recent years MRA has been used increasingly for analyzing plant disease epidemics but with varying success (Royle, 1973). In studying the epidemic of the hop mildew (*Pseudoperonospora humuli*) Royle (1973) used the following set of independent variables: duration of leaf wetness, duration of leaf wetness with a rainfall contribution, quantity of rainfall, duration of rainfall, mean temperature, mean vapor pressure deficit, duration of sunshine, airborne spore density, and funnel-trap spore catch. It appeared from the MRA that dew itself did not permit substantial infection even in combination with the required temperature. Only moderate levels of infection appeared at leaf wetness caused by dew when sufficient inoculum was available. Usually, but not always, heavy infection occurred after a period of leaf wetness due to rain. This was explained in a laboratory experiment. Infection was severe after an infection period in the light, when zoospores of *P. humuli* settled near the stomata. Relatively little infection followed in the dark, because zoospores settled at random on the leaf surface and most of them subsequently did not enter the stomatal pores. Rain often occurs during the day, whereas dew forms exclusively at night, in the dark. In the case described by Royle, it appeared that although free water on the leaf surface was provided equally by rain and dew, they differed in time of diurnal occurrence. This scheduling was decisive for settling and penetration of zoospores. In this example, MRA guided a well-planned experiment after which the infection process was much better understood. MRA indicated the choice of the situational factors (free water, temperature), the independent variable (light), and the response (settling and penetration of the zoospores).

   Kranz (1974) mentioned a number of equations proposed to estimate and predict effects of environmental factors on the disease process (Table III). In some of these equations relatively large values of the partial regression coefficients were attributed to moisture variables. There was one

## TABLE III

### Some MRA Equations[a] Proposed to Estimate and Predict Effects of Environmental Factors

| Disease | Host | Equation | Terms | Authors |
|---------|------|----------|-------|---------|
| Blister blight | Tea | $y = 33 + 0.3145x_1 - 0.03725x_1x_2$ | $y$ = number of blisters per 100 shoots about 3 weeks ahead, $x_1$ = spores caught per unit volume of air, $x_2$ = mean daily sunshine | Kerr and Rodrigo (1967a) |
| Leaf rust | Wheat | $y = -3.3998 + 0.0606x_1 + 0.7675x_2$ $+ 0.4003x_3 + 0.0077x_4$ | $y = 1 \log$ g% leaf rust severity 14 days after date of prediction (DP), $x_1 = 1 \log$ g% leaf rust severity on DP, $x_2 = 1 \log$ g% leaf rust severity on DP, $x_3$ = growth stage of wheat in integers from 1–9, $x_4$ = Schrödter's temperature equivalence | Burleigh et al. (1972) |
| Apple scab | Apple | $y = 0.08017 + 0.009W_{i-b}(sk)$ $+ 0.00211H_{2(t-10)}$ | $y$ = rate of change in disease severity; $W_{i-b}(sk) = \Sigma_{i=1}^{n} f(x_i)$ = spore germination function at date $i$ minus the incubation period $b$ in days, $n$ number of hours with leaf wetness and $f(x_i)$ the temperature equivalent by Schröder, $H_{2(t-10)}$ = frequency of hours with temp. 18–20°C 10 days before disease reading | Analytis (1973) |

[a] References fully given by Kranz (1974). A selection of equations from a table given by Kranz (1974) in "Epidemics of Plant Diseases" (Springer-Verlag).

exception, the regression model of Kerr and Rodrigo (1967a,b) for blister blight of tea, where only two independent variables appeared in the equation: spore catch per unit volume of air and mean daily sunshine. But duration of sunshine is used here instead of duration of wetness, both being highly complementary in their situation. Measurements of sunshine are taken routinely in many posts in Sri Lanka. Here, for practical reasons, sunshine duration was used instead of wetness duration, an uncommon measurement.

MRA is a relatively simple mathematical technique, and in many standard programs elaborate calculations are avoided. Although functional relationships cannot be described by a regression equation, Massie and Nelson (1973) used regression models in subroutines of a simulator. Regression models are empirical by nature—the product of experiment and observation. The models make no provision for effects of untested factors such as changes in weather patterns, pathogen populations, and systems of crop management. MRA is valuable, however, in quantitatively evaluating variables with respect to their response. In epidemiology, MRA has a bridging function, and there is no good reason to omit it during routine analysis of an epidemic. As in the example from Royle's study, MRA showed the way to further work on the infection process. For such reasons MRA is already useful.

### 3. Simulation

Simulation—mathematical simulation as it is called by Waggoner (1974)—is a logical assembly of information on the interrelations among pathogen, host, and environment, gathered to calculate patterns of epidemics. In this context a simulator is *in concreto* a computer program written in a language that fits its structure and purpose. Simulation is intriguing because of the challenge it presents the model builder to organize in an orderly fashion the relations he thought were important in the pattern of the epidemic.

The purpose of simulation, and model making in general, is to describe an epidemic in space and in time. The logical application of this information is prediction of an epidemic. Here again the purpose should be solely to predict absolute numbers of infections or diseases in space and time. At present, however, this seems difficult. When crop loss–disease relations are incorporated in decisions of pest management, such epidemiological findings are essential.

The simulator transforms data on weather, edaphic factors, crop, and pathogen into disease progress, and this is done in a structured way. The well-known "apparent infection rate" in the differential equation is "split" into various components that can be indicated as constituting the rate quantitatively. In contradistinction to MRA the results can in principle be applied in a variety of years and places.

The simulator is in a way a notebook, in which increases of levels of disease and changes of rates of infection are recorded. It enables quantitative evaluation of the consequences of changes somewhere in the epidemic with little hindrance imposed by the complexity of the processes. But the simulator is much more than a notebook. It can be made to help follow stepwise one event after another.

*a. Structure of the Simulator.* Essential information for building a simulator can be found in sections on etiology in books on plant pathology. The life cycle of the pathogen is first translated into a flow diagram by the programer. The next step is to determine the impact of environment on each stage of the life cycle. The simulator, like the epidemic itself, is mainly concerned with rates of processes in the infection cycle (germination, sporulation). By setting initial values and by accumulation of rates over a certain period, the levels are obtained, indicating the state of the epidemic at that time. Waggoner (1974) stated that at the beginning one can concentrate on the known relations between crop, environment, pathogen, and time, without worrying about relations that are not obvious.

*b. Characteristic Time.* The simulator is built to describe the pattern of an epidemic in time. Characteristic times in the components of the program that are to be assembled, must be considered early. From all possible time periods, i.e., fractions of a second to millenia, the present author believes the 24-hr day should be considered the main unit of time in steps in the program. This accounts to ca. 100 steps for a growing season. One day is not considered a separate step, but rather one in a sequence. The state of the system at the end of one day determines also the events during the next day. In some cases spore germination leads to infection when two successive days provide suitable weather for the pathogen. Hence the sequential events of these two days are decisive, rather than the conditions during one day (Bashi and Rotem, 1974). If sufficient inoculum is present, but the day has an insufficient number of hours of leaf wetness for completion of the infection cycle, and that day is followed by another one with the necessary additional number of hours of leaf wetness, infection occurs. Each day considered separately and screened for length of period of leaf wetness would appear insignificant for infection, but in combination such days are effective.

The processes within the day determine the "value" of that day for the pattern of the disease. The hour is a practical and adequate time unit for processes within the day. In experiments the hour as a time unit means at least half-hourly records; hence up to 48 measurements a day are involved. The duration of leaf wetness must be measured at least every 30 min. If the hourly values of temperature and light should reflect averages, two measurements on nonintegrating sensors may be too few for determining rates of various processes. If information on their temperature dependence permits a more accurate measurement, and the increased accuracy is necessary, measurements should be taken more frequently. Half-hourly measurements do not imply 48 steps in the program each day. Data reduction comes after measurement. Duration of leaf wetness, for example, can be expressed in only one value.

A workable division of a program that simulates an epidemic is one that balances predictions of importance of various processes with actual experimental data. Units in the program, having clearly defined inputs and outputs, should reflect units in the experimentation. Symmetry in the treatment of experiments and the units in the simulator facilitates experimental checks on relations within simulator units. In the simulator, time and place of infection in the growing season, i.e., suitable infection days, should be defined in relation to inoculum, moisture, temperature, light, and the presence of susceptible crop plants.

The length of the symptomless period, i.e., the time from infection to appearance of initial symptoms, is considered in simulators that predict disease incidence at a given time after recorded "infective days." This period is unimportant for multiplication of the pathogen, but it is important for studies on injury to the plant or crop. The length of the symptomless period usually is a function of temperature, race–cultivar combination, and sometimes leaf wetness periods, with the latter possibly mediated by leaf water potential, and hence indirectly by growth of the pathogen (Cook and Papendick, 1972; Stavely and Slana, 1975).

To a large degree the latent period, the time between infection and first sporulation, controls the pattern of an epidemic. It generally is a function of temperature, race–cultivar combination, and sometimes light and leaf wetness duration. The latent period of *Leptosphaeria nodorum* became shorter with increasing duration of the period of leaf wetness (Shearer and Zadoks, 1972, 1974). The shortest time between infection and appearance of ripe pycnidia was ca. 6 days at 15°C with continuous leaf wetness (24 hr per day). This was similar to the time from inoculation to appearance of pycnidia on wheat meal agar at similar temperature and light conditions (Van der Wal and Luuring, 1976).

Sporulation intensity and duration are greatly influenced by leaf water potential, air humidity, temperature, light, and race–cultivar combinations (Van der Wal and Zadoks, 1976). Repeated cycles of sporulation with alternating dry and wet periods have been demonstrated for *Leptosphaeria nodorum* (Scharen, 1966).

At the present time dispersal figures in the field are merely guesses. Rational rules for removal of spores and their lifting in wind and rain have been incorporated in simulators, but both rules and parameters are uncertain though important in the calculus of absolute numbers of infections. Waggoner (1974) stated that "until the requisite biological and meteorological knowledge of dispersal is acquired, we shall continue encountering difficulties in comparing the output of a simulator to the observation of the absolute infection, and Bourke (1970) can justly continue complaining that simulation successes are mere predictions of the general rise of infection with time." The area with disease symptoms, however, can be related to spore production. The density of infecting spores on the leaf surface and the resulting extent of symptoms can be determined as well. In a well-chosen experiment these can be bridged for various pathogens in the field. Both initial relations, i.e., symptoms and their production of spores, and spore density causing symptoms, are highly specific for a cultivar–race combination. Dispersal, however, is specific only for the species. It is reemphasized that dispersal must be quantified, because a lack of specific information greatly limits the usefulness of simulators in crop protection.

For details on structure of simulators, the reader is referred to Waggoner (1974) and Krause and Massie (1975). A discussion on units of multiplication of spores or infected leaf areas, overlapping of infection cycles, resulting in a more or less smooth pattern of the epidemic, and stochastic processes is beyond the scope of this chapter. Because moisture controls various processes in many epidemics, its source, distribution during the growing season, and scheduling within the day should be known and incorporated when using simulators.

## IV. MOISTURE AS A FACTOR IN FORECASTING

The incidence of plant disease varies from season to season and place to place, because of differences in amount of initial inoculum and in environmental conditions. Effective control measures may be worthwhile in a year of severe disease but unnecessary in another, and reliable forecasts of likely incidence of plant disease thus save growers a great deal of money, provided that forecasts can be made early enough to organize control measures (Tarr, 1972), and provided also that the grower has enough confidence in the forecast to use it in planning control measures.

Epidemics of unimportant diseases are worth studying because of their possible contribution to better understanding of epidemics in general. Forecasting must be concentrated on epidemics that are relevant to the problem of world hunger. A forecasting organization is not only a model, but also includes mechanisms for systematic collection of data (the input for the model), and for distributing results to users, advisory and plant protection services, farmers, etc.

Forecasts can be effective if control of disease is possible. Interest in forecasting is increasing together with development of control measures. The possible type of control is reflected in the purpose of forecasting. If disease control is only preventive, plants must be sprayed before infection is expected. Forecasting the expected time of infection is then necessary. Many chemicals used in plant protection are more preventive than curative, and therefore prediction of infection must be specified as to time and place.

Prediction of plant diseases has a relatively long history; forecasting of some diseases has been studied for 40 years, including rice blast (*P. oryzae*) (Tarr, 1972), late blight of potato (*P. infestans*) (Bourke, 1970), and grape mildew (*Plasmopara viticola*) in Europe (Tarr, 1972). In the next sections a few forecasting systems will be briefly reviewed. The reader is referred to Tarr (1972) for more details about forecasting systems in operation.

## A. INITIAL INOCULUM OR WEATHER PRIMARILY DETERMINES THE EPIDEMIC

Systems of disease forecasting need not be complicated. In some cases, such as bunt and smut in cereals (*Tilletia* and *Ustilago* spp.), there is hardly any effect of weather on disease. The spores are transmitted from one season to another with the seed or via the seed that contains the pathogen. Lack of resistance of the plant or absence of effective seed dressing will lead to a fair infection. Outbreaks of such diseases are common in areas where seed dressing is neglected or not possible.

The amount of inoculum does not seem to limit the disease potential. Some soil moisture is necessary for spore germination, as it is for seed germination. According to Chester (1943) the incidence of leaf rust (*P. recondita*) can be forecast by using the degree of infection in the young wheat crop in Oklahoma after overwintering. Since weather in the area studied was always favorable to development of rust, the end result of the epidemic was predicted once the initial conditions were known in the period from March to April. The weather from December to March was a key factor. The fungus does not increase to any extent at temperatures below 10°C, since uredospore germination and penetration are slow below this temperature. In Oklahoma, March is the first month in which the temperature approximates 10°C, and therefore is critical. Subsequently the fungus multiplies, requiring several consecutive generations, each of about 10 days. The 10°C monthly mean can be used in selecting places on earth for the critical time of the initial inoculum level, for example, January in India, May in Canada, and the beginning of April in New York State (Tarr, 1972).

Potato blight had been intensively studied in Europe, the British Isles, the United States, and elsewhere. The first successful warning system was in the Netherlands (Tarr, 1972), where the four Dutch rules were formulated some 40 years ago. The appearence of blight depended on (1) a night temperature below dew point for at least 4 hr, that is, dew for at least that amount of time, (2) a minimum temperature of 10°C or higher, (3) mean cloudiness on the next day of at least 0.8, and (4) at least 0.1 mm of rain during the next 24 hr.

This method was considered very reliable in The Netherlands but in the southwest of England it was found that the four rules could be reduced to two, namely, a minimum temperature of 10°C and a relative humidity not falling below 75% for at least 2 days (Beaumont period). When applied in the southwest of England, the Dutch rules tended to forecast blight earlier than it actually appeared. Blight symptoms could be expected 2–3 weeks after a Beaumont period.

In other cases, such as forecasts of outbreaks of grape fruit mildew (*Plasmopara viticola*), warnings are based entirely on weather data; the amount of inoculum probably is not limited in the range that is of interest to the grower. In the Perugia areas of Italy, outbreaks are likely when the dew point is higher than 12°C, the minimum temperature for spore germination. The already present inoculum then becomes active (Tarr, 1972). In France the first outbreaks are predicted from the presence of active zoospores, i.e., at temperatures above 11°C, and precipitation that keeps the soil surface wet for several days. Secondary spread depends on rain and temperatures higher than 8°C, and especially on retention of moisture on leaves for at least 6 hr at temperatures between 11° and 20°C. Epidemics of this disease can be expected during warm humid nights with heavy dew or mists, or in rainy weather, particularly if the absence of sunshine prolongs the persistence of moisture on leaf surfaces. In some years forecasting reduced the number of spray applications from five or six to one or two (Tarr, 1972).

Forecasting of *Phytophtora phaseoli* on Lima bean is based on rainfall–temperature relations. A day (the characteristic time unit previously mentioned) is considered favorable for the disease when the 5-day moving temperature is less than 26°C, with a minimum of 7°C, and 10-day total rainfall is 30.5 mm or more. The disease is then likely to appear after approximately 8 consecutive favorable days. The spread of leaf spot of groundnut caused by *Mycosphaerella berkeleyii* and *M. arachidecola* in Georgia, United States, is favored by diurnal periods of 10 hr or longer of relative humidity of 95% or higher, and a temperature during these periods of approximately 21°C (Tarr, 1972).

In Germany observations on blight incidence have been correlated since 1958 with weather conditions at fifteen trial sites in representative potato-growing areas of West Germany to provide more material for the forecasting system as described by Bourke (1970). This system considers the following four states in the growth and life cycle of *P. infestans:* (1) sporulation, which requires 10 hr or more of moist conditions, i.e., hours during which there is either measurable precipitation or at least 90% humidity; (2) germination of sporangia and infection of the plant, requiring moist conditions of at least 4 hr; (3) mycelium growth depending on temperature but not on atmospheric moisture; and (4) suppression of spread of the pathogen during dry periods when RH is less than 70%.

Data on RH, temperature, and rainfall are recorded hourly at some fifty stations, codified on a weekly basis, and fed into a computer which weighs the observations with the appropriate parameters and calculates the weekly weather rating. A current total weather rating is obtained each week by adding the weekly rating, starting from a known average date of emergence of the early potato crop. A total rating of less than 150 warrants a negative forecast and, on the day on which a rating of 150 is reached (first critical date) growers are advised to beware of blight, which is likely to appear 10–40 days later. Immediate control measures should be taken when the total weather rating reaches 270, the disease normally appearing between 15 days before and 15 days after the critical day (Tarr, 1972).

Among the plant diseases of major importance, potato blight (*P. infestans*) is undoubtedly the one whose appearance and spread are most completely under direct control of environmental factors (Bourke, 1959). In most potato-growing countries the fungus is almost invariably present to such a degree as to give rise to an epidemic whenever suitable weather occurs and persists. No phenological observations other than that the plants are above ground are required to supplement weather data. Weather effects on growth of potato plants themselves do not play a major role in blight epidemiology. Hence it is reasonable to concentrate entirely on environmental influences on propagation and spread of the fungus. Krause *et al.* (1975) combined the Hyre and the Wallin system into one computer program (Blitecast) based on rainfall, RH, and temperature.

## B. Initial Inoculum Level and Weather Together Determining the Epidemic

In the case of rice blast (*p. oryzae*) in Japan the prediction of first infection is based largely on overwintering of the fungus, the time of appearance of conidia in the field, and weather as related to development of the rice plant (Tarr, 1972). Fields, sown with susceptible early cultivars, which tend to become severely infected before the disease appears in other cultivars, are examined for appearance of the disease. The annual incidence of foot rot in wheat, *Cercosporella herpotrichoides,* varies with the amount of inoculum available, prevailing weather conditions, and methods of wheat culture (Rowe and Powelson, 1973a,b). Frequent rains are necessary to assure inoculum dispersal and the high humidity required for sporulation and infection. In areas where foot rot is a serious problem, abundant moisture usually is present from late autumn to early spring. Temperature has the greatest influence on sporulation, which can occur for up to 50 days. Based on the temperature sporulation curve for *C. herpotrichoides*, sporulation periods can be identified from field temperature data and used to assess the seasonal epidemic potential.

In analyzing the seasonal progress of apple scab (*Venturia inaequalis*) weather data must be supplemented by observations on the initial level of inoculum, and on spore ripening and dispersal (Bourke, 1959). In France the warning threshold for *Venturia pirina* on pear was about 1000–1500 ascospores per hour on traps in the orchards when trees were in full bloom and fruit swell. In New York fungicidal sprayings are based on temperature–moisture relations since perithecia discharge only when thoroughly wetted and infection is likely after 30 hr of leaf wetness at 5°C to 9.5 hr at 15°C (Tarr, 1972; Table IV).

## C. Prediction of Weather Suitable for Infection

The relation between the air masses passing over the surface of a location and the actual environment the fungus experiences is influenced

TABLE IV
DEFINITION OF THE MILLS PERIOD[a]

| Temperature (°C) | 5.6 | 7.2 | 10.0 | 12.8 | 14.4 | 16.7 |
|---|---|---|---|---|---|---|
| Hours of leaf wetness | 30 | 20 | 14 | 11 | 10 | 9 |

[a] As given by the British Federation of Plant Pathologists: The period of apple leaf wetness satisfying requirements of leaf infection by *Venturia inaequalis* ascospores at a stated temperature. From Anonymous (1973).

by exposure of the location to sunshine, wind, and altitude, as discussed in relation to epidemics of rice blast disease (Suzuki, 1975).

When passing air masses carry moisture as well as inoculum (Nagarajan *et al.*, 1976), the deposition of rain loaded with spores is an accurate index of infection, because inoculum and favorable conditions are combined. Satellite photographs showing the trajectory of cloud systems are modern tools in predicting infection time and place, i.e., forecasting the first infection in areas where the inoculum is imported annually. Three upper air synoptic rules for the 700-mbar level have been formulated (Nagarajan and Singh, 1975).

Synoptic weather maps have been used to predict infection periods (Bourke, 1970). Correlations were made between periods of favorable blight infection and surface weather charts for those periods. In general, two synoptic situations giving rise to potato blight were identified. When waves of maritime tropical air and stagnant or slowly moving air giving lengthy periods of wet overcast weather were developing, spraying advice was issued. Wallin and Riley (1960) related daily surface weather to the 5-day mean surface weather and upper air maps. Wind direction at the surface and at 10,000 ft elevation carrying moving air masses with moisture suitable for the fungus appeared useful in 5-day forecasts of favorable weather. Movement of atmospheric pressure systems to target areas was also studied for 30-day forecasts. Bourke and Wallin outlined the use of synoptic and prognostic weather maps for forecasting (Krause and Massie, 1975). Weather maps can be used in forecasting air masses that are likely to pass over a target area, creating conditions favorable for infection. They can be used to predict certain weather situations in large areas (many thousands of $km^2$), but they are not very useful in locating infection in small areas, such as those covered by a few farms, because of differences between weather and microclimates.

## D. MRA in Prediction

The value of MRA in this application lies in its capability to assemble predictive equations rather than to analyze relationships between variables. Tea blister blight (*Exobasidium vexans*) proved to be a convenient disease for an epidemiological field study in Ceylon (Kerr and Rodrigo, 1967a). One of the advantages of studying tea diseases is that the crop is harvested by frequent handplucking of young shoots with two or three leaves and a bud. By examining the harvested tissues an accurate measure of disease attributed to infection can be obtained within a relatively short time. In a sense the harvested shoot tips are used as traps. Two multiple regression equations were assembled by Kerr and Rodrigo

(1967b) to explain variation in the number of blisters per 100 shoots in terms of daily spore concentration and daily duration of either surface wetness or sunshine, and the infection periods 15–25 days prior to each measurement of disease. They described a simple calculator so that tea planters could easily use the forecast method, and economize in applying fungicides. They concluded their forecasting procedure was one of the best that is based on MRA (Fig. 5). The elegance of the technique is clear. It is simple, adapted to wide application, and has true confidence of growers.

## E. EVALUATION OF FORECASTING

Most forecasting systems currently in operation are actually late-warning disease prediction systems rather than ones that predict infection. They forecast disease (symptom) appearance *after* the weather suitable for infection has been recorded. We still cannot accurately predict the environmental parameters necessary for infection prediction (Krause

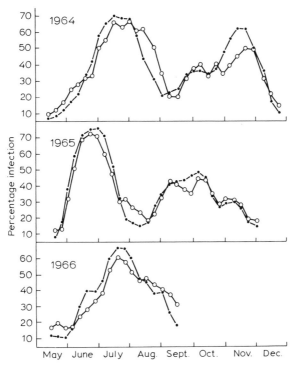

Fig. 5. Predicted (o———o), by using MRA, and measured (•———•) disease incidence of blister blight of tea in Sri Lanka, in 1964, 1965, and 1966. (From Kerr and Rodrigo, 1967b.)

and Massie, 1975). This situation must change, as long as control measures remain primarily preventive.

The usefulness of synoptic weather maps can be increased when the relation between the edaphic factors (microclimate) and their effect on the incoming air masses can be established. Dew point, wind speed, wind direction, and temperature must be translated into a.o. hours of duration of leaf wetness. Simulation in forecasting has not yet developed sufficiently to evaluate its merit in crop protection, although in various areas of Europe and the United States simulators are being designed and tested (Waggoner, 1974).

The weakest points in quantifying relations between crop, pathogen, environment, and time are still determination of the amount of inoculum in time and space, and dispersion, i.e., the relation between spore production and amount of new infection in successive cycles.

The role of the various classes of moisture in epidemiology and forecasting has been discussed in this chapter. This knowledge can be made to work especially in simulation, and to a lesser extent in MRA. Although both approaches still face problems of quantification, they seem promising, and their rate of development is high enough to postulate that weather prediction may become the limiting factor in accurately forecasting infection. Epidemiologists should urge the appropriate institutions and people to develop more reliable weather forecasts for at least 4 days, especially with respect to the intriguing role that moisture plays in epidemiology and forecasting of plant diseases.

## REFERENCES

Analytis, S. (1973). Methodik der analyse von Epidemien dargestellt am Apfelschorft [Venturia inaequalis (Cooke) Aderh.]. *Acta Phytomed.* **1**, 1–75.
Anonymous. (1973). A guide to the use of terms in plant pathology. *Phytopathol. Pap.* **17**, 1–55.
Baily, N. T. J. (1957). "The Mathematical Theory of Epidemics." Griffin, London.
Bashi, E., and Rotem, J. (1974). Adaptation of four pathogens to semi-arid habitats as conditioned by penetration rate and germinating spore survival. *Phytopathology* **64**, 1035–1039.
Bourke, P. M. A. (1959). Meteorology and the timing of fungicide applications against potato blight. *Int. J. Bioclimatol. Biometeorol.* **3**, Part II, Sect. E, 1–8.
Bourke, P. M. A. (1970). Use of weather information in the prediction of plant disease epiphytotics. *Annu. Rev. Phytopathol.* **8**, 345–370.
Box, J. E. (1965). Design and calibration of a thermocouple psychrometer which uses the Peltier effect. *In* "Humidity and Moisture" (A. Wexler, ed.), Vol. I, pp. 110–121. Van Nostrand-Reinhold, Princeton, New Jersey.
Burleigh, J. R., Eversmeyer, M. G., and Roelfs, A. P. (1972). Development of linear equations for predicting wheat leaf rust. *Phytopathology* **62**, 947–953.

Burrage, S. W. (1972). Dew on wheat. *Agric. Meteorol.* **10**, 3–12.

Butt, D. J., and Royle, D. J. (1974). Multiple regression analysis in the epidemiology of plant diseases. *In* "Epidemics of Plant Diseases" (J. Kranz, ed.), Ecol. Ser. No. 13, pp. 78–114. Springer Verlag, Berlin and New York.

Chester, K. S. (1943). The decisive influence of later winter weather on wheat leaf rust epiphytotics. *Plant Dis. Rep., Suppl.* **143**, 133–144.

Cook, R. J., and Christen, A. A. (1976). Growth of cereal root-rot fungi as affected by temperature-water potential interactions. *Phytopathology* **66**, 193–197.

Cook, R. J., and Papendick, R. I. (1972). Influence of water potential of soils and plants on root diseases. *Annu. Rev. Phytopathol.* **10**, 349–374.

De Weille, G. A. (1969). A climatological model typifying the days on which onion crops get infected by downy mildew conidia. *Biometeorol, Proc. Int. Biometeorol. Congr. 5th, 1969* Vol. 4, Part II, p. 198.

Duvdevani, S., Reichert, I, and Palti, J. (1946). The development of downy and powdery mildews of cucumbers as related to dew and other environmental factors. *Palest. J. Bot.*, Rehovot Ser. **5**, 127–151.

Eyal, Z. (1971). The kinetics of pyconospore liberation in Septoria tritici. *Can. J. Bot.* **49**, 1095–1099.

Fournet, J. (1969). Propriétés et rôle du cirrhe du Septoria nodorum Berk. *Ann. Phytopathol.* **1**, 87–94.

Gregory, P. H. (1973). "The Microbiology of the Atmosphere." Plant Sci. Monogr., 2nd ed., Leonard Hill, London.

Hine, R. B. (1976). Epidemiology of pink disease of pineapple fruit. *Phytopathology* **66**, 323–327.

Hirst, J. M., Stedman, O. J., and Hogg, W. H. (1967). Long distance spore transport: methods of measurement, vertical spore profiles and the detection of immigrant spores. *J. Gen. Microbiol.* **48**, 329–355.

Hogg, W. H., Hounnam, C. E., Mallik, A. K., and Zadoks, J. C. (1969). Meteorological factors affecting the epidemiology of wheat rusts. W.M.O., *Tech. Rep. Ser.* **238**, T.P. 130, Tech. Note 99, 1/143.

Hunter, J. E., and Kunimoto, R. K. (1974). Dispersal of Phytophthora palmivora Sporangia be wind-blown rain. *Phytopathology* **64**, 202–206.

Jowett, D., Browning, J. A., and Cournoyer Haning, B. (1974). Non-linear disease progress curves. *In* "Epidemics of Plant Diseases" (J. Kranz, ed), Ecol. Stud. No. 13, pp. 115–136. Springer-Verlag, Berlin and New York.

Kempthorne, D., Bancroft, T. A., Gowen, J. W., and Lush, J. L. (eds). (1954). "Statistics and Mathematics in Biology." Iowa State Coll. Press, Ames.

Kerr, A., and Rodrigo, W. R. F. (1967a). Epidemiology of tea blister blight (Exobasidium vexans). III. Spore deposition and disease prediction. *Trans. Br. Mycol. Soc.* **50**, 49–55.

Kerr, A., and Rodrigo, W. R. F. (1967b). Epidemiology of tea blister blight (Exobasidium vexans). IV. Disease forecasting. *Trnas. Br. Mycol. Soc.* **50**, 609–614.

Kranz, J. (1974). The role and scope of mathematical analysis and modeling in epidemiology. *In* "Epidemics of Plant Diseases" (J. Kranz, ed.), Ecol. Stud. No. 13, pp. 7–54. Springer-Verlag, Berlin and New York.

Krause, R. A., and Massie, L. B. (1975). Predictive systems: Modern approaches to disease control. *Annu. Rev. Phytopathol.* **13**, 31–47.

Lapwood, D. H., and Adams, M. J. (1973). The effect of a few days of rain on the distribution of common scab (Streptomyces scabies) on young potato tubers. *Ann. Appl. Biol.* **73**, 277–283.

Lapwood, D. H., and Lewis, B. G. (1967). Observations on the timing of irrigation and the incidence of potato common scabe (Streptomyces scabies). *Plant Pathol.* **16**, 131–135.

Lapwood, D. H., Wellings, L. W., and Rosser, W. R. (1970). The control of common scab of potatoes by irrigation. *Ann. Appl. Biol.* **66**, 397–405.

Massie, L. B., and Nelson, R. R. (1973). The use of regression analysis in epidemiological studies of southern corn leaf blight. *Phytopathology* **63**, 205.

May, K. R. (1967). Physical aspects of sampling air-borne microbes. *Symp. Soc. Gen. Microbiol.* **17**, 60–80.

Meredith, D. S. (1973). Epidemiological consideration of plant diseases in the tropical environment. *Phytopathology* **63**, 1446–1454.

Monteith. J. L. (1957). "Dew." *Q. J. R. Meteorol. Soc.* **83**, 322–341.

Monteith, J. L. (1972). "Survey of instruments for Micrometeorology." I. B. P. Handb. No. 22. Blackwell, Oxford.

Nagarajan, S., and Singh, H. (1975). The Indian stem rust rules—an epidemiological concept on the spread of wheat stem rust. *Plant Dis. Rep.* **59**, 133–136.

Nagarajan, S., Singh, H., Joshi, L. J., and Saari, E. E. (1976). Meteorological conditions associated with long distance dissemination and deposition of *Puccinia graminis* tritici uredospores in India. *Phytopathology* **66**, 198–203.

Noffsinger, T. L. (1965). Survey of techniques for measuring dew. *In* "Humidity and Moisture" (A. Wexler, ed.), Vol. 2, pp. 523–531. Van Nostrand-Reinhold, Princeton, New Jersey.

Pielou, E. C. (1969). "An Introduction to Mathematical Ecology." Wiley, New York.

Politowski, K., and Browning, J. A. (1975). Effects of temperature, light, and dew duration on relative numbers of infection structures of Puccinia coronata avenae. *Phytopathology* **65**, 1400–1404.

Post, J. J. (1959). De thermohygrograph als bladnatschrijver. *Versl. Kon Ned. Meteorol. Inst.* V-51.

Rapilly, F., and Skajennikoff, M. (1974). Etudes sur l'inoculum de *Septoria nodorum* Berk., agent de la septoriose du blé. II. Les pycnidiospores. *Ann. Phytopathol.* **6**, 71–82.

Rapilly, F., Foucault, B., and Delacasedieux, J. (1973). Etudes sur l'inoculum de *Septoria nodorum* Berk. (*Leptosphearia nodorum* Müller), agent de la septoriose du blé. I. Les ascospores. *Ann. Phytopathol.* **5**, 131–141.

Rawlins, S. L. (1976). Measurement of water content and the state of water in soils. *In* "Water Deficits and Plant Growth" (T. T. Kozlowski, ed.), Vol. 4, pp. 1–55. Academic Press, New York.

Rotem, J., and Reichert, I. (1964). Dew—a principal moisture factor enabling early blight epidemics in a semi-arid region of Israel. *Plant Dis. Rep.* **48**, 211–215.

Rowe, R. C., and Powelson, R. L. (1973a). Epidemiology of *Cercosporella* Footrot of wheat: Spore production. *Phytopathology* **63**, 981–984.

Rowe, R. A., and Powelson, R. L. (1973b). Epidemiology of *Cercosporella* Footrot of wheat: Disease spread. *Phytopathology* **63**, 984–988.

Royle, D. J. (1973). Quantitative relationships between infection by the hop downy mildew pathogen, *Pseudoperonospora humuli*, and weather and inoculum factors. *Ann. Appl. Biol.* **73**, 19–30.

Scharen, A. L. (1966). Cyclic production of pycnidia and spores in dead wheat tissue by *Septoria nodorum* (under favorable weather conditions). *Phytopathology* **56**, 580–581.

Schurer, K., and Van der Wal, A. F. (1972). An electronic leaf wetness recorder. *Neth. J. Plant Pathol.* **78**, 29–32.

Shearer, B. L., and Zadoks, J. C. (1972). The latent period of *Septoria nodorum* in wheat. I. The effect of temperature and moisture treatments under controlled conditions. *Neth. J. Plant Pathol.* **78**, 231–241.

Shearer, B. L., and Zadoks, J. C. (1974). The latent period of *Septoria nodorum* in wheat. II. The effect of temperature and moisture treatments under field conditions. *Neth. J. Plant Pathol.* **80**, 48–60.

Smith, W. H. (1970). "Tree Pathology." Academic Press, New York.

Snedecor, G. W., and Cochran, W. G. (1972). "Statistical Methods", 6th ed. Iowa State Univ. Press, Ames.

Stavely, J. R., and Slana, L. J. (1975). Relation of postinoculation leaf wetness to initiation of Tobacco Brown spot. *Phytopathology* **65**, 897–901.

Suzuki, H. (1975). Meteorological factors in the epidemiology of rice blast. *Annu. Rev. Phytopathol.* **13**, 239–256.

Tarr, S. A. J. (1972). "The Principles of Plant Pathology." Macmillan, New York.

Taylor, C. F. (1956). A device for recording the duration of dew deposits. *Plant Dis. Rep.* **40**, 1025–1028.

Tu, J. C., and Hendrix, J. W. (1970). The summer biology of *Puccinia striiformis* in southeastern Washington. II. Natural infection during the summer. *Plant Dis. Rep.* **54**, 384–386.

Van der Plank, J. E. (1963). "Plant Diseases: Epidemics and Control." Academic Press, New York.

Van der Wal, A. F., and Cowan, M. C. (1974). An ecophysiological approach to crop losses, exemplified in the system wheat, leaf rust and glume blotch. II. Development, growth, and transpiration of uninfected plants and plants infected with Puccinia recondita f.sp. triticina and/or Septoria nodorum in a climate chamber experiment. *Neth. J. Plant Pathol.* **80**, 192–214.

Van der Wal, A. F., and Luuring, B. B. (1976). Leptosphaeria nodorum Müller (Septoria nodorum Berk.). *Sticht. Ned. Graan Cent., Tech. Ber.* **22**, 1–78.

Van der Wal, A. F., and Zadoks, J. C. (1976). Towards mass production of uredospores of brown rust on wheat (Puccinia recondita f.sp. tritici). *Cereal Rusts Bull.* **4**, 9–13.

Van der Wal, A. F., Smeitink, H., and Maan, G. C. (1975). An ecophysiological approach to crop losses exemplified in the system wheat, leaf rust and glume blotch. III. Effects of soil-water potential on development, growth, transpiration, symptoms, and spore production of leaf rust-infected wheat. *Neth. J. Plant Pathol.* **81**, 1–13.

Waggoner, P. E. (1974). Simulation of epidemics. *In* "Epidemics of Plant Diseases" (J. Kranz, ed.), pp. 137–160. Springer-Verlag, Berlin and New York.

Wallin, J. R., and Riley, J. A. (1960). Weather map analysis—an aid in forecasting potato late blight. *Plant Dis. Rep.* **44**, 227–234.

Wexler, A. ed. (1965). "Humidity and Moisture", Vols. I and II. Van Nostrand-Reinhold Princeton, New Jersey.

Yarwood, C. E. (1959). Microclimate and infection. *In* Plant Pathology 1908–1958, pp. 548–556. (Holton, C. S., et al., eds), Univ. of Wisconsin Press, Madison.

# AUTHOR INDEX

Numbers in italics refer to the pages on which the complete references are listed.

# SUBJECT INDEX

## A

ABA, *see* Abscisic acid
*Abies balsamea,* 232
Abiotic disease, 101–115
Abscisic acid, 28, 42, 47, 68, 69, 86, 94
Abscission, 27, 28, 103, 106, 111, 113
Abscission zone, 28
Absorption
  ions, 12
  minerals, 68, 107, 185
  phosphorus, 93
  potassium, 5, 32
  solutes, 36, 42
  water, 8, 20, 21, 26, 31, 35, 37–42, 47, 48,
    51, 63, 64, 66, 82, 107, 110, 128, 229,
    260
Accumulation
  potassium, 48
  solutes, 51
*Acer* sp., 131
*Acer saccharum,* 239
Acervuli, 157
Acetic acid bacteria, 271
*Acetomonas* sp., 271
Actinomycetes, 110, 182, 191
Active transport, 48
Adventitious root, 13, 103–105
Aecia, 126, 136
Aeciospores, 120, 126, 136, 145, 149
Aeration, 64, 106, 109
*Aethalium septicum,* 148
Aflatoxin, 200, 206
Agaricaceae, 225
Agaricales, 223
Age, 2, 80
Aging, 69
Air spaces, 41
Alfalfa, *see Medicago sativa*
*Allium cepa,* 160

Almond, *see Prunus amygdalus*
*Alternaria solani,* 144
*Alternaria tenuis,* 147, 148
*Alternaria* sp., 137, 202–205
Aluminum, 108
Aluminum oxide, 259
Amino acids, 6, *see also* specific amino acids
Ammonia, 110
Anaerobic storage, 240
Anaerobiosis, 64, 103, 106, 107, 110, 206
*Ananas comosus* sp., 271
Angular leaf spot, 5
Antagonism, 67, 189, 191
Anthesis, 40
Anthracnose, 133, 157–159, 164
Antibiotics, 6, 186
*Antirrhinum majus,* 106
*Aphanomyces cochliodes,* 11
*Aphanomyces euteiches,* 179, 188
Aphyllophorales, 223, 225
Apical meristem, 12
*Apium graveolens,* 110, 115
Apothecia, 188, 244
Apple, *see Malus* sp.
Apple scab, 121, 281, 289
Appressoria, 260, 275
Arabitol, 180
*Armillaria mellea,* 65
Asci, 119, 120, 124, 125
Ascomycetes, 43, 119–124, 185, 188, 189,
  224–226, 231
Ascospores, 120–123, 138, 144, 148, 155,
  188
Ash, *see Fraxinus* sp.
Aspen, *see Populus* sp.
*Aspergillus brevicale,* 148
*Aspergillus candidus,* 209
*Aspergillus chevalieri,* 187
*Aspergillus flavus,* 186, 189, 200, 204, 206,
  209